GREENLAND

ICELAND

Sanlúcar de
Barrameda

Route of
the *Trinidad*
(1522)

Bermuda

Azores

Canaries

MEXICO

Cuba

Cape Verde
Islands

Pacific

PANAMA

Ocean

Cape St. Augustin

Tiburones

BRAZIL

Puka Puka

Santa Lucia Bay

Valparaiso

PATAGONIA

Rio de la Plata

Port
Saint
Julian

Strait
of Magellan

– – – Tracks of the ships
of Magellan's armada

• • • • • the pope's line of
demarcation (1494) with
equivalent longitudinal
line in the Eastern
Hemisphere

© 2003 David Cain

Over the Edge of the World

ALSO BY LAURENCE BERGREEN

Voyage to Mars:
NASA's Search for Life Beyond Earth

Louis Armstrong:
An Extravagant Life

Capone:
The Man and the Era

As Thousands Cheer:
The Life of Irving Berlin

James Agee:
A Life

Look Now, Pay Later:
The Rise of Network Broadcasting

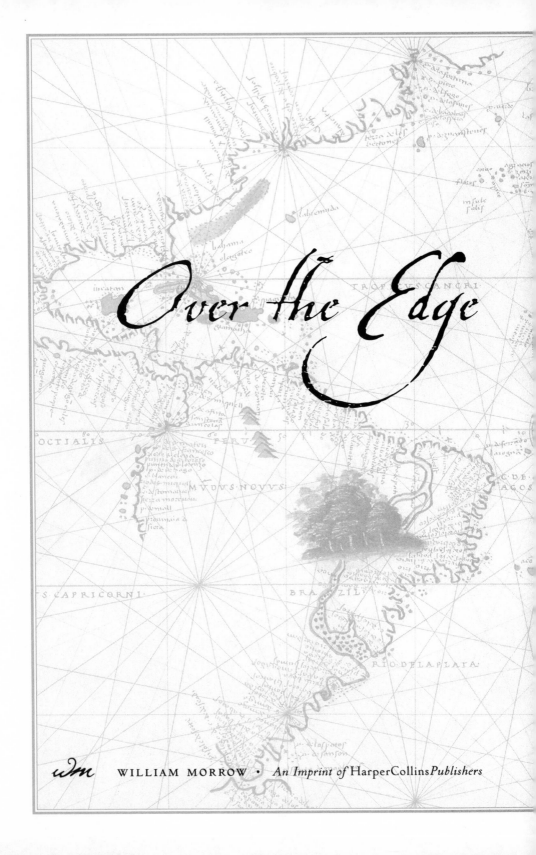

Over the Edge

WILLIAM MORROW · *An Imprint of* HarperCollins*Publishers*

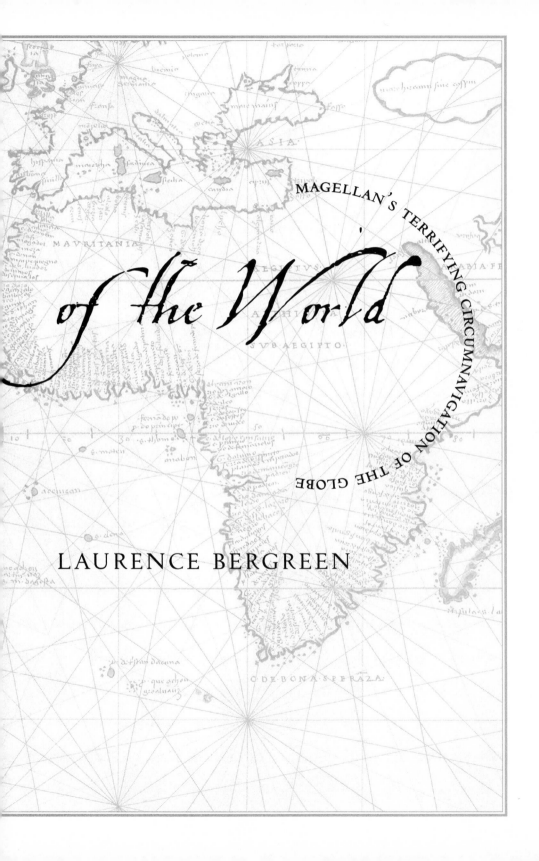

MAGELLAN'S TERRIFYING CIRCUMNAVIGATION OF THE GLOBE

of the World

LAURENCE BERGREEN

ILLUSTRATION CREDITS

Insert page 1 Mariners' Museum, Newport News, VA; page 2 (top) Museo de America, Madrid, Spain/Bridgeman Art Library, (bottom) Scala/ Art Resource, NY; page 3 (top) Réunion des Musées Nationaux/Art Resource, NY, (bottom) Historical Picture Archive/CORBIS; page 4 (top) Giraudon/Art Resource, NY, (bottom) Archivo Iconografico, S.A./CORBIS; page 5 (top) Royalty-Free/CORBIS, (bottom) Scala/ Art Resource, NY; page 6 (top) Réunion des Musées Nationaux/Art Resource, NY, (bottom) Courtesy of the John Carter Brown Library at Brown University; page 7 (top) Mariners' Museum, Newport News, VA, (bottom) Historical Picture Archive/CORBIS; page 8 (top and bottom) Jacques Descloitres, MODIS Land Rapid Response Team/ NASA; page 9 (top) author's photograph, (bottom) Jon Diamond; page 10 (top) Collection Kharbine-Tapabor, Paris, France/Bridgeman Art Library, (bottom) Mary Evans Picture Library; page 11 (top) Mary Evans Picture Library, (bottom) Bibliothèque Nationale, Paris, France/Bridgeman Art Library; page 12 (top) Bruce Dale/National Geographic Image Collection, (bottom) Lindsay Hebberd/ CORBIS; page 13 Stapleton Collection/CORBIS; page 14 (top and bottom) Beinecke Rare Book and Manuscript Library, Yale University; page 15 Collection Kharbine-Tapabor, Paris, France/Bridgeman Art Library; page 16 (top) Courtesy of the John Carter Brown Library at Brown University, (bottom) Private Collection/Bridgeman Art Library

HarperCollins books may be purchased for educational, business, or sales promotional use. For information please write: Special Markets Department, HarperCollins Publishers Inc., 10 East 53rd Street, New York, NY 10022.

FIRST EDITION

Designed by Betty Lew
Endpaper map © 2003 by David Cain

Printed on acid-free paper

Library of Congress Cataloging-in-Publication Data
Bergreen, Laurence.
    Over the edge of the world: Magellan's terrifying circumnavigation of the globe/Laurence Bergreen.—1st ed.
    p. cm.
    ISBN 0-06-621173-5
    1. Magalhães, Fernão de, d. 1521—Journeys.   2. Explorers—Portugal—Biography.   3. Voyages around the world. I. Title.
G420.M2B47 2003
910'.92–dc21
[B]   2003050143

03 04 05 06 07 WBC/RRD 10 9 8 7 6 5 4 3 2 1

In memory of my brother and father

How a Ship having passed the Line was driven by
storms to the cold Country towards the South Pole;
and how from thence she made her course to the
tropical Latitude of the Great Pacific Ocean; and
of the strange things that befell; and in what manner
the Ancyent Marinere came back to his own Country.

—Samuel Taylor Coleridge,
"The Rime of the Ancient Mariner"

Contents

Book Three: Back from the Dead

Principal Characters

King Charles I (later Charles V, emperor of the Holy Roman Empire)
King Manuel (king of Portugal)
Juan de Aranda
Juan Rodríguez de Fonseca (bishop of Burgos)
Cristóbal de Haro (financier)
Ruy Faleiro (cosmographer)
Beatriz Barbosa (Magellan's wife)
Diogo Barbosa (Magellan's father-in-law)

The Armada de Molucca
(at the time of departure from Seville)

Trinidad
Ferdinand Magellan (Captain General)
Estêvão Gomes (pilot major)
Gonzalo Gómez de Espinosa (*alguacil,* or master-at-arms)
Francisco Albo (pilot)
Pedro de Valderrama (chaplain)
Ginés de Mafra (seaman)
Enrique de Malacca (interpreter)
Duarte Barbosa (supernumerary)

Álvaro de Mesquita (Magellan's relative, supernumerary)

Antonio Pigafetta (chronicler, supernumerary)

Cristóvão Rebêlo (Magellan's illegitimate son, supernumerary)

San Antonio

Juan de Cartagena (captain and inspector general)

Antonio de Coca (fleet accountant)

Andrés de San Martín (astrologer and pilot)

Juan de Elorriaga (master)

Gerónimo Guerra (clerk)

Bernard de Calmette, also known as

Pero Sánchez de la Reina (chaplain)

Concepción

Gaspar de Quesada (captain)

João Lopes Carvalho (pilot)

Juan Sebastián Elcano (master)

Juan de Acurio (mate)

Hernando Bustamente (barber)

Joãozito Carvalho (cabin boy)

Martin de Magalhães (supernumerary)

Victoria

Luis de Mendoza (captain)

Vasco Gomes Gallego (pilot)

Antonio Salamón (master)

Miguel de Rodas (mate)

Santiago

Juan Rodríguez Serrano (captain)

Baltasar Palla (master)

Bartolomé Prieur (mate)

A Note on Dates

$\mathcal{D}$ates are given in the Julian calendar, in effect since the time of Julius Caesar. With modifications, this calendar was adopted by Christian churches around the world, including those in Spain.

Sixty years after the completion of Magellan's voyage, in 1582, Spain, France, and other European countries migrated to the Gregorian calendar, decreed by Pope Gregory XIII and designed to correct incremental errors in the Julian system. It took more than two centuries to complete the transition to the new calendar throughout Europe, since Protestant nations resisted the change. To correct for accumulated errors, ten days were omitted, so that October 5, 1582, in the Julian calendar suddenly became October 15, 1582, in the Gregorian.

In addition to this calendar shift, Magellan's voyage had its own record-keeping issues. The dates of various events recorded by the two official chroniclers of the expedition, Antonio Pigafetta and Francisco Albo, occasionally diverge by one day. The discrepancy may be due to human error, and it may also have been caused by the way each diarist reckoned the day. Albo, a pilot, followed the custom of ships' logs, which began the day at noon rather than at midnight. In contrast, Pigafetta used a nonnautical frame of reference in his diary. Thus, an event occurring on a given morning might have been put down a day apart in the records maintained by the two.

Finally, the international date line did not exist before Magellan's voyage. (It now extends west from the island of Guam, in the Pacific Ocean.) As Albo and Pigafetta neared the completion of their cir-cumnavigation, they were astonished to note that their calculations were off, and their voyage around the world actually took one day longer than they had thought.

Measurements

One fathom equals six feet.
One Spanish league (*legua*) equals approximately four miles.

One *bahar* (of cloves) equals 406 pounds.
One *quintal* equals 100 pounds.
One *cati* (a Chinese measurement) equals 1.75 pounds.

One *braza* (of cloth) equals about five and a half feet.

One *maravedí* equals approximately 12 modern cents.

Over the Edge of the World

PROLOGUE

A Ghostly Apparition

Oh! dream of joy! is this indeed
The light-house top I see?
Is this the hill? is this the kirk?
Is this mine own countree?

On September 6, 1522, a battered ship appeared on the horizon near the port of Sanlúcar de Barrameda, Spain.

As the ship came closer, those who gathered onshore noticed that her tattered sails flailed in the breeze, her rigging had rotted away, the sun had bleached her colors, and storms had gouged her sides. A small pilot boat was dispatched to lead the strange ship over the reefs to the harbor. Those aboard the pilot boat found themselves looking into the face of every sailor's nightmare. The vessel they were guiding into the harbor was manned by a skeleton crew of just eighteen sailors and three captives, all of them severely malnourished. Most lacked the strength to walk or even to speak. Their tongues were swollen, and their bodies were covered with painful boils. Their captain was dead, as were the officers, the boatswains, and the pilots; in fact, nearly the entire crew had perished.

The pilot boat gradually coaxed the battered vessel past the natural hazards guarding the harbor, and the ship, *Victoria*, slowly began to make her way along the gently winding Guadalquivir River to Seville, the city from which she had departed three years earlier. No one knew what had become of her since then, and her appearance came as a surprise to those who watched the horizon for sails. *Victoria* was a ship of mystery, and every gaunt face on her deck was filled with the dark secrets of a prolonged voyage to unknown lands. Despite the journey's hardships, *Victoria* and her diminished crew accomplished what no other ship had ever done before. By sailing west until they reached the East, and then sailing on in the same direction, they had fulfilled an ambition as old as the human imagination, the first circumnavigation of the globe.

*T*hree years earlier, *Victoria* had belonged to a fleet of five vessels with about 260 sailors, all under the command of Fernão de Magalhães, whom we know as Ferdinand Magellan. A Portuguese nobleman and navigator, he had left his homeland to sail for Spain with a charter to explore undiscovered parts of the world and claim them for the Spanish crown. The expedition he led was among the largest and best equipped in the Age of Discovery. Now *Victoria* and her ravaged little crew were all that was left, a ghost ship haunted by the memory of more than two hundred absent sailors. Many had died an excruciating death, some from scurvy, others by torture, and a few by drowning. Worse, Magellan, the Captain General, had been brutally killed. Despite her brave name, *Victoria* was not a ship of triumph, she was a vessel of desolation and anguish.

And yet, what a story those few survivors had to tell—a tale of mutiny, of orgies on distant shores, and of the exploration of the entire globe. A story that changed the course of history and the way we look at the world. In the Age of Discovery, many expeditions ended in disaster and were quickly forgotten, yet this one, despite the misfortunes that befell it, became the most important maritime voyage ever undertaken.

This circumnavigation forever altered the Western world's ideas about cosmology—the study of the universe and our place in it—as well as geography. It demonstrated, among other things, that the earth was round, that the Americas were not part of India but were actually a separate continent, and that oceans covered most of the earth's surface. The voyage conclusively demonstrated that the earth is, after all, one world. But it also demonstrated that it was a world of unceasing conflict, both natural and human. The cost of these discoveries in terms of loss of life and suffering was greater than anyone could have anticipated at the start of the expedition. They had survived an expedition to the ends of the earth, but more than that, they had endured a voyage into the darkest recesses of the human soul.

Book One

In Search of Empire

CHAPTER I

The Quest

He holds him with his skinny hand,
"There was a ship," quoth he.
"Hold off! unhand me, grey-beard loon!"
Eftsoons his hand dropt he.

*O*n June 7, 1494, Pope Alexander VI divided the world in half, bestowing the western portion on Spain, and the eastern on Portugal.

Matters might have turned out differently if the pontiff had not been a Spaniard—Rodrigo de Borja, born near Valencia—but he was. A lawyer by training, he assumed the Borgia name when his maternal uncle, Alfonso Borgia, began his brief reign as Pope Callistus III. As his lineage suggests, Alexander VI was a rather secular pope, among the wealthiest and most ambitious men in Europe, fond of his many mistresses and his illegitimate offspring, and endowed with sufficient energy and ability to indulge his worldly passions.

He brought the full weight of his authority to bear on the appeals of King Ferdinand and Queen Isabella, the "Catholic Monarchs" of Spain who had instituted the Inquisition in 1492 to purge Spain of Jews and Moors. They exerted considerable influence over the papacy, and they had every reason to expect a sympathetic hear-

ing in Rome. Ferdinand and Isabella wanted the pope's blessing to
protect the recent discoveries made by Christopher Columbus, the
Genoese navigator who claimed a new world for Spain. Portugal,
Spain's chief rival for control of world trade, threatened to assert its
own claim to the newly discovered lands, as did England and France.

Ferdinand and Isabella implored Pope Alexander VI to support
Spain's title to the New World. He responded by issuing papal
bulls—solemn edicts—establishing a line of demarcation between
Spanish and Portuguese territories around the globe. The line extended from the North Pole to the South Pole. It was located one
hundred leagues (about four hundred miles) west of an obscure
archipelago known as the Cape Verde Islands, located in the
Atlantic Ocean off the coast of North Africa. Antonio and Bar-
tolomeo da Noli, Genoese navigators sailing for Portugal, had discovered them in 1460, and ever since, the islands had served as an
outpost in the Portuguese slave trade.

The papal bulls granted Spain exclusive rights to those parts of
the globe that lay to the west of the line; the Portuguese, naturally,
were supposed to keep to the east. And if either kingdom happened
to discover a land ruled by a Christian ruler, neither would be able
to claim it. Rather than settling disputes between Portugal and Spain,
this arrangement touched off a furious race between the nations to
claim new lands and to control the world's trade routes even as they
attempted to shift the line of demarcation to favor one side or the
other. The bickering over the line's location continued as diplomats
from both countries convened in the little town of Tordesillas, in
northwestern Spain, to work out a compromise.

*I*n Tordesillas, the Spanish and Portuguese representatives agreed
to abide by the idea of a papal division, which seemed to protect the
interests of both parties. At the same time, the Portuguese prevailed
on the Spanish representatives to move the line 270 leagues west;
now it lay 370 leagues west of the Cape Verde Islands, at approxi-

mately 46° 30' W, according to modern calculations. This change placed the boundary in the middle of the Atlantic, roughly halfway between the Cape Verde Islands and the Caribbean island of Hispaniola. The new boundary gave the Portuguese ample access to the African continent by water and, even more important, allowed the Portuguese to claim the newly discovered land of Brazil. But the debate over the line—and the claims for empire that depended on its placement—dragged on for years. Pope Alexander VI died in 1503, and he was succeeded by Pope Julius II, who in 1506 agreed to the changes, and the Treaty of Tordesillas achieved its final form.

The result of endless compromises, the treaty created more problems than it solved. It was impossible to fix the line's location because cosmologists did not yet know how to determine longitude—nor would they for another two hundred years. To further complicate matters, the treaty failed to specify whether the line of demarcation extended all the way around the globe or bisected just the Western Hemisphere. Finally, not much was known about the location of oceans and continents. Even if the world was round, and men of science and learning agreed that it was, the maps of 1494 depicted a very different planet from the one we know today. They mixed geography with mythology, adding phantom continents while neglecting real ones, and the result was an image of a world that never was. Until Copernicus, it was generally assumed that the earth was at the absolute center of the universe, with the perfectly circular planets—including the sun—revolving around it in perfectly circular, fixed orbits; it is best to conceive of the earth as nested in the center of all these orbits.

Even the most sophisticated maps revealed the limitations of the era's cosmology. In the Age of Discovery, cosmology was a specialized, academic field that concerned itself with describing the image of the world, including the study of oceans and land, as well as the world's place in the cosmos. Cosmologists occupied prestigious chairs at universities, and were held in high regard by the thrones of Europe. Although many were skilled mathematicians, they often concerned themselves with astrology, believed to be a legitimate branch of astronomy, a practice that endeared them to insecure rulers in search of reassurance in an uncertain world. And it was changing faster than cosmologists realized. Throughout the six-teenth century, the calculations and theories of the ancient Greek and Egyptian mathematicians and astronomers served as the basis of cosmology, even as new discoveries undermined time-honored assumptions. Rather than acknowledge that a true scientific revolu-tion was at hand, cosmologists responded to the challenge by trying to modify or bend classical schemes, especially the system codified by Claudius Ptolemy, the Greco-Egyptian astronomer and mathe-matician who lived in the second century A.D.

Ptolemy's massive compendium of mathematical and astronomi-cal calculations had been rediscovered in 1410, after centuries of neglect. The revival of classical learning pushed aside medieval notions of the world based on a literal—yet magical—interpreta-tion of the Bible, but even though Ptolemy's rigorous approach to mathematics was more sophisticated than monkish fantasies of the cosmos, his depiction of the globe contained significant gaps and errors. Following Ptolemy's example, European cosmologists disregarded the Pacific Ocean, which covers a third of the world's surface, from their maps, and they presented incomplete renditions of the Amer-ican continent based on reports and rumors rather than direct observations. Ptolemy's omissions inadvertently encouraged explo-ration because he made the world seem smaller and more navigable than it really was. If he had correctly estimated the size of the world, the Age of Discovery might never have occurred.

Amid the confusion, two kinds of maps evolved: simple but accu-rate "portolan" charts based on the actual observations of pilots, and far more elaborate concoctions of cosmographers. The charts simply showed how to sail from point to point; the cosmographers tried to include the entire cosmos in their schemes. The cosmographers relied primarily on mathematics for their depictions, but the pilots relied on experience and observation. The pilots' charts covered harbors and shorelines; the cosmographers' maps of the world, filled with beguiling speculation, were often useless for actual navigation. Neither approach successfully applied the terms of the Treaty of Tordesillas to the real world.

Although it might be expected that pilots worked closely with cosmologists, that was far from the case. Pilots were hired hands who occupied a lower social stratum. Many of them were illiterate and relied on simple charts that delineated familiar coastlines and harbors, as well as on their own instincts regarding wind and water. The cosmologists looked down on pilots as "coarse men" who possessed "little understanding." The pilots, who risked their lives at sea, were inclined to regard cosmologists as impractical dreamers. Explorers setting out on ocean voyages to distant lands needed the skills of both; they took their inspiration from cosmologists, but they relied on pilots for execution.

Although the Treaty of Tordesillas was destined to collapse under the weight of its faulty assumptions, it challenged the old cosmological ways. On the basis of this fiction, based on a profound misunderstanding of the world, Spain and Portugal competed to establish their global empires. The Treaty of Tordesillas was not even a line drawn in the sand; it was written in water.

*E*mboldened by the Treaty of Tordesillas, Ferdinand and Isabella looked for ways to exploit the portion of the globe granted to Spain. Success proved elusive: Christopher Columbus's voyages to the New World all failed to find a water route to the Indies. A generation after Columbus, King Charles I resumed the quest to establish a global Spanish empire. He, or his advisers, recognized that the Indies could provide priceless merchandise, and the most precious commodity of all was spices.

Spices have played an essential economic role in civilizations since antiquity. Like oil today, the European quest for spices drove the world's economy and influenced global politics, and like oil today, spices became inextricably intertwined with exploration, conquest, imperialism. But spices evoked a glamour and aura all their own. The mere mention of their names—white and black pepper, myrrh, frankincense, nutmeg, cinnamon, cassia, mace, and cloves, to name a few—evoked the wonders of the Orient and the mysterious East.

Arab merchants traded in spices across land routes reaching across Asia and became adept at boosting prices by concealing the origins of the cinnamon, pepper, cloves, and nutmeg with which they enriched themselves. The merchants maintained a virtual monopoly by insisting these precious items came from Africa, when in fact they grew in various places in India, and China, and especially throughout Southeast Asia. Europeans came to believe that spices came from Africa, when in fact they only changed hands there. To protect their monopoly, Arab spice merchants invented all sorts of monsters and myths to conceal the ordinary process of harvesting spices, making it sound impossibly dangerous to acquire them.

The spice trade was central to the Arab way of life. Muhammad, the prophet of Islam, belonged to a family of prominent merchants, and for many years traded in myrrh and frankincense, among other spices, in Mecca. Arabs developed sophisticated methods of extracting essential oils from aromatic spices used for medical and other therapeutic purposes. They formulated elixirs and syrups derived from spices, including *julāb,* from which the word "julep" derives. During the Middle Ages, Arab knowledge of spices spread across western Europe, where apothecaries developed a brisk trade in concoctions made from cloves, pepper, nutmeg, and mace. In a Europe starved for gold (much of it controlled by the Arabs), spices became more valuable than ever, a major component of European economies.

Despite the overwhelming importance of spices to their economy, Europeans remained dependent on Arab merchants for their supply. They knew the European climate could not sustain these exotic spices. In the sixteenth century, the Iberian peninsula was far too cold—colder than it is now, in the grip of the Little Ice

Age—and too dry to cultivate cinnamon, cloves, and pepper. An Indonesian ruler was said to have boasted to a trader who wanted to grow spices in Europe, "You may be able to take our plants, but you will never be able to take our rain."

Under the traditional system, spices, along with damasks, diamonds, opiates, pearls, and other goods from Asia, reached Europe by slow, costly, and indirect routes over land and sea, across China and the Indian Ocean, through the Middle East and Persian Gulf. Merchants received them in Europe, usually in Italy or the south of France, and shipped them overland to their final destination. Along the way, spices went through as many as twelve different hands, and every time they did, their prices shot up. Spices were the ultimate cash crop.

The global spice trade underwent an upheaval in 1453, when Constantinople fell to the Turks, and the time-honored overland spice routes between Asia and Europe were severed. The prospect of establishing a spice trade via an ocean route opened up new economic possibilities for any European nation able to master the seas. For those willing to assume the risks, the rewards of an oceanic spice trade, combined with control over the world's economy, were irresistible.

The lure of spices impelled sober, cautious financiers to back highly risky expeditions to unknown parts of the globe, and enticed young men to risk their lives. In Spain, the best and perhaps the only reason to risk going to sea was the prospect of getting rich in the Spice Islands, wherever they were. If a sailor devoted years of his life to getting there and back, and if he managed to bring home a small sack stuffed with spices such as cloves or nutmeg, legitimately or not, he could sell it for enough to buy a small house; he could live off the proceeds for the rest of his life. An ordinary seaman might attain a modest degree of wealth, but a captain had a right to expect much more than that in the Age of Discovery—not only vast riches and fame, but titles to pass on to his heirs and foreign lands to rule.

*P*ortugal was the first European nation to exploit the sea for spices and the global empire that went along with them. The quest began as early as 1419, when Prince Henry, the third son of João I and his English wife, Philippa, established his court at Sagres, a stark outcropping of rock at the southernmost edge of Portugal. Known as Prince Henry the Navigator, he rarely went to sea himself; instead, he inspired others to conquer the ocean. Portuguese ships faced obstacles so overwhelming, so shrouded in ignorance and superstition, that only extraordinarily confident and accomplished mariners dared to venture into the Ocean Sea, as the Atlantic Ocean was then known.

As a young soldier, Prince Henry had fought against Arabs, and he was determined to drive them from the Iberian peninsula and from North Africa. At the same time, he learned much from his avowed enemy: their trade routes, their science and mapmaking, and most of all, their navigational techniques. When Prince Henry came to Sagres, Europeans knew little about the ocean beyond latitude 27°N, marked by Cape Bojador in West Africa. It was believed that the waters south of this point teemed with monsters, that their storms made them too violent to navigate, and that inescapable fogs would envelop wayward ships. In the face of all these dangers, Prince Henry offered a bold reply, "You cannot find a peril so great that the hope of reward will not be greater."

In pursuit of his goal, he attracted navigators, shipwrights, astronomers, pilots, cosmographers, and cartographers, both Christians and Jews, to the academy at Sagres, where they cooperated in the enterprise of exploring the world, under Henry's direction. They designed a new type of ship, the small, maneuverable caravel, distinguished by her triangular lateen sail (the name *lateen* came from the word "Latin"), borrowed from Arab vessels. Until this time, European vessels such as galleys relied on oarsmen or fixed sails for power. With their shallow draught and movable sails, Henry's caravels could set a course close to the wind, and they could tack, that is, shift their course to take advantage of the wind from one direction and then from another, zigzagging against the wind toward a fixed point. With their maneuverable sails and impressive seaworthiness, caravels became the vessels of choice for exploration.

Even so, the ocean proved extremely hazardous. Prince Henry sent no less than fourteen expeditions to Cape Bojador within twelve years, and they all failed. He convinced Gil Eannes, a Portuguese explorer, to try once more, and in 1434, Eannes finally accomplished what so many had said was impossible: He sailed safely past Cape Bojador. The following year Eannes, together with Alfonso Gonçalves Baldaya, returned to Cape Bojador; fifty leagues past the cape, they explored a large bay and came upon a caravan of men and camels. Baldaya sailed farther south and collected thousands of sealskins; this was the first commercial cargo brought back to Europe from that part of Africa. On subsequent voyages, Portuguese ships brought gold, animal hides, elephant tusks—and slaves.

Every captain sponsored by Prince Henry was under orders to record the tides, the currents, and the winds, and to compile accurate charts of the coastlines. Voyage by voyage, these charts added to the Portuguese knowledge of the oceans and of the world beyond the Iberian peninsula.

Although Portugal was celebrated for leading Europe into the Age of Discovery, Portuguese kings often frustrated their heroic mariners. In 1488, during the reign of João II, Bartolomeu Dias reached the southernmost point of Africa and rounded what is now known as the Cape of Good Hope; his voyage opened new possibilities for Portuguese trade and conquest. On his return, Dias attempted to claim a reward for his feat, but received practically none. Ten years later, when King Manuel I had succeeded to the throne, Vasco da Gama retraced Dias's route around the tip of Africa and reached Mozambique on the southeastern coast; there he replenished his supplies and sailed farther east to establish an ocean route to India. Da Gama received a royal appointment as viceroy of India, and King Manuel anointed himself "Lord of Guinea and of the navigation and commerce of Ethiopia, Arabia, Persia, and India"—all of it thanks to Vasco da Gama. Across Europe, other monarchs disparaged Manuel as "The Grocer King," and Vasco da Gama came to believe that he had been inadequately rewarded for his service to the crown. In time he joined the ranks of explorers who became estranged from this vain ruler.

King Manuel's indifference to those who had risked their lives to advance the cause of the Portuguese empire had much to do with his ingrained fear of rivals within Portugal. Ever since the start of his reign in 1495, he had enjoyed great commercial success as the wealth of the Indies flowed into the royal coffers, thanks to the exploits of da Gama and other Portuguese explorers, all of which the king took as his due. But King Manuel was no adventurer, and he lacked an appreciation beyond the strictly commercial aspects of what his explorers had done for the Portuguese empire. Rather than doing battle himself, he preferred to remain in his palace, faithful to his wife and to the Church, and tending to Portugal's domestic issues.

Manuel's harshest policies concerned the Jews of Portugal, who distinguished themselves as scientists, artisans, merchants, scholars, doctors, and cosmographers. In 1496, when King Manuel wished to take the daughter of Ferdinand and Isabella as his wife, he was told that he could do so only on condition that he "purify" Portugal by expelling the Jews, as Spain had done four years earlier. Rather than lose this valuable segment of the population, Manuel encouraged conversions to Christianity—forced conversions, in many cases. As "new Christians" (the title fooled no one), Portuguese Jews continued to occupy high positions in the government, and received royal trading concessions, in Brazil especially. Despite these accommodations, anti-Semitism in Portugal led to a massacre of Jews in Lisbon in 1506. Manuel punished those responsible, but the legacy of bitterness lingered, and many Jews left the country for the Netherlands.

*T*hroughout all the turmoil, Portugal retained its ambition to wrest control of the spice trade from the Arabs, and to reach the Spice Islands. In pursuit of this goal, daring, even reckless mariners pre-sented themselves to the king to seek backing for their journeys of exploration to these exotic and dangerous new worlds. Most met with frustration, for the Portuguese court was a place of intrigue, suspicion, double-dealing, and envy.

Among the most persistent supplicants was a minor nobleman with a long and checkered history in the service of the Portuguese empire in Africa: Fernão de Magalhães, or Ferdinand Magellan. According to most accounts, he was born in 1480, in the remote mountain parish of Sabrosa, the seat of the family homestead. He spent his childhood in northwestern Portugal, within sight of the pounding surf of the Atlantic. His father, Rodrigo de Magalhães, traced his lineage back to an eleventh-century French crusader, De Magalhãis, who distinguished himself sufficiently to be rewarded with a grant of land from the duke of Burgundy. Rodrigo himself qualified as minor Portuguese nobility, and served as a sheriff of the port of Aveiro.

Less is known about Magellan's mother, Alda de Mesquita, and there is room for intriguing speculation. The name *Mesquita*, mean-ing mosque, was a common name among Portuguese *conversos* who sought to disguise their Jewish origins. It is possible that she had Jew-ish ancestry, and if she did, Ferdinand was also Jewish, according to Jewish law. Nevertheless, the family considered itself Christian, and Ferdinand Magellan never thought of himself as anything other than a devout Catholic.

Even these basic outlines about Magellan's ancestry are in doubt. In 1567, his heirs began squabbling over his estate, and questions arose over his exact place in the Magalhães family tree. The diffi-culties in tracing Magellan's ancestry arise from the idiosyncrasies of Portuguese genealogy. For example, until the eighteenth century, males usually assumed their father's last name, but the females often chose other surnames for themselves. They took on their father's name, or their mother's, or even a saint's name. And some children assumed a grandfather's name, or their mother's last name, or still other family names. Ferdinand Magellan's brother Diogo took on the name de Sousa, from his paternal grandmother's family. The irregularities make it difficult to determine even today exactly which branch of the Magalhães family tree can rightfully claim the explorer.

*A*t twelve years of age, Ferdinand Magellan and his brother Diogo moved to Lisbon, where they became pages at the royal court; there Ferdinand took advantage of the most advanced education in Portugal, and he was exposed to topics as varied as religion, writing, mathematics, music and dance, horsemanship, martial arts, and, thanks to the legacy of Prince Henry the Navigator, algebra, geometry, astronomy, and navigation. Through his privileged position at court, Ferdinand came of age hearing about Portuguese and Spanish discoveries in the Indies, and he was privy to the secrets of the Portuguese exploration of the ocean. He even assisted with preparing fleets leaving for India, familiarizing himself with provisions, rigging, and arms.

Magellan seemed destined to become a captain himself, but in 1495, his patron, King João, the leader of a faction with only tenuous claims to the throne, suddenly died. João's successor, Manuel I, mistrusted young Magellan, who had, after all, been allied with the opposition. As a result, the fast-rising courtier found his career stymied. Although he retained his modest position at court, the prospect of leading a major expedition for Portugal seemed to vanish.

Finally, in 1505, after a decade of anonymous service at the palace, Ferdinand and Diogo Magellan received dual assignments aboard a mammoth fleet consisting of twenty-two ships bound for India, all of them under the command of Francisco de Almeida. Ferdinand Magellan spent the next eight years trying to establish a permanent Portuguese presence in India, dashing from one trading post to another, and from one battle to the next; he survived multiple wounds and, if nothing else, learned to stay alive in a hostile environment.

In this, the first phase of his career abroad, Magellan had displayed remarkable bravery and toughness, but in the end his foreign service proved a mixed adventure. He invested most of his fortune with a merchant who soon died; in the ensuing confusion, Magellan lost most of his assets. He petitioned King Manuel for restitution, but the king refused the request. After all those years of service abroad to the crown, all the dangers he had experienced, and the wounds he had received, his relationship to the court was no better than it had been when he left home years before.

*R*eturning to Lisbon, Magellan, still bristling with ambition, commenced a new phase in his career. Seeking to make himself useful to the crown, he involved himself with the Portuguese struggle to dominate North Africa. In 1513, he seemed to find an ideal opportunity to demonstrate his loyalty and usefulness to the crown when the city of Azamor, in Morocco, suddenly refused to pay its annual tribute to Portugal. The Moroccan governor, Muley Zayam, ringed the city with a powerful, well-equipped army. King Manuel responded to the challenge by sending the largest seaborne force ever to sail for his kingdom: five hundred ships, fifteen thousand soldiers, the entire military might of this small nation.

Among the hordes of soldiers sent to defend the honor of Portugal was Ferdinand Magellan, along with an aging steed, the only mount he could afford on his drastically reduced budget. He rode courageously into battle, only to lose his horse to the Arabs. What started so bravely for Magellan turned into a near disaster, as he barely escaped from the siege with his life. The larger picture was more favorable, as Portugal reclaimed the city, but Magellan remained indignant. He had lost his horse in the service of his country and king! And the Portuguese army was offering him only a fraction of what he considered to be his mount's true value as compensation.

Displaying a hotheadedness and tactlessness that bedeviled his entire career, Magellan wrote directly to King Manuel, insulting numerous ministers by circumventing their jealously guarded authority, and insisted on receiving full compensation for the horse. Manuel proved no more generous than he had been on the occasion of Magellan's previous demand for compensation of his lost investment. The new request was swiftly dismissed as a minor nuisance.

Magellan's reaction was telling; rather than quitting the field of battle in disgust, he stubbornly remained at his post, somehow acquired a new horse, and participated in skirmishes with the Arabs who swooped out of the desert wastes to harass Portuguese soldiers guarding Azamor. Magellan showed himself to be a fearless warrior, engaging in hand-to-hand combat with the enemy on a daily basis. In one confrontation, he received a serious wound from an Arab lance, which left him with a shattered knee and a lifelong limp; it also ended his career as a soldier. With his irrational idealism and loyalty, his wounds, and his unquenchable thirst for battle and righting perceived wrongs, Magellan came to seem like a real-life Don Quixote.

At last, he received a taste of recognition he craved when his service in battle and war wounds earned him a promotion to the rank of quartermaster. The position entitled him to a share of the spoils of war, which proved to be his undoing. In a subsequent battle, Arabs surrendered a immense herd of livestock, over 200,000 goats, camels, and horses. Magellan was among the officers responsible for distributing the spoils in an equitable fashion, and he decided to pay off tribal allies with some of the captured animals. As a result of this transaction, Magellan and another officer were indicted for selling four hundred goats to the enemy and keeping the proceeds for personal gain.

The charges were, on their face, preposterous. Magellan, as a quartermaster, was entitled to his spoils of war, and it was not clear that he received any. He failed to respond to the charges, and with-out authorization, left Morocco for Lisbon, where he appeared before King Manuel. Magellan did not apologize for his conduct in Morocco, but demanded an increase in the allowance he received as a member of the royal household, his *moradia*. Making a bad situa-tion even worse, he lectured the king, reminding him that he, Ferdi-nand Magellan, was a nobleman, and had rendered lifelong service to the crown, and had the wounds to show for it. Nothing but a more generous *moradia* would suffice to acknowledge his stature, his sense of honor, and his idealism. Jealous rivals whispered behind Magellan's back that his limp was merely an act designed to elicit sympathy.

King Manuel's judgment, when it came, was swift and sure: The insolent and foolish Magellan was to return to Morocco immedi-ately to face charges for treason, corruption, and leaving the army without authorization. This he did. After investigating the evidence, a tribunal in Morocco dismissed all charges against him, and he returned to Lisbon clutching a letter of recommendation from his commanding officer. Displaying superhuman stubbornness, Magel-lan went back to his sovereign king to demand the increased *moradia* with more vehemence than ever.

Once more, the king refused.

*M*agellan was entering middle age, with a bad leg and an un-fairly tarnished reputation. Short and dark, and teetering on the brink of poverty, he looked nothing like the aristocrat he thought himself to be. And he still yearned to distinguish himself in the ser-vice of Portugal, to make a name for himself that would rank with the important figures of the day, the explorers who opened new trade routes for Portugal in the Indies, and in the process became rich themselves. It seemed that Magellan was merely compounding his folly by asking the king who had refused to increase his *moradia* to back an entire expedition, but the would-be explorer saw matters differently. He was offering the king a scheme, admittedly a bit vague and risky, to fill the royal coffers with the wealth of the Indies.

Acknowledging that he needed help to persuade King Manuel, Magellan brought a prominent personage with him: Ruy Faleiro, a mathematician, astronomer, and nautical scholar. He was, in short, that quintessential Renaissance man, a cosmologist. Documents of the era always refer to Faleiro as a *bachiller*, in other words, a student (and perhaps also a teacher) at a university. Born in Covilhã, a town in mountainous eastern Portugal, Faleiro was a brilliant but unstable man who impressed his colleagues with his demonic personality; like many scholars of the day, he may have been a *converso*. He often worked closely with his brother Francisco, an influential scholar in his own right and the author of a well-regarded study of navigation, and it was likely that each Faleiro brother planned to play a major role in the expedition.

Despite the impressive credentials he brought to the venture, Ruy Faleiro had his own troubled history with King Manuel. The king had refused Faleiro's application to become a "Judiciary Astron- omer," and worse, had appointed a rival to a new chair in astron- omy at the University of Coimbra. So when Magellan and Faleiro presented themselves at court with their plan, the king was already prejudiced against both men, the stubborn, defiant Magellan and the mercurial Faleiro, men whose requests he had refused in the past.

*B*y the time Magellan made his case for a voyage, King Manuel, then fifty-one years old, had entered the throes of a personal crisis. He believed that his long reign was coming to an end; his adored wife had recently died in childbirth, and he decided to abdicate in favor of his son. But when the young man proved ungrateful, Manuel abruptly changed his mind, decided to remain on the throne, and arranged to marry his son's fiancée, Leonor, the twenty-

year-old sister of King Charles of Spain. All the while, it was
rumored, she continued to have relations with the boy, Prince João,
a situation that caused no end of scandal and derision at court. So
the Portuguese sovereign whom Magellan petitioned with his ambitious plan was a deeply suspicious, unhappy, and conflicted man—a
man determined to keep others from attaining fame and power.

Three times, Magellan asked for royal authorization for a voyage
to the Indies to discover a water route to the fabled but little known
Spice Islands. Three times, the king, who had disliked and mistrusted Magellan for more than twenty years, refused.

Finally, in September of 1517, Magellan asked if he could offer
his services elsewhere, and, to his astonishment, the king replied
that Magellan was free to do as he pleased. And when Magellan
knelt to kiss the king's hands, as custom dictated, King Manuel
concealed them behind his cloak and turned his back on his
petitioner.

*T*he humiliating rejection proved to be the making of Ferdinand
Magellan.

After he received the final rebuff from the Portuguese king, he
suddenly found direction in his life, and he moved quickly, carried
along by his own ambitions and by the tides of history. By October
20, 1517, he had arrived in Seville, the largest city in Andalusia, in
southwestern Spain. Ruy Faleiro, and possibly Francisco, joined him
there in December, the three of them forming a close-knit team of
Portuguese expatriates seeking their fortune in a rowdy and ener-
getic new land. Within days of his arrival, Magellan signed documents formally making him a subject of Castile and its young king,
Charles I. No longer was he Fernão de Magalhães; in Spain, he
became known as Hernando de Magallanes.

There was ample precedent for Magellan's emigration to Spain.
His boyhood hero, Christopher Columbus, had come to Spain from

Genoa to seek backing for the discovery of a route to the Indies, and, after years of delay and frustration, had finally won it from King Charles's grandparents, Ferdinand and Isabella. Magellan believed that he could do what the Genoese navigator had claimed to do but had never actually accomplished: reach the fabulous Indies by sailing westward across the ocean.

Tensions ran so high between Spain and Portugal that an international incident could result from an expedition following this route. Portugal had long been notoriously secretive about its empire, almost as secretive as the Arabs had been. By an edict of the Portuguese king, announced on November 13, 1504, anyone revealing discoveries or plans for missions of exploration could be executed. From 1500 to about 1550, not one book concerning Portuguese discoveries was published, at least in Portugal itself. Private individuals, during most of the sixteenth century, were not allowed to possess materials pertaining to the India trade and related subjects. Portuguese charts and maps were regarded as classified information and treated as state secrets. Had Magellan sailed on behalf of his homeland, his voyage around the world might have been lost to history.

Fortunately, the Spanish developed a different approach to empire building. Obsessive record keepers, given to documenting everything—laws, lineage, finances—they applied the same scrutiny to Magellan's voyage. Unlike the Portuguese or the Arabs, the Spanish proclaimed their exploits to stake their claim to various parts of the world. Furthermore, the Age of Discovery coincided with the discovery of movable type and the spread of printed books and pamphlets across Europe, supplemented by influential handwritten presentation copies compiled by professional scribes for the libraries of nobility. All of these accounts helped to spread the news of the discovery of the New World, and to reshape maps and popular conceptions of the globe.

Magellan brought with him many of Portugal's most precious and sensitive secrets: information about secret expeditions, a familiarity with Portuguese activity in the Indies, and an acquaintance with Portuguese navigational knowledge of the world beyond Europe. He was a rare breed, an explorer, schooled in the royal tradition established by Prince Henry the Navigator. But he needed a sponsor.

At eighteen, Charles I, king of Castile, Aragon, and Leon, was keenly aware of his august antecedents. He had preceded Magellan to Spain by only a year, and was as much of a stranger, or more. A Hapsburg, he had come of age in Flanders, drinking beer and speaking Flemish. He was now trying to learn the Spanish language and Spanish customs as quickly as he could. Endowed with a classic Hapsburg physique—tall, fair, with a hugely prominent chin—he towered above most of his subjects. He was trying to grow a beard to cover the broad expanse of his chin, and was on his way to becoming an accomplished horseman. It was said that he even participated in bullfights to display his valor.

His thirst for fame and glory had become apparent as soon as he arrived in Spain, and it was encouraged by his advisers, most of them highly placed officials of the Church who had been in power since the days of Ferdinand and Isabella and who saw in the young king the perfect vehicle for advancing their own ambitions. Less than a year after coming to Spain, Charles was elected king of the Romans, thanks to a great deal of string-pulling by the members of his family. The appointment meant that he would eventually become crowned King Charles V, emperor of the Holy Roman Empire, but to receive the title he would have to pay vast sums of money, essentially bribes, to the electors, who were based in Germany, and he looked to the Indies and the New World as a source of revenue to advance his personal ambitions. Explorers such as Magellan could be very useful to a young king in search of glory and in need of money.

*T*he timing of Magellan's arrival in Spain was auspicious, but his overall prospects were decidedly mixed. Although he possessed specialized knowledge and experience of the vast yet secret Portuguese empire, he was an unknown quantity to the Spanish court and ministers. He spoke Spanish haltingly, and relied on scribes for written communication in that language. He had renounced his loyalty to Portugal, but he remained an outsider in Spain, on probation and under suspicion. In these difficult circumstances, getting the financial backing for his proposed voyage would require a superhuman expenditure of effort and cunning, as well as a generous amount of luck. Spain, in this era, remained a feudal society ruled by a powerful, feared, and corrupt clergy. Bishops' illegitimate children, often referred to as "nephews" and "nieces," played prominent roles in public life. Cruelty, hypocrisy, and tyranny imbued the social order in which Magellan now found himself, but for the moment, he prospered by appealing to the Spanish court's longing to dominate world trade, and by insinuating himself into the country's power structure.

Soon after arriving in Seville, Magellan became acquainted with Diogo Barbosa, another Portuguese expatriate who had settled in this city fourteen years earlier and was now Knight Commander of the Order of Santiago, among his other distinctions. Diogo's nephew, Duarte, had sailed for Portugal, and Magellan was probably influenced by Duarte's accounts of his journey. At the same time, Magellan began to woo Diogo's daughter Beatriz; the relationship developed very quickly, and they married before the year was out. Suddenly, Magellan had an important sponsor in Seville, as well as a financial stake, because Beatriz brought with her a dowry of 600,000 *maravedís*. She might have been pregnant at the time of their marriage; the child, named Rodrigo, was born the following year.

*G*uided by the Barbosa family, Ferdinand Magellan prepared to persuade the powerful Casa de Contratación, or House of Commerce, to allow him to undertake his audacious voyage. Founded in Seville on January 20, 1503, by Queen Isabella, the Casa managed expeditions to the New World on behalf of the crown, and carried out its administrative chores with the bureaucratic zeal for which the Spanish were famous. At the time of its founding, the Casa de Contratación was housed near the Seville shipyards, in the Atarazanas, or arsenal, but to emphasize its authority, Queen Isabella moved it to the royal palace itself, the Alcázar Real. The Casa's role quickly expanded from collecting taxes and duties to administering all aspects of exploration, including registering cargoes and proclaiming rules for the outfitting of ships and their weapons. Within a few years of its founding, the Casa began giving instructions to captains, and imposing punishments for smuggling, which was ever present. Soon the Casa functioned as a maritime court, adjudicating contract disputes and insurance claims for all voyages to the New World. The Casa even administered cosmography, maintaining and updating the *padrón real*, or royal chart, which served as a master copy for charts distributed to all ships leaving Spain. By 1508, the Casa acquired a *piloto mayor*, pilot major or chief pilot, who administered a school of navigation to train navigators and sailors who wished to advance themselves. (The very first *piloto mayor* was Amerigo Vespucci, who gave his name to the Americas.)

The Casa de Contratación was controlled by one man, who was neither a navigator nor an explorer. Juan Rodríguez de Fonseca, the bishop of Burgos, had served as Queen Isabella's chaplain and had managed Columbus's expeditions even before the Casa came into existence. A cold, manipulative bureaucrat who jealously guarded his power, Fonseca made himself essential to all Spanish expeditions to the New World. Anyone wishing Spain's backing would have to obtain Fonseca's blessing—which, as legions of explorers would testify, was also a curse.

Columbus and Fonseca despised one another and fought bitterly. Fonseca was forever trying to get Ferdinand and Isabella and their successors to ignore the claims of independent entrepreneurs such as Columbus and to exert complete control over the expeditions that Spain sent to the New World. This meant, of course, that Fonseca would control the expeditions, and reap the full benefits from their trading. In the midst of the dispute, Columbus physically attacked Fonseca's accountant, kicking and assaulting him as a proxy for Fonseca himself. Nevertheless, Fonseca gradually exerted his will over Columbus, and by the time Magellan appeared on the scene, the balance of power over trading privileges had shifted decisively from the explorer to the crown. Magellan, and others in his position, would have to settle for what the crown granted them—still a fortune beyond imagining—rather than establishing their own foreign trading empires. There could be no expedition to the Spice Islands without the backing of Fonseca and his Casa de Contratación.

When Magellan approached representatives of the Casa de Contratación and declared that he believed that the Spice Islands were located within the Spanish hemisphere, he was telling them exactly what they wanted, indeed, *needed* to hear. Peter Martyr, a chronicler with access to the highest circles of the Casa, could barely conceal their gloating. "If the affair has a favorable outcome, we will seize from the Orientals and the King of Portugal the trade in spices and precious stones."

Still, the provisions of the Treaty of Tordesillas posed serious obstacles for the proposed expedition. Members of the Casa failed to see how Magellan could avoid trespassing on Portuguese interests by sailing west until he reached the East. Anticipating this objection, Magellan referred the distinguished members of the Casa to a clause in the Treaty of Tordesillas that allowed Spain or Portugal the freedom of the seas to reach lands belonging to one empire or the other. Such a clause was open to many interpretations, and Magellan might sail into conflict with Portugal if he attempted to take advantage of it.

Then there was the question of Magellan's nationality. The prospect of a Portuguese leading a Spanish expedition through Portuguese waters made nearly everyone at the Casa de Contratación uneasy; if the Portuguese became aware of the expedition, relations between the two countries might be strained to the breaking point. Yet the Casa's newest member looked at matters quite differently. Juan de Aranda, an ambitious merchant, took the Portuguese navigator aside and offered to lobby on behalf of the expedition in exchange for 20 percent of the profits. Privately, Magellan resented Aranda's intrusion into his scheme, but the merchant held out the best hope for keeping the expedition's prospects alive. And so Magellan agreed to cooperate.

Aranda wrote enthusiastically on behalf of Magellan, only to be reprimanded by the Casa de Contratación, which reminded him that he was not entitled to negotiate the terms of the expedition on his own. Worse, Magellan's comrade Ruy Faleiro was outraged to hear that Aranda had insinuated himself into the expedition and flew into a tirade so severe that it caused Magellan to back down. There was more to Faleiro's rage than simple indignation; it was a symptom of his growing mental instability. Aranda, for his part, attempted to apologize to Faleiro, and, despite the violent disagreement, contrived to obtain an audience for Magellan with King Charles in the city of Valladolid in north central Spain. It was here that Ferdinand and Isabella were married, and where Christopher Columbus died, and it now served as the capital city of Castile. And on January 20, 1518, Magellan, along with Ruy Faleiro and Ruy's brother Francisco, set out from Seville for Valladolid.

*M*agellan's arrival in the capital city coincided with a period of instability within the innermost circles of the royal court. Castile's regent, Cardinal Ximenes de Cisneiros, had just died on his way to assist the inexperienced King Charles, and poisoning was suspected. More than anyone else in Spain, the cardinal had ensured the safety of the newly arrived king, providing 32,000 soldiers to preserve order, but now he was gone, and young Charles sorely missed the

prelate's guiding hand. Instead, he relied on the advice of a group of
Flemish ministers for every decision. Guillaume de Croy, Seigneur
de Chièvres, perhaps the most able of the lot, had long served as
Charles's tutor, schooled him in the exercise of power, and jealously
guarded his authority over the lad. The young king's inner circle
also included Ximenes's successor, Chancellor Sauvage, and Cardinal Adrian of Utrecht. Despite his subsequent elevation to the
papacy as Adrian VI, this cardinal seems to have earned no one's
admiration. Wrote one nineteenth-century historian of Adrian: "Of
low extraction, and a person of weak character, his advancement
must always be regarded with wonder." Such were the men on whom
an immature king from a foreign culture, speaking a foreign tongue,
depended to make decisions concerning affairs of state.

Aranda obtained a meeting for Magellan with the king's Flemish
ministers to consider a proposal to assemble an expedition for the
Spice Islands. And Magellan came well armed for what would be the
most important meeting of his life. To begin, he offered tantalizing
letters from his friend Francisco Serrão, a Portuguese explorer,
describing the riches of the Spice Islands.

Serrão's odyssey began in 1511, when he assumed command of
one of three ships, dispatched by the Portuguese viceroy of India
and bound for the Spice Islands, using an easterly route. Surviving
shipwrecks and pirates, Serrão and several companions arrived at
Ternate, in the Spice Islands, the following year. In all likelihood,
they were the first Europeans to visit these fabled islands. Serrão
carefully cultivated Ternate's small ruling class, especially its king,
and tried to promote trade between Ternate and Portugal, but the
brisk transoceanic trade that he expected was slow to materialize.
Rather than giving up, Serrão stayed on. Surrounded by the scent of
drying cloves, soothed by the attentions of his newly acquired island
wife, he wrote beguiling letters to Magellan describing the extravagant beauty and wealth of the Spice Islands and inviting his friend
to visit and see for himself. "I have found here a new world richer

and greater than that of Vasco da Gama," he wrote. "I beg you to join me here, that you may sample for yourself the delights that surround me."

Magellan had every intention of visiting Serrão in the island paradise: "God willing, I will soon be seeing you, whether by way of Portugal or Castile, for that is the way my affairs have been leaning: you must wait for me there, because we already know it will be some time before we can expect things to get better for us." And when Magellan made a promise, he did everything in his power to keep it.

Significantly, Serrão's letters placed the Spice Islands far to the east of their true position; he located them squarely within the Spanish hemisphere, as defined in the Treaty of Tordesillas. This error might have been intentional, to disguise the Spice Islands' location from outsiders, but in any event his geographical legerdemain alleviated Spain's principal anxiety: Magellan's expedition to the Spice Islands must not violate the treaty.

To dramatize his mission, Magellan then displayed his slave of long standing, Enrique, who was believed to be a native of the Spice Islands. (This was not quite accurate, but in any event, Enrique could act as an interpreter.) According to one account, Magellan also brought another slave from the Indies, an attractive female from Sumatra who spoke many languages.

After presenting the slaves, Magellan spoke excitedly of his intention to sail along the eastern coast of what is now called South America until the land ended and he would be able to turn west toward the Spice Islands; he invoked the seven years he had spent in the service of Portugal, administering its empire and flourishing spice trade; and, to clinch his argument, he displayed a map or a globe (the wording in the original documents is ambiguous) depicting the route he planned to take. A crucial part of the map was obscured, however: the part that showed a waterway extending through South America toward the Spice Islands. Although Magellan, in his zeal to persuade the king's ministers to back the expedition, all but gave the strait's location away, he remained fearful that someone would steal his map and his strategy and launch a rival expedition before he could organize his own.

"Magellan had a well-painted globe in which the entire world was depicted," wrote Bartolomé de las Casas, a historian and missionary who took part in the meeting. "And on it he indicated the route he proposed to take." Reliable information about trade routes was so sensitive and precious that governments zealously guarded all maps and charts, which were essential to national security, and for Magellan to display a map likely purloined from Portugal was the equivalent of selling nuclear secrets at the height of the Cold War.

Magellan's conception of the world he planned to explore was fatally inaccurate. Like most explorers of the Age of Discovery, his ideas about the size of the globe, and location of landmasses, were inspired by Ptolemy. Had Magellan comprehended the size of the Pacific, its currents, storms, and reefs, it is unlikely that he would have dared to mount an expedition. But without the Pacific Ocean to inform his calculations, the estimated length of his route came to only half the actual distance. Magellan confidently predicted that it would take him at most two years to reach the Spice Islands and return to Spain with ships bulging with precious cargo. All he would have to do was find a way to get around or through South America, and he would be at the doorstep of the Indies. This was nearly the same mistake that Columbus had made over and over, during his four voyages. And it was a mistake that would be corrected only at the cost of great suffering and of many lives during the voyage Magellan now proposed.

After making his presentation to the ministers, Magellan was invited to discuss the proposed expedition in greater detail with Fonseca and Las Casas.

"I asked him what way he planned to take," the historian wrote, "and he answered that he intended to go by Cape Saint Mary, which we call the Río de la Plata, and from thence to follow the coast until he hit the strait."

Las Casas remained skeptical of Magellan's belief in the strait. "But suppose you do not find any strait by which you can go into the other sea?" he asked. Magellan told him that if he could not locate the strait, he "would go the way the Portuguese took." Although Magellan sounded ready to contravene the terms of the Treaty of Tordesillas, the king and his advisers were too intrigued to turn him away. "This Ferdinand Magellan must have been a man of courage and valiant in his thoughts for undertaking great things," Las Casas marveled, "although he was not of an imposing presence because he was small in stature and did not appear in himself to be much, so that people thought they could put it over him for want of prudence and courage."

In Magellan's case, appearances were deceiving. His ideas were big enough, and promised to be lucrative enough, to convince King Charles and his powerful advisers to back them.

*I*mmediately after the meetings at Valladolid, the potential co-leaders of the expedition presented a list of demands to the crown; they were couched in respectful language, but they were demands nonetheless. They included an exclusive franchise on the Spice Islands for a full ten years, 5 percent of the rent and proceeds "of all such lands that we would discover," and the privilege of trading for their own accounts, so long as they paid taxes to the king. They asked to keep any "islands" they discovered for themselves, if they discovered more than six, as well as permission to pass the newly discovered lands on to "our heirs and successors."

Magellan's insistence on a ten-year exclusive on voyages to the Spice Islands appeared preposterous in a fast-changing world, but he was concerned that Spain would send duplicate expeditions as soon as his was out of sight of land, expeditions guided by his theories and secrets, expeditions that might succeed if he failed. Magellan was right to insist on this point, although he was powerless to enforce it.

*O*n March 22, 1518, King Charles, from his royal seat in Valladolid, offered Magellan and Faleiro a contract "regarding the discovery of the Spice Islands." The document was a charter to discover a new world on behalf of Spain. "Inasmuch much as you, Bachelor Ruy Faleiro and Ferdinand Magellan, gentlemen born in the Kingdom of Portugal, wishing to render us a distinguished service, oblige yourselves to find in the domains that belong to us and are ours in the area in the Ocean Sea, within the limits of our demarcation, islands, mainlands, rich spices," it began, "we order that the following contract with you be recorded."

In the first clause, King Charles appeared to accede to Magellan's insistence on a ten-year exclusive: "Since it would be unjust that others should cross your path, and since you take the labors of this undertaking upon yourselves, it is therefore my wish and will, and I promise that, during the next ten years, I will give no one permission to go on discoveries along the same regions as yourselves." Nevertheless, he did not honor this promise. Just as Magellan feared, King Charles dispatched a follow-up expedition to the Spice Islands only six years after Magellan's departure from Spain. The Spice Islands were too valuable to entrust to the luck and skill of a single explorer.

King Charles enjoined Magellan and Faleiro to respect Portugal's territorial rights under the terms of the Treaty of Tordesillas: "You must so conduct this voyage of discovery that you do not encroach upon the demarcation and boundaries of the Most Serene King of Portugal, my very dear Uncle and Brother, or otherwise prejudice his interests, except within the limits of our demarcation." He reminded Magellan of the delicate diplomatic and family situation complicating the rivalry of Spain and Portugal for mastery of the seas and of world trade. Portugal's sovereign, King Manuel, had married not just one but *two* of Charles's aunts, first Isabel and then María. And now he was planning to marry

Charles's sister, Leonor, within a matter of weeks. The family ties, with their complex web of sentiment and formality, kept Spain and Portugal from all-out war with each other, but they did not extinguish the intense rivalry between the two nations; they drove it underground or into the diplomatic realm, where it was no less fierce.

Charles I had every intention of overtaking the elderly king of Portugal. No matter what the language of the contract seemed to say, the impatient young king wished to bend the terms of the Treaty of Tordesillas to Spain's advantage by insisting that the Spice Islands lay within the Spanish hemisphere. And if it was impossible to prove this point, then it was equally impossible to disprove it. To succeed, Magellan's expedition need only give Spain a reasonable argument for claiming the Spice Islands.

From Magellan's standpoint, this was a remarkable contract because it gave him nearly everything he had wanted. The grant of lands, for instance, proved more generous than Magellan had any right to expect. "It is our wish and our will that of all the lands and islands that you shall discover, to grant to you . . . the twentieth part, and shall besides receive the title of Lieutenant and Governors of the said lands and islands for yourselves and your sons and heirs in freehold for all times, provided that the supreme [authority] shall remain with Us and with the Kings who come after Us." Magellan would find his name all over the newly redrawn maps of the world, maps depicting lands that he not only discovered, but *possessed:* Magellan islands, Magellan lands, entire realms belonging to Ferdinand Magellan and his legitimate male heirs. The world, or at least a significant part of it, could be his.

From Fonseca's point of view, this was hardly an advantageous contract, for it gave Magellan too much power over the expedition. It would take Fonseca months, but eventually he would have his revenge on Magellan, and exert the control over the expedition that had been denied him in the royal contract.

*K*ing Charles also promised Magellan five ships: "two each of 130 tons, two each of 90 tons, and one of sixty tons, equipped with crew, food, and artillery, to wit that the said ships are to go supplied for two years, and the other people necessary." The fleet would be called the Armada de Molucca, after the Indonesian name for the Spice Islands.

The ships were mostly black—pitch black. They derived their blackness, and their ominous aura, from the tar covering the hull, masts, and rigging, practically every exposed surface of the ship except for the sails. Their sterns rose high out of the water, towering as much as thirty feet over the waves, so high that a man standing on the stern deck seemed to rule the sea itself. Their height exaggerated movement; even in relatively calm water, they tossed the men about like toy figures.

The ships were among the most complicated machines of their day, wonders of Renaissance technology, and the product of thousands of hours of labor by skilled artisans working at their specialized trades. They were relatively small, out of necessity. One of Seville's limitations as a port was the shallowness of the Guadalquivir River; ships had to be sufficiently small and light to negotiate the narrow waterway to the Atlantic. Thus Magellan's flagship, *Trinidad*, weighed 100 tons; *San Antonio*, which carried many of the provisions, weighed 120 tons; *Concepción* 90 tons, *Victoria* 85 tons; and *Santiago*, to be used for reconnaissance, weighed just 75 tons.

With the exception of *Santiago*, a caravel, the ships were all classified as *naos*, a term that simply meant ships. No illustrations of them have survived, so it is difficult to determine exactly how they were configured, but accounts from Magellan's time mention their daunting stern castles, their multiple decks, and the profusion of *obras muertas*, or "dead wood," to ornament the officers' quarters. Each ship had three masts, one of which carried a lateen sail.

*A*lthough King Charles was supposed to pay for Magellan's ships, according to the contract, he was deeply in debt. To cover the expedition's cost, the Casa de Contratación turned to a familiar presence in financial circles, Cristóbal de Haro, who represented the House of Fugger, an influential banking dynasty based in Augsburg, Germany. Haro's name derived from the city of Haro, in north central Spain. Haro (the city) flourished as a center of winemaking, and it also sheltered a community of Jewish goldsmiths and bankers until a civil war broke out in the fourteenth century and drove the Jews from their homes. Many of the persecuted Jews adapted by becoming *conversos*, adopting Christian-sounding names, Cristóbal de Haro's ancestors among them.

For years, Haro served as the Fuggers' man in Lisbon, trading in spices, lending money for secret Portuguese expeditions, and forging friendships with many of the great explorers of the era, including Bartolomeu Dias. His familiarity with secret Portuguese expeditions, or with tantalizing rumors about their findings, gave him privileged information concerning the existence of a strait leading through the American landmass to the Indies—the same possibility that animated Magellan's furious desire to explore the East. Following a bitter dispute with King Manuel, Haro left Lisbon for Seville, where he renewed his acquaintanceship with Magellan, and combined their enthusiasm for a search for the strait.

For an explorer in need of financial backing, Cristóbal de Haro was the ideal friend; the House of Fugger, for which he worked, had enough money to finance ten expeditions, or more; indeed, it had more money than King Charles. By bringing in Haro, the king and his advisers would be giving up a significant amount of the profits. Given the hazards of the spice trade, and the uncertainty of long ocean voyages, financiers like Haro could be induced to risk their capital on such ventures for only one reason: the lure of extraordinary profits. If successful, or even partly successful, a fleet returning

from the Indies could yield a profit of 400 percent; the more pragmatic Haro estimated that Magellan's expedition could yield a
profit of 250 percent. Meanwhile, he advanced money at an interest
rate of 14 percent.

The official accounting of the expedition put the cost at
8,751,125 *maravedís*, including the five ships, provisions, salaries
paid out in advance, and fittings for the ships. Magellan's pay came
to 50,000 *maravedís*, and an additional 8,000 *maravedís* each month.
By royal order, his monthly salary went directly to his wife, Beatriz.

Of the overall cost of the expedition, the king's share came to
6,454,209 *maravedís*, much of it provided at high interest by Haro.
Although royal documents place Haro's contribution to the great
enterprise at a modest 1,616,781 *maravedís*, that number is deceptive. Because his backers, the House of Fugger, also financed expeditions for Portugal, they probably concealed the full extent of their
contribution by loaning additional money to the king.

In a final piece of official business, King Charles conferred the
title of captain on both Magellan and Faleiro. Given the hazards of
exploration, it was not unusual for expeditions in the Age of Discovery to have co-captains, but in this case, the arrangement unintentionally sowed the seeds of bitter disputes at sea. The powers
granted to the pair were sweeping and unequivocal. "We order the
master and boatswains, pilots, seamen, ship boys and pages, and any
other persons and officials there may be in the said fleet, whatever
persons who are and reside in the said lands and islands to be discovered ... that they shall regard, accept, and consider you as our
Captains of the said fleet. As such, they shall obey you and comply
with your orders, under the penalty or penalties which, in our
name, you shall impose." As the language made clear, Magellan
and Faleiro had absolute authority at sea. "We authorize you to execute sentence on their persons and goods. ... If during the voyage
of the said fleet there should arise any disputes and conflicts, at
sea as well ashore, you shall deliver, determine and render justice

with respect to them, summarily and without hesitation nor question of law."

Magellan could only have marveled at the speed with which his plan to reach the Spice Islands had come together. King Charles risked Spain's authority and reputation on the expedition, and the backers risked their capital, but Magellan would risk even more: his very life.

CHAPTER II

The Man Without a Country

The Sun came up upon the left,
Out of the sea came he!
And he shone bright, and on the right
Went down into the sea.

When word of Magellan's spectacular commission reached Portugal, King Manuel reacted with alarm. The navigator had betrayed them all, and the members of the royal court were at a loss to understand why he had done so. The Portuguese court historian João de Barros, who had a passing acquaintance with Magellan, contended that a demonic force had possessed the navigator: "Since the devil always maneuvers so that the souls of men entertain evil deeds in whose undertaking he shall perish, he prepared this occasion for this Ferdinand Magellan to become estranged from his king and his kingdom, and to go astray." No one in Portugal dared to admit the actual reason for Magellan's behavior, that King Manuel had refused to back the navigator, humiliating him over and over again.

King Manuel did what he could to ruin Magellan's name while, at the same time, trying to lure Magellan and Faleiro back to Portugal. He involved the Portuguese ambassador to King Charles's court,

Álvaro da Costa, who sought out the two exiles, promising that King Manuel would reconsider their request for an expedition. Da Costa was explicit about the dire consequences that would befall the two if they continued with their plan to sail for Spain; they would offend God, King Manuel, and relinquish all personal honor. Nor would matters end there; their families and heirs would suffer, and they would upset the delicate truce between Spain and Portugal at the very moment that King Manuel was planning to marry King Charles's sister, Leonor.

Magellan refused to be swayed by the ambassador's entreaties. He suspected that if he returned to Portugal he would be thrown into jail, tried for treason, and executed. Summoning all his meager diplomatic skills, Magellan replied that he had formally renounced his allegiance to King Manuel and given his loyalty to King Charles. He had no obligation to serve anyone else.

 Frustrated by Magellan's stubbornness, Álvaro da Costa appealed to King Charles himself. "Your Highness has plenty of vassals for discoveries without having to turn to those malcontents," he argued. Uncertain about how to handle the matter, King Charles turned to his advisers for guidance, and they reiterated their position that the Spice Islands lay in the Spanish hemisphere, and Magellan's expedition would not violate the Treaty of Tordesillas. King Charles followed the advice, and Magellan and Faleiro retained his backing in spite of pressure from Portugal.

Da Costa tried to put the best face on his failed attempt at diplomacy. He wrote to King Manuel that Magellan and Faleiro actually wished to return to Portugal, but King Charles prevented them from doing so. Da Costa probably believed his letter would remain confidential, but its contents became known, much to the outrage of King Charles. Ultimately, da Costa's false claims hurt Portugal's cause, and hardened King Charles's determination to stand by his two embattled explorers. Portugal's attempt to attack

Magellan confirmed the belief of King Charles's advisers that they had hit on a scheme of great strategic value. Yet relations between the two neighboring countries were more complicated than they appeared. Despite all the tension between them, King Manuel proceeded with his plans to marry Charles I's sister, Leonor, according to a contract dated July 16, 1518. In so doing, rivals for the control of world trade would be yoked by marriage. Instead of ending the strife, the impending union pushed the conflict off-shore. Rather than competing head to head on the Iberian penin-sula, Spain and Portugal would grapple for control of trade routes around the world. They remained simultaneously rivals and allies, as affairs of state and matters of the heart alternated in rapid succession.

Four days after King Manuel completed his nuptial arrange-ments, the Spanish monarch instructed the Casa de Contratación to proceed with Magellan's expedition to the Spice Islands without delay. Magellan and Faleiro were to receive money to begin their preparations, and they were ordered to Seville to outfit their ships.

City of Gold. City of Water. City of Faiths. "*Quien no ha visto Sevilla*," runs a saying, "*no ha visto maravilla.*" "Who has not seen Seville, has not seen wonder." For centuries, Seville, the preeminent city of Andalusia, has held Spain in its thrall. "I have placed Seville, or rather God has placed her, as the mother of all the cities and cen-ter of the glory and excellences of that territory," wrote an early his-torian of the city, "for it is the most populous and greatest of her capitals." Now, at the height of the Age of Discovery, Seville hov-ered at the apex of its prosperity and influence. The city straddling the Quadalquivir River was an amalgam of Roman, Visigoth, Mus-lim, Jewish, and Christian cultures. Its fame reverberated through-out the known world, borne on ships to destinations only vaguely located on maps. Throughout Europe, only Venice, Naples, and Paris were larger; Seville, with a population of about 100,000, was on a par with Genoa and Milan, each of them a thriving trading center; London, the largest city in Britain, claimed only half as many inhabitants as boisterous Seville.

Above all, Seville was a commercial center, "well adapted to every profitable undertaking, and as much was brought there to sell as was bought, because there are merchants for everything," in the words of a sixteenth-century observer. "It is the common homeland, the endless globe, the mother of orphans, and the cloak of sinners, where everything is a necessity and no one has it." Only Seville was capable of providing Magellan with the technology, the labor, and the financial resources to travel halfway around the world in search of lands to claim and spices to bring back to Europe.

It was also a city of faith, the home of the third largest church in the world, after Saint Peter's in Rome and Saint Paul's in London. Work on Seville's cathedral continued for well over a hundred years, until 1519, the year Magellan set sail for the Spice Islands. With its bell tower, vaults, chambers, and fantastic amalgam of Gothic, Greco-Roman, and Arabic architecture, the cathedral became the expression of Seville's striving, a world unto itself. The flame of the Catholic faith burned most brightly in Seville during Semana Santa, Holy Week, lasting from Palm Sunday to Easter Sunday, when solemn, almost frightening processions of religious penitents coursed through the city's narrow, winding streets and capacious squares. The penitents walked barefoot over the sharp stones and splinters embedded in the streets, bearing a wooden cross, their feet bleeding, displaying their wounds in emulation of Jesus. This was an act of piety straight out of the Middle Ages, a demonstration of blind obedience to an omnipotent Lord, a recognition of and mastery over mortal suffering, and an acknowledgment of humankind's sinful state. As such, it served as good practice for the rigors and pains of a voyage of discovery.

*A*s Magellan and Faleiro arrived in Seville to commence preparations in earnest for the voyage, the ill will between Spain and Portugal led to rumors that the lives of the Portuguese co-commanders
were in danger. It was said that Bishop Vasconcellos, a confidant of
King Manuel, inspired an assassination attempt. Magellan was
inclined to ignore the death threats, but King Charles took the
intimidation so seriously that he provided bodyguards for Magellan
and Faleiro, granted them another audience, proclaimed them
Knights of the Order of Santiago, and reaffirmed the terms of their
original commission. Having done all he could to demonstrate his
support of the two Portuguese, King Charles urged them to begin
their expedition as soon as possible. Time was short, and an empire
was at stake.

*S*omething has come up," Magellan wrote to King Charles on
Saturday, October 23, 1518, in the midst of outfitting the fleet for
the voyage. Unlike many captains, Magellan involved himself in the
day-to-day preparations, even loading goods onto the ships as if he
were an ordinary seaman, not the Captain General, and that was
how the trouble started. Despite his close interaction with the
sailors and dockworkers, or perhaps because of it, Magellan believed
he did not receive the cooperation and respect to which he was
accustomed. In desperation, he appealed to the one individual who
could restore order.

It is possible that Magellan's problems stemmed at least in part
from his inadequate Spanish; time and again, he had to rely on
translators, and his inability to communicate underscored his outsider status. Even now, writing to King Charles, he had to rely on a
scribe, "because I still do not know how to write in Spanish as well I
should." He proceeded to explain the matter. "I had to haul one of
the ships to shore because there was an ebb tide. I got up at three in

the morning to make sure that the riggings were in place and when it was time to work I ordered the men to put up four flags with my coat of arms on the mast where those of the captain are customarily placed, while those of Your Majesty were to be placed on top of *Trinidad,* which is the name of the ship." The unusual juxtaposition of signs, emphasizing that a Portuguese captain was sailing for Spain, attracted a large, gossipy crowd of onlookers. "Because in this world there is never any lack of envious people, they began to talk. They said I had done wrong in putting up my coat of arms on the cap- stan." The crowd thought that the four flags containing Magellan's coat of arms signified those of the king of Portugal. Resentment boiled along until a functionary ordered Magellan to remove the offending flags. "I approached him and told him that those flags were not of the King of Portugal but my own, and that I was a vassal of Your Majesty." And he refused to remove the flags. Another Spanish official approached Magellan with the same demand. No, Magellan explained, he would not take down the flags.

As Magellan was explaining all this to the official, the man who had first approached him, "without any warning and without any authority to do so...came up the steps calling the people to seize the Portuguese captain who had put up the flags of the King of Por- tugal." He demanded to know why Magellan chose to display these flags, and Magellan, not surprisingly, refused to explain.

At that moment, chaos erupted. The insolent functionary "called military officers to seize me and laid his hands on me, shouting that they would seize me and my men." Worse, "There were some who showed their intentions to harm my men rather than to help us do what was for the service of Your Majesty." At that point, the two officials who had challenged Magellan got into a fight with each other over how to treat Magellan. The workmen outfitting *Trinidad* quickly fled, as did a number of the sailors, further exasperating Magellan, who stood by helplessly as he watched the local officials disarm the sailors, and even arrest several of them and lead them away to prison. In the struggle, one of Magellan's pilots was stabbed as he was going about his work. Although Magellan escaped harm, his dignity and authority had suffered a blow. To make matters worse, the fight had occurred in the open, under the watchful eye of a Portuguese spy, who would carry news of the brawl back to Lisbon.

"Because I believe that Your Majesty does not approve of mal-treating men who leave their kingdom and their own kind to come and serve Your Majesty," Magellan wrote, "I ask you most humbly to decide what would be best for your service. Whatever Your Majesty orders would give me utmost satisfaction because I con-sider the affront done me not an affront done to one Ferdinand Magellan, but done to a captain of Your Majesty."

Magellan's fury at the incident was understandable. An exile, he enjoyed the protection of King Charles, but in reality he was at the mercy of the mob and self-appointed busybodies. If he could not maintain order here on the quay at Seville, how would he lead men on the perilous journey across an uncharted ocean to the Spice Islands? And if there was another uprising on a distant shore, where it would be impossible for him to summon the king's help, how would matters stand?

Within days of receiving Magellan's letter, King Charles demonstrated his loyalty and punished the offenders—those who had boarded *Trinidad*, stabbed the pilot, seized Magellan him-self—and arrested the sailors. Preparations for the voyage contin-ued, but the flag incident served as a warning to Magellan that his men, especially the Spaniards, posed a danger as great as the sea itself.

On April 6, 1519, the king sent orders to another officer, Juan de Cartagena—orders that became the most controversial aspect of the entire expedition—to serve as the inspector general of the fleet under the command of the two Portuguese commanders. Yet his salary was considerably more than Magellan's, the highest of any in the fleet: 110,000 *maravedís*. Essentially, Cartagena was to have the final say over all commercial aspects of the expedition; he was the chief accountant and representative of the king's treasury. "You must see to it that a book is kept in which you will make entry of all that is loaded in the holds. These things must be marked with your mark, each different class of merchandise being by itself; and you must designate particularly what belongs to each person, because, as will be seen later, the profits must be allotted at so much to the pound, in order that there may be no fraud."

That was not all. It would be Cartagena's job to "see to it carefully that the bartering and trading of said fleet is done to the greatest possible advantage to our estates." He would have to check every entry in every book, and once it met his approval, sign off on it. At every step, he was to exercise "much care and vigilance." And it was certainly in his interest to do so, because he had invested his own funds in the expedition. This provision made Magellan responsible to Cartagena for all commercial decisions. The wording ("to the greatest possible advantage to our estates") enabled Cartagena to step in at any moment and prevent Magellan from enriching himself, even if he believed he was entitled to do so under his own contract with King Charles. Implicitly, these new instructions to Cartagena took precedence over the prior arrangement.

There was more. Cartagena was to function as the eyes and ears of the king throughout the voyage. "You will advise us fully and specifically of the manner in which our instructions and mandates are complied with in said lands; of our justices; of the treatment of natives of said lands; . . . [and] how said captains and officers"—meaning, especially, Magellan and Faleiro—"observe our instructions, and other matters of our service." If the co-commanders were negligent in any way, Cartagena was to report their behavior to the Casa de Contratación in writing. The instructions were so thorough that a Spaniard predisposed to mistrust Magellan and Faleiro could conclude that he, and not they, had the final say on the conduct of the entire voyage.

And that was exactly the conclusion to which Cartagena came.

*E*ven as he undercut Magellan's authority, King Charles remained concerned about the possibility of open conflict between Spain and Portugal and tried his hand at personal diplomacy. Writing to Manuel from Barcelona in February 28, 1519, Charles confessed, "I have been informed by letters which I have received by persons near you that you entertain some fear that the fleet which we are dispatching to the Indies, under the command of Ferdinand Magellan and Ruy Faleiro, might be prejudicial to what pertains to you in those parts of the Indies"—which was putting it mildly. Charles continued. "In order that your mind may be freed from anxiety, I thought to write to you to inform you that our wish has always been, and is, duly to respect everything concerning the line of demarcation which was settled and agreed upon with the Catholic king and queen my sovereigns and grandparents." And he vowed that "our first charge and order to the said commanders is to respect the line of demarcation and not to touch in any way, under heavy penalties, any regions of either lands or seas which were assigned to and belong to you by the line of demarcation."

*O*n May 8, 1519, amid frenzied preparations for departure, King Charles delivered his final instructions for the voyage to Magellan and Faleiro, instructions so detailed that the two commanders might have thought the king would be coming along with them in one of the ships.

Magellan and Faleiro were ordered to record every landfall and landmark they attained, and if they came across any inhabited lands, they were to "try and ascertain if there is anything in that land that will be to our interest." They were also to treat humanely any indigenous peoples they happen to find, if only to make it possible for the fleet to assure its supply of food and water. Magellan could seize any Arabs he found in the Portuguese hemisphere—an implicit admission that he might violate the Treaty of Tordesillas, after all—and, if he wished, sell them for slaves. In contrast, if Magellan came across Arabs in the Spanish hemisphere, he was to treat them well and to make treaties with their leaders. Only if they were belligerent could Magellan subject them to punishment, as a warning. But this was not in any sense a slaving expedition. Magellan was to go in search of spices and lands, and nothing else, and when he reached the Spice Islands, his instructions were to "make a treaty of peace or commerce with the king or lord of that land" before he attempted to load the goods onto his ships.

Although the king warned Magellan to be careful in his dealings with Indians ("Beware you do not trust the natives because sometimes, on account of going unarmed, disasters happen"), the orders insisted that Magellan treat them fairly. "You shall not cheat them in any way, and . . . you shall not break [the deal]. . . . You shall not consent in any manner that any wrong or harm be done to them . . . ; rather, you shall punish those who do harm."

And on a sensitive point, Magellan and Faleiro had to see to it that the crew members had no contact with local women. "You shall never consent to have anyone touch a woman . . . the reason being that in all those parts the people, on account of this thing over and above all, might rebel and do harm." That order would prove impossible to enforce, as would another clause prohibiting the use of firearms; members of the expedition were forbidden to discharge them in newly found lands lest they terrify the Indians on whose goodwill they would depend. It was another well-intended but impractical edict; if the men had weapons, they would use them.

The orders spelled out what Magellan should do in the event that one of the ships became separated from the fleet: "They should wait a month at the place agreed before and leave a sign which will consist of five rocks put on the ground forming a cross on both sides of the river, and another cross of sticks. You will also leave something written in a receptacle buried in the ground indicating the time and date the ship came by."

The orders also covered minor, but important matters. Sailors, for instance, had permission to write whatever they wished in their letters home; fortunately for future historians of this voyage, there was to be no censorship. Blasphemy, on the other hand, was for-bidden aboard ship (Magellan found it impossible to enforce this directive), as were card and dice players. Since these items were ubiquitous aboard the ships of discovery, it is unlikely that Magel-lan bothered to discourage games of chance, but he may have suc-ceeded in preventing professional gamblers and cardsharps from enlisting in the crew and fleecing the other crew members during the voyage.

In addition to Magellan and Faleiro, the co-commanders of the expedition, a copy of these orders was sent to Cartagena, in his capacity of inspector general. And it is possible that Cartagena, with his exalted impression of his role, took these orders to mean that the king of Spain now considered him equal in rank to Magellan.

King Manuel made one more attempt to subvert the Armada de Molucca. He sent his agent, Sebastián Álvares, a factor, to Seville, with orders to undo Magellan's resolve. On July 18, 1519, Álvares secretly reported that the officials at the Casa de Contratación "can-not stomach" Magellan. The artful spy spoke vaguely but provoca-tively of disputes between the Casa de Contratación and Magellan concerning the sailors' salaries. And he recounted his own efforts to persuade Magellan to call off the expedition: "I went to the lodgings of Magellan, where I found him arranging baskets and boxes with victuals of preserves and other things"—delicacies to be enjoyed by the expedition's leaders rather than the crew. Álvares began a care-fully rehearsed argument to persuade Magellan to abandon his nefarious scheme. I wished to recall to his memory how many times, as a good Portuguese and his friend, I had spoken to him, and opposed the great error he was committing. . . . I always told him . . . that he should see that this road has as many dangers as a St. Catherine's Wheel." According to the oft-repeated legend, Emperor Maxentius, a fierce pagan, captured a young convert to Christianity named Catherine in A.D. 305. It was said that fifty philosophers tried to persuade her that her belief in Christianity was foolish, but Catherine, despite her youth, confounded their arguments and converted them to the faith. Maxentius ordered those unlucky philosophers put to death, and Catherine was sent to prison, where the emperor's wife visited her and also converted to Christianity. At that, the emperor decided that Catherine herself must die. He ordered a wheel embedded with razors to be constructed; Catherine was bound to its rim, but instead of slicing her to pieces, the wheel shattered, and its splinters and razors injured the onlookers. In despair, the emperor finally ordered Catherine beheaded. If Magellan did not desire to suffer the fate of Catherine, Álvares urged him to "return to his native country and the favor of your Highness, where he would always receive benefits."

Magellan replied that he was committed to Spain, and nothing could change his mind.

Álvares's practiced reply would have unnerved a weaker soul than Magellan. "I said to him, that to acquire honor unduly, and when acquired by such infamy, was neither wisdom nor honor . . . for he might be certain that the chief Castilians of this city, when speaking of him, held him to be a vile man, of low blood, since to the disservice of his true king and lord he accepted such an enterprise." Furthermore, "He might be sure that he was held to be a traitor in going against the State of your Highness." Every term of opprobrium that Álvares hurled at Magellan strengthened the mariner's resolve to carry out his mission. Even Álvares was impressed by Magellan's conviction. "It seemed to me that his heart was true as to what befitted his honor and conscience."

Despite his resolve, Magellan suffered pangs of conscience over his decision to abandon his homeland. "He made a great lamenta-tion," Álvares observed, "but that he did not know of anything by means of which he could reasonably leave a king who had shown him so much favor. I told him . . . that he should weigh his coming to Portugal."

Leaving Magellan to his torment, Álvares tried to persuade him-self and King Manuel that the expedition would never come to pass. He counted on the once brilliant Faleiro's deteriorating mental state to aid the Portuguese skulduggery. "I spoke to Ruy Faleiro on two occasions," Álvares reported to his sovereign. "It seems to me that he is like a man deranged in his senses. . . . It seems to me that, if Fer-dinand Magellan were removed, Ruy Faleiro would follow whatever Magellan did."

If the fleet somehow managed to depart, Álvares advised that the five ships were barely seaworthy. "They are very old and patched up, for I saw them when they were beached for repairs. It is eleven months since they were repaired, and they are now afloat, and they are caulking them in the water. I went on board [one of them] a few times, and I assure Your Highness that I should be ill inclined to sail in them to the Canaries." These islands were only a few days' sail from the Iberian coast, and if the ships could not be trusted to sail that far, how could they possibly reach the Indies?

Álvares went on to boast that he knew what course the fleet planned to follow. Once the ships crossed the Atlantic, if they crossed it, Brazil would remain "on their right hand" as they sailed to the line of demarcation dividing the Spanish and Portuguese halves of the world. He erroneously informed the king that the fleet would then sail across open water west and northwest to the Spice Islands. "There are no lands laid down in the maps which they carry with them," Álvares noted with malicious glee. "Please God the

Almighty that they make such voyage as did the Côrte Reals"—
Portuguese explorers whose fleet had sunk without a trace.

Of all the problems Álvares recounted, the most serious was Ruy
Faleiro's fragile mental state. Since leaving Portugal, and perhaps
even before, the brilliant cosmographer had exhibited signs of instability. One acquaintance said that Faleiro "sleeps very little and wanders around almost out of his mind." Others remarked on his
irritability or simply stated that he had lost his mind. The evidence,
fragmentary though it is, suggests that Faleiro was suffering from
bipolar disorder or some form of extreme depression. Magellan
remained silent on the subject of his colleague's condition, but all
around him Spanish officials commented on the danger of taking
Faleiro in his unstable condition on a long and trying voyage. What
if he went mad, misused his authority as co-admiral, and endangered the entire expedition?

Even King Charles took note of Faleiro's condition, and on July
26, 1519, issued a royal certificate declaring that Faleiro would not
sail with Magellan. Instead, the cosmographer would remain in
Seville to prepare for another expedition that would follow in Magellan's wake. This violation of the ten-year exclusive King Charles
had granted Magellan was more likely a face-saving gesture
designed to preserve what little dignity Faleiro had left, for he never
went to sea.

Magellan seemed relieved to rid himself of the unstable Faleiro;
he agreed to the removal as long as the fleet could keep the cosmographer's precious, state-of-the-art navigational instruments, which is
exactly what occurred. Faleiro's trove consisted of thirty-five compasses, supplemented by an additional fifteen devices that Magellan
purchased in Seville; a wooden astrolabe constructed by Faleiro himself; six metal astrolabes of a more common variety; twenty-one
wooden quadrants; and eighteen hourglasses, some of which Magellan purchased himself. Then there were the charts, twenty-four in
all, most of them top secret, all of them extremely valuable. An

unauthorized individual caught with a chart could be punished severely, even with death. They were kept under lock and key, and under armed guard. Of the total number of charts, six had been drafted by Faleiro. Eighteen others were the work of the cosmographer Nuño Garcia, seven of these under the direction of Faleiro, and eleven more under the direction of Magellan. All of these precious items remained with the armada, at Magellan's disposal. The fleet also carried quantities of prepared blank parchment, as well as dried skins to make still more parchment, if necessary, for additional maps.

So the team of Magellan and Faleiro, the driving force behind the expedition since their days in Lisbon, was sundered. In reality, the architect of Faleiro's removal was probably Fonseca rather than King Charles. As head of the Casa de Contratación, Fonseca had long been looking for a way to alter the arrangement whereby two Portuguese commanded the expedition, and Faleiro's illness provided just the excuse he needed. There is a story that Fonseca artfully provoked a quarrel between the two Portuguese comrades by entrusting the royal standard to Faleiro, indicating that he, not Magellan, would be the Captain General of the fleet. Magellan was said to have become so incensed that he requested Faleiro's removal from the enterprise, and Fonseca was only too glad to comply.

Fonseca replaced the would-be explorer with Andrés de San Martín, a well-connected Spanish cosmographer and astrologer who had long sought a lofty role for himself in the Casa's affair. San Martín occupied a prestigious position in the roster, and commanded a generous salary—an advance of 30,000 *maravedís,* plus an additional 7,500 to cover expenses—but he did not hold Faleiro's exalted rank. Faleiro had dazzled the Spanish with his brilliance, passion, and his aura of mysticism. San Martín, in contrast, was a fully qualified astronomer and astrologer who enjoyed the respect of the Spanish authorities, and nothing more.

The removal of Faleiro opened the way for Cartagena, the

inspector general, to take his place. From Fonseca's point of view, the promotion contained a certain numerical logic because the expedition would now have one Spanish and one Portuguese leader, but Magellan did not view matters that way. He considered himself the sole Captain General, and Cartagena simply the inspector general, not a co-admiral. Archbishop Fonseca clearly had another idea, for he appointed Cartagena as Faleiro's replacement, specifying that he was *"persona conjunta."* The exact meaning of this title was subject to varying interpretations, but at the minimum it meant that Magellan was supposed to consult with Cartagena in all matters. At the maximum, it meant that the two were co-commanders, with Cartagena, as inspector general, having a slight edge in his capacity as Magellan's official supervisor.

Although he had no experience at sea, Juan de Cartagena found himself leading one of the largest maritime expeditions mounted by Spain. This bizarre situation had much to do with his relationship with the man who appointed him, Archbishop Fonseca. Cartagena was considered Fonseca's nephew, but as everyone realized, that term was a euphemism: In reality, Cartagena was Fonseca's illegitimate son. Nor was he the only example of this peculiar brand of nepotism. The fleet's accountant, Antonio de Coca, was the "nephew" of Fonseca's brother. Not only that, but Fonseca appointed two close "friends" and "servants" of his as captains of two of the ships; these were Luis de Mendoza, who assumed command of *Victoria,* and Gaspar de Quesada, of *Concepción.* Not surprisingly, all three captains appointed by Fonseca—Cartagena, Quesada, and Mendoza—despised and looked down on Magellan from the moment they came on board.

Here, at last, was Fonseca's revenge on Magellan. No matter what the contract said, Fonseca had managed to stifle Magellan's authority, and, potentially, his share of the proceeds of the expedition, by appointing his natural son and his close allies to virtually all the important positions in the armada. Collectively, they, and not Magellan, would have the final say over the disposition of the fleet and its finances. They, and not Magellan, would decide the allocation of personnel and resources. Magellan still held the rank of Captain General, it was true, but it was reduced in power; from Fonseca's point of view, Magellan served at the pleasure of his Castilian captains, rather than the other way around. The arrangement made it impossible for Magellan and his captains to make decisions in the best of circumstances, even if they felt goodwill toward one another. And if they lacked mutual trust and respect, which was far more likely to be the case, it set the stage for endless challenges to Magellan's authority, in other words, for mutiny.

*N*ot content with the removal of Faleiro, the archbishop turned his malign attention to Juan de Aranda, who had first introduced Magellan to the Castilian court. Fonseca launched an investigation in Juan de Aranda's business arrangements with Magellan and Faleiro; all three were interrogated separately. Under oath, Magellan described the fees Aranda had received for the services he rendered to the explorers, and the signed agreement to distribute a portion of the proceeds to Aranda. On June 15, 1519, Aranda himself went before the Supreme Council of the Indies, and by all accounts acquitted himself well. He had served the interests of the Spanish crown in his dealings with Magellan and Faleiro, and as for his personal stake in the expedition, it was the custom of the era.

Despite these favorable indications, the Supreme Council censured Aranda for his actions, declaring that he had committed a criminal act by receiving money from Magellan; the judgment was signed by the council's president, who just happened to be Fonseca. Two weeks later, the Spanish crown took up the council's charges against Aranda and removed him from any further involvement with the expedition. He was, in short, disgraced. Fonseca could have tarred Magellan and Faleiro with the same brush, but they were not the targets of the inquiry, which concluded that these men were innocent of scandal. With the purge of the unstable Faleiro and now the acquisitive Aranda, Magellan could only have felt a sense of relief mingled with fear of what the all-powerful Fonseca might do next to the Armada de Molucca.

*A*s the date of departure approached, Magellan turned his attention to the complicated and hugely expensive matter of provisioning the ships. During the long months of preparation, Magellan's five ships were tied up at a dock known as the Puerto de las Muelas, because it was paved with millstones. It was here, at Millstone Dock, that the ships took on all the sailing gear, arms, provisions, and furnishings that they would bring on the voyage. It was the only dock where wine, an essential part of the sailors' diet, was permitted to be loaded. The dock, and the area around it, throbbed with activity, the waters constantly stirred by small craft coming and going, the streets packed with carts bearing supplies, all of them checked by customs inspectors who made certain that the merchants paid their tariffs—and their payoffs—to the proper authorities.

Magellan approached the task of provisioning with as much attention to detail as he did the outfitting of the ships, and with good reason. The food represented a considerable investment: 1,252,909 *maravedís*, nearly as much as the cost of the entire fleet, and that figure covered just enough food to see them through the first leg or two of the voyage. It was expected that the sailors would be looking for additional food at almost every port, and in the ocean itself.

Of the food that Magellan took on at Seville, nearly four-fifths consisted of just two items, wine and hardtack. Wine was considered the most important; it was tax free, and an official was required to come aboard and make certain it had not soured or become contaminated. The wine was stored in casks, which were carefully maintained, and in pipes sealed with a cork and with pitch. These were meticulously stowed on board the ships according to a plan designed to maximize the use of the limited space below deck.

Hardtack, the other staple of the sailor's execrable diet, consisted of coarse wheat flour, including the husk, kneaded with hot water (never cold), and cooked twice. The result, a tough, brittle biscuit known as *biscocho,* was stored for up to a month before it was sold. Inevitably, the hardtack degraded in the humid conditions at sea, and when it became soft, and rotten, and inedible, it was called *mazamorra*; the sailors boiled the stuff until it turned into a mush known as *calandra,* said to be so vile that even starving sailors refused it.

The ships also held flour stored in wooden barrels, to be kneaded with seawater and then grilled as a kind of tortilla, as well as meat, usually pork, bacon, ham, and especially salted beef. And some meat came on the hoof. The fleet carried seven cows and three pigs; they were slaughtered just before or just after departure; otherwise, they would have eaten their way through a considerable amount of valuable food. Their presence turned the ship into a floating barn, with an odor to match. Barrels of cheese, almonds in the shell, mustard, and casks of figs were also loaded on board the ships. As unlikely as it sounds, Magellan's fleet carried fish—sardines, cod, anchovies, and tuna—all of it dried and salted. In expectation of catching fresh fish along their route, the ships' holds included a generous amount of fishing line and a plentiful supply of hooks. There was little in the way of fresh vegetables; instead, the sailors consumed chickpeas, beans, rice, garlic, almonds, lentils. All fruit was preserved. Raisins, a particular favorite of the sailors, came in two varieties, "sun raisins," dried in the open air, and "lye raisins," boiled in a mild lye solution. Magellan also carried with him jelly and jam preserves, including a cider jam known as "diacitron." The officers brought with them a delicacy in the form of preserved quince, *carne de membrillo,* a jam made from the small, hard, applelike fruit. As the voyage wore on, quince jelly would play a crucial role in the lives of the sailors, and Magellan's as well.

There were casks filled with vinegar, which was used as a disinfectant both for the ships and for contaminated water. On rare occasions, starving sailors would add vinegar to rotting hardtack. Sugar and salt also had a place in the list of provisions. Salt was plentiful, and was used for preserving meat and fish through the voyage, while sugar was scarce. It was administered to sailors who had fallen ill, but not used in food. Honey, far cheaper, served as the universal sweetener.

These provisions made for an unhealthy diet, high in salt, low in protein, and lacking vitamins that sailors needed to protect themselves against the rigors of the sea. Given the inadequacy and volatility of his food supply, it was no surprise that Magellan's first thought on arriving at ports of call was replenishing his stock, and, along with it, his sailors' health and morale.

*D*isputes over the crew's composition and pay bedeviled Magellan until the moment of the fleet's departure from Seville. Three of his Spanish pilots demanded to be paid as much as the more experienced Portuguese pilots Magellan had retained, but King Charles refused, reminding them that they had already been richly rewarded with a full year's pay in advance, free lodgings in Seville, and the prospect of knighthood.

The composition of the crew engendered greater controversy. Magellan was suspected of packing the roster with his countrymen, but the reality was that experienced Spanish seamen willing to enlist on the voyage were scarce, and so he was forced to include many foreigners. The Casa de Contratación decreed that Magellan must limit his entire crew to 235 men, including cabin boys. If he did not obey this constraint, the Casa sternly warned, the resulting "scandal or damage" would be blamed on him, "as it would any person who chooses to disobey a royal command." When the armada's roster, bloated with well-connected Spaniards, exceeded this number, the Casa stopped short of halting preparations, but it placed Magellan on warning. And when he hired no less than seventeen apprentice sailors, or *grumetes*, he was forced to let them go. He was reminded that key positions such as bookkeepers and bursars must be filled by

Spaniards. Magellan protested that he had retained the services of only two Portuguese bursars, and he pleaded in writing with the Casa to allow the men he had enlisted to board the ships, regardless of nationality. If he could not have the crew he wanted, he insisted that he would abandon the expedition.

The Casa would not let matters rest there. On the day before the fleet's departure from Seville, August 9, 1519, Magellan was summoned from his frantic last-minute preparations to testify that he had made every effort to hire Spanish officers and crew members rather than foreigners. He had, in fact, gone to great lengths to comply, and he swelled with pride as he delivered his sworn statement. "I proclaimed [through a town crier] in this city [Seville], in squares and markets and busy places and along the river that anyone— sailors, cabin boys, caulkers, carpenters, and other officers—who wished to join the Armada should contact me, the captain, or talk to the masters of the ship, and I also mentioned the salaries stipulated by the king. Sailors will receive 1,200 *maravedís*, cabin boys 800 *maravedís*, and pages 500 *maravedís* every month, and carpenters and caulkers five ducats every month. None of the villagers born here wanted to join the Armada." And that was the truth. Qualified sailors were rare in Seville, and qualified sailors willing to risk their lives on a voyage to the Spice Islands rarer still.

Desperate to recruit qualified crew members for the expedition, Magellan cast his net even further. He sent his master-at-arms to Málaga with a letter from the Casa de Contratación indicating the salaries and benefits those joining the Armada de Molucca would receive. Other officers fanned out to popular seaports such as Cádiz in search of willing hands, but those willing to risk their lives on a voyage into the unknown proved scarce. "I couldn't find enough people," Magellan explained, "so I accepted all the foreigners we needed, foreigners such as Greeks, and people from Venice, Genoa, Sicily, and France." Although he did not say so, few Spanish seamen wanted to sail under a Portuguese captain.

As matters stood at the time of departure, Magellan had official permission to hire only a dozen Portuguese; in reality, he was taking nearly forty with him. At the last minute, he sacrificed three relatives whom he had quietly enlisted, one of whom was a pilot approved by the Casa, but he kept berths for at least two others: Álvaro de Mesquita, a relative on his mother's side, and Cristóvão Rebêlo, his illegitimate son.

Magellan's last-minute compromises on the composition of the crew placated the Casa, and the Captain General received final permission to proceed with his expedition. To guarantee this voyage, he had sacrificed his allegiance to his homeland, his partnership with Ruy Faleiro, and a considerable amount of his authority as Captain General, but he had kept his essential mission intact. After twelve months of painstaking preparation, the Armada de Molucca was at last ready to conquer the ocean.

Just before departure, the officers and crew of the five ships comprising the fleet attended a mass at Santa María de la Victoria, located in Triana, the sailors' district.

During the ceremony, King Charles's representative, Sancho Martínez de Leiva, presented Magellan with the royal flag as the Captain General knelt before a representation of the Virgin. This marked the first occasion that Charles had bestowed the royal colors on a non-Castilian. Magellan could only have felt that he was now invested with the king's full trust.

Still kneeling, his head bent, Magellan swore that he was the king's faithful servant, that he would fulfill all his obligations to guarantee the success of the expedition, and when he was finished, the captains repeated the oath and swore to obey Magellan and to follow him on his route, wherever it might lead.

Among those in attendance at Santa María de la Victoria that day was a Venetian scholar named Antonio Pigafetta who had spent

long years in the service of Andrea Chiericati, an emissary of Pope
Leo X. When the pope appointed Chiericati ambassador to King
Charles, Pigafetta, who was about thirty years old at the time, followed the diplomat to Spain. By his own description, Pigafetta was a
man of learning (he boasted of having "read many books") and religious conviction, but he also had a thirst for adventure, or, as he put
it, "a craving for experience and glory."

Learning of Magellan's expedition to the Spice Islands, he felt
destiny calling, and excused himself from the diplomatic circles to
seek out the renowned navigator, arriving in Seville in May 1519,
in the midst of feverish preparations for the expedition. During the
next several months, he helped to gather navigational instruments
and ingratiated himself into Magellan's trust. Pigafetta quickly
came to idolize the Captain General, despite their differing nation-
alities, and was awestricken by the ambitiousness and danger of the
mission. Nevertheless, Pigafetta decided he had to go along.
Although he lacked experience at sea, he did have funds and
impeccable papal credentials to recommend him. Accepting a
salary of just 1,000 *maravedís,* he joined the roster as a *sobrasaliente,*
a supernumerary, receiving four months of his modest pay in
advance.

Magellan, who left nothing to accident, had an assignment for
Pigafetta; the young Italian diplomat was to keep a record of the
voyage, not the dry, factual pilot's log, but a more personal, anecdotal, and free-flowing account in the tradition of other popular travel works of the day; these included books by Magellan's
brother-in-law, Duarte Barbosa; Ludovico di Varthema, another Italian visitor to the Indies; and Marco Polo, the most celebrated Italian
traveler of them all. Making no secret of his ambition to take his
place in letters beside them, Pigafetta readily accepted the assignment. His loyalties belonged to Magellan alone, not to Cartagena or
to any of the other officers. For Pigafetta, the Armada de Molucca
was the tangible result of Magellan's daring, and if the expedition

succeeded, it would be the result of Magellan's skill and God's will—of that Pigafetta was quite certain.

From the moment the fleet left Seville, Pigafetta kept a diary of events that gradually evolved from a routine account of life at sea to a shockingly graphic and candid diary that serves as the best record of the voyage. He took his role as the expedition's official chronicler seriously, and his account is bursting with botanical, linguistic, and anthropological detail. It is also a humane and compassionate record written in a distinctive voice, naïve yet cultivated, pious yet bawdy. Of the handful of genuine chronicles of foreign lands available at the time, only Pigafetta's preserved moments of self-deprecation and humor; only his betrayed the realistic fears, joys, and ambivalence felt by the crew. His narrative anticipates a modern sensibility, in which self-doubt and revelation play roles. If Magellan was the expedition's hero, its Don Quixote, a knight wandering the world in a foolish, vain, yet magnificent quest, Pigafetta can be considered its antihero, its Sancho Panza, steadfastly loyal to his master while casting a skeptical, mordant eye on the proceedings. His hunger for experience makes it possible to experience Magellan's voyage as the sailors themselves experienced it, and to watch this extraordinary navigator straining against the limits of knowledge, his men's loyalty, and his own stubborn nature.

Pigafetta was not the only diarist on the voyage. Francisco Albo, *Trinidad*'s pilot, kept a logbook, and some of the surviving sailors gave extensive interviews and depositions on their return to Spain, or wrote their own accounts from memory. The plethora of firsthand impressions of the voyage, combined with the fantastically detailed Spanish records, make it possible to re-create and understand it from a variety of perspectives, ranging from the deeply personal and casually anecdotal to the official and legalistic; royalty and ordinary seamen alike have their voices in this epic of discovery.

An important limitation governed all the accounts, varied as they are. They provide only the European perspective on a voyage that affected nations and cultures around the world, often profoundly. There is no testimony from the individuals whom Magellan's fleet would visit. Occasionally, we can glean disturbing hints of the reactions of those whom the armada would visit, and what they thought of the intruders in their black ships, the men who had come from a great distance, men bearing gifts but also guns.

*M*agellan's departure deeply affected the fortunes of those he left behind. His wife, Beatriz, pregnant with their second child, lived quietly in the city under the protection of her father. She received a monthly stipend, as specified in Magellan's contract, but she was, in fact, a hostage to the Spanish authorities. If word should reach Seville that Magellan had done anything untoward during the expedition, or exhibited disloyalty to King Charles, she would be the first person the king's agents would seek out.

Although it seemed Magellan had placed his pregnant wife and young child at risk in the hostile environment of Seville, he did take elaborate precautions to ensure their future—and his own posthumous glory—in his will, dated August 24, 1519. Magellan knew from experience the risks of embarking on his voyage of discovery. He knew that each day of the voyage he would be at the mercy of forces he could scarcely contemplate, forces that only his fervent belief in God and unswerving loyalty to King Charles would be able to help him surmount. Although he coveted the renown and rewards of a successful voyage, he realized he might die far from home, in a part of the world that was still a blank on European maps. This knowledge imparted to his will a special weight and urgency.

In the will, Magellan left thousands of *maravedís* to various churches and religious orders, all of them in Seville, which he designated as his permanent home in this life and the next: "I desire that if I die in this city of Seville my body may be buried in the monastery of Santa María de la Victoria in Triana—ward and precinct of the city of Seville—in the grave set apart for me; and if I die on this said voyage, I desire that my body may be buried in a church dedicated to Our Lady, in the nearest spot at which death seize me and I die." He proposed very specific and pious plans for his funeral rites: "And I desire that on the said day of my burial thirty masses may be said over my body—two *cantadas* and twenty-eight *rezadas*, and that they shall offer for me the offering of the bread and wine and candles that my executors desire; And I desire that in the said monastery of Santa María de la Victoria a thirty-day mass may be said for my soul, and that the accustomed alms may be given therefor; and I desire that on the said day of my burial three poor men may be clothed—such as I have indicated to my executors—and that to each may be given a cloak of gray stuff, a shirt, and a pair of shoes, that they may pray to God for my soul; and I also desire that upon the said day of my burial food may be given to the said three paupers, and to twelve others, that they may pray for my soul."

Magellan made certain that all acknowledged family members and retainers would be well taken care of. He specified that Beatriz's entire dowry of 600,000 *maravedís* be returned to her; that his illegitimate son, Cristóvão Rebêlo, whom he called "my page," receive a legacy of 30,000 *maravedís*; and that his slave, Enrique, be freed. Since Enrique, like Cristóvão, would accompany Magellan on the voyage to the Spice Islands, the terms of his freedom were of particular interest: "I declare and ordain as free and quit of every obligation of captivity, subjection, and slavery, my captured slave Enrique, mulatto, native of the city of Malacca, of the age of twenty-six years more or less, that from the day of my death thenceforward forever the said Enrique may be free and manumitted, and quit, exempt, and relieved of every obligation of slavery and subjection, and that he may act as he desires and thinks fit."

All that, plus 10,000 *maravedís*.

Magellan envisioned leaving behind a great empire. He left to

Rodrigo, "my legitimate son," along with any other legitimate male heirs that he might have with Beatriz Barbosa, all the rights and titles King Charles had granted to him for the voyage to the Spice Islands; in other words, these children might grow up to find themselves the rulers of distant lands, administered by Spain, and very wealthy rulers at that. All that Magellan asked was that they give a portion of their income to their mother, making her a wealthy widow. And even if she remarried after Magellan's death, "I desire that there be given and paid to her the sum of two thousand Spanish doubloons."

The will covered every eventuality that might befall a great explorer such as Magellan—except what would actually occur once he set forth from Seville.

The Portuguese reacted bitterly to the imminent departure of the Armada de Molucca. King Manuel ordered the harassment of Magellan's relatives who remained in their homeland. To make his dishonor public, vandals were sent to the family estate in Sabrosa; they tore the Magellan escutcheon from the gates and smashed it to the ground. Even young relatives of Magellan found themselves the object of derision and were stoned. Fearing for their lives, they fled the country. Francisco de Silva Téllez, who claimed to be Magellan's nephew, eventually sought refuge in the Portuguese colony of Brazil, where he dictated instructions that suggest the depth of shame stirred by Magellan's betrayal: "I order all my relatives and heirs to put no other stone nor shield of arms in my house ... in Sabrosa because I want them forever effaced, in the same condition that our lord and King prescribed, as punishment of Ferdinand Magellan's crime of moving to Castile." Should others take up Magellan's mantle, his nephew warned that he would refuse to acknowledge them "should I learn that they had entertained feelings and designs so base and ruinous to their families as befell my father and me, who felt compelled to leave our house out of shame and fear that our neighbors would attack us, as they justly could not suffer him who went against Portugal, his motherland, to serve the Castilians, our natural enemies."

Abandoned, the Sabrosa estate fell into disrepair, and another house rose on the site. The stone that once held the Magellan escutcheon met with a special fate: It was covered with excrement.

CHAPTER III

Neverlands

And now the STORM-BLAST came, and he
Was tyrannous and strong:
He struck with his o'ertaking wings,
And chased us south along.

On the tenth of August," Antonio Pigafetta recorded in his diary, "the fleet, having been furnished with all that was necessary for it and having in the five ships people of divers nations to the number of two hundred and thirty-seven in all, was ready to depart from... Seville, and firing all the artillery we set sail with the staysail only." Pigafetta's head count probably omitted about twenty crew members also on board the ships. Only Magellan remained behind, making last-minute provisioning arrangements; he would join the fleet shortly before its final departure from Spain.

To reach the Atlantic, the five ships, their colors set, negotiated the sinuous Guadalquivir River, whose hazards immediately tested the pilots' abilities. Fed by rainwater in winter and melting snows in spring and summer, the Guadalquivir empties into the Gulf of Cádiz. The last forty miles, traversing a seemingly endless stretch of tidal marches known as Las Marismas, presented special perils. Hidden sandbanks, the hulls of shipwrecks, and shallow areas lurked beneath the river's turbid waters, and occasionally these obstacles visited disaster on an expedition even before it reached the open sea. Pigafetta, new to the problems of navigation, suddenly became alert to the dangers of the Guadalquivir. "There was a bridge over the river by which one went to Seville, which bridge was in ruins, although two columns remained at the bottom of the water. Wherefore you must have practiced and expert men of the country to point out the proper channel for passing safely between these two columns, for fear of striking on them."

Although defeated and driven from Spain, the Moors had left their indelible marks on the Spanish psyche, landscape, and bloodlines. "Going by this river we passed a place named Gioan de Farax, where there was a great settlement of Moors," Pigafetta noted of one encampment. The Guadalquivir derived its very name from the Arabic original, Wadi al-Kabir, meaning "great river," as the Arab rulers of the region designated it. And, as everyone aboard these ships knew, Moorish pirates still patrolled the Spanish coast, looking for ships laden with precious resources and, most of all, with weapons—ships like those of the Armada de Molucca.

A week after leaving Seville, the fleet reached the snug coastal town of Sanlúcar de Barrameda, the final point of departure for the Ocean Sea. "You enter it on the west wind and depart from it on the east wind," said Pigafetta, repeating the lore he recently learned. On arrival, the crew found a windswept seaport, seemingly poised on the edge of the world, and reverberating with a sense of adventure. Over the centuries, Sanlúcar de Barrameda had witnessed a succession of conquerors, from Roman to Arab to, most recently, King Alfonso X, who claimed it in 1264. In 1498, Christopher Columbus chose it as the departure point for his third voyage to the New World, and Magellan might have chosen the same port to announce that he planned to build on and outdo Columbus's accomplishments.

Beyond the huddled town lay the churning waters of the

Atlantic. To Magellan and his crew, the body of water was known simply as the Ocean Sea, believed to girdle the globe. At the sight of these seething green waters, every sailor's pulse quickened; their lives depended on conquering this element. Many ships had departed from the Sanlúcar de Barrameda, and some had been fortunate enough to return from distant ports and newly discovered lands, but none had circumnavigated the entire world.

*M*agellan took command of his fleet just before departure, and made sure that his sailors led a pious existence during what might be their last days on land. "A few days after, the Captain General went along the said river in his boat and the masters of other ships with him," wrote Pigafetta, "and we remained for some days at the port to hear mass on land at a church named Our Lady of Barrameda near Sanlúcar, where the Captain General ordered all those of the fleet to confess themselves before going farther. In which he showed the way to others. Moreover he would not allow any woman, whoever she might be, to come onto the fleet and to the ships, for many good reasons."

Magellan's autocratic style extended beyond religious observance. To stifle dissent, Pigafetta writes, Magellan concealed the ultimate goal of the expedition from his rank-and-file sailors. "He did not wholly declare the voyage that he wished to make, lest the people from astonishment and fear refuse to accompany him on so long a voyage as he had in mind to undertake, in view of the great and violent storms of the Ocean Sea whither he would go." The assertion needs clarification. As a Portuguese mariner, Magellan was accustomed to secrecy when it came to voyages of discovery; that was the Portuguese way. Yet everyone realized the fleet was bound for the Spice Islands; it was even called the Armada de Molucca. Perhaps Pigafetta meant that Magellan wished to keep his plan to find a strait—a waterway leading to the East—to himself until it was too late for disloyal crew members to desert. Inevitably, the plan meant trouble, because once the fleet encountered storms, then uncharted waters, and finally a search for an unknown strait, the men whom he had hoodwinked into coming along were likely to rise up in rebellion against him.

In the pages of his diary, Pigafetta confided another and far more troubling reason for Magellan's unusual secrecy: "The masters and captains of the other ships of his company loved him not. I do not know the reason, unless it be that he, the Captain General, was Portuguese, and they were Spaniards and Castilians, which peoples have long borne ill-will and malevolence toward one another."

To assert his authority over his resentful and contentious captains, Magellan gave strict sailing orders designed to reinforce his unquestioned authority. They were "good and honorable regulations," in Pigafetta's words, and consistent with procedures followed by other fleets of the era. "First, the said Captain General desired that his ship should go before the other ships and that the others should follow him; and to this end he carried by night on the poop of his ship a torch or burning fagot of wood, which they called a *farol*, that his ships should not lose him from sight. Sometimes he put a lantern, at other times a thick cord of lighted rushes, called a *trenche*, which was made of rushes soaked in water and beaten, then dried by the sun or by smoke." If the flagship, *Trinidad*, signaled, the others were to reply; that way, Magellan could tell if his fleet was following him. "And when he wished to change course because the weather changed, or the wind was contrary, or he wanted to reduce speed, he had two lights shown. And if he wanted others to haul in a bonnet (which is a part of the sail attached to the mainsail), he showed three lights. Thus by three lights, even if the weather was good for sailing faster, he meant that the said bonnet be brought in, so that the mainsail could be sooner and more easily struck and furled when bad weather came on suddenly." Four lights on *Trinidad* signaled that the other ships should strike sail. If the watchman suddenly discovered land, or even a reef, Magellan would display lights or fire a mortar.

Magellan set a traditional system of watches, an essential precaution. There were to be three: "the first at the beginning of the night, the second at midnight, and the third toward daybreak. . . . And every night the said watches were changed, that is to say, he who had made the first watch made on the morrow the second, and he who had made the second then made the third. And after this manner they changed every night. Then the Captain [General] ordered that his regulations, both for signals and watches, be strictly observed, that their voyage be made with greater safety."

Magellan's strict procedures demanded discipline from an inexperienced crew lacking respect for the Captain General. The most innocuous aspect of his standing orders—the requirement that all ships report to *Trinidad* at dusk—rankled the most because it demonstrated that Magellan, and no one else, served as the leader of the Armada de Molucca.

Leaving the mouth of the Guadalquivir River on September 20, 1519, the five ships of the armada plunged into the Atlantic Ocean. Juan de Escalante de Mendoza, an experienced Spanish seaman, described the exhilaration and frenzy of sailing past Sanlúcar de Barrameda into the Atlantic. "When the hour had arrived in which they had to make sail," he wrote, "the pilot ordered the men to raise all but one of the anchors and to attach the cable on the last anchor to the capstan . . . and with the yards and sails aloft, he ordered two apprentices to climb the foremast and stand ready to unfurl the sails when and as they were ordered and directed." Amid the intricately choreographed flurry of activity aboard the ships, officers shouted orders, but their words at this crucial moment sounded more like prayers than commands. "And if the special pilot for the sandbar said that it was time to make sail, the ship's pilot would call out the following to the two men aloft on the yard: 'Ease the rope of the foresail, in the name of the Holy Trinity, Father, and Son, and Holy Spirit, three persons in one single true God, that they may be with us and give us good and safe voyage, and carry us and return us safely to our homes!'" With those words ringing in their ears, the sailors hauled the hemp ropes holding anchors, set the sails, and felt the breeze freshen against their faces. The ships picked up speed, and the coastline began to recede; there was no turning back now. It would sustain them all, or it would destroy them all. To reach his goal, Magellan would have to master both the great Ocean Sea and a sea of ignorance.

*I*t was a dream as old as the imagination: a voyage to the ends of the earth. Yet until the Age of Discovery, it remained only a dream. At the time, Europe was deeply ignorant of the world at large. Magellan undertook his ambitious voyage in a world ruled by super-stition, populated with strange and demonic creatures, and rever-berating with a longing for religious redemption. To the average person, the world beyond Europe resembled the fantastic realms depicted in *The Thousand and One Nights,* a collection of tales including "The Seven Voyages of Sindbad the Sailor." Going to sea was the most adventurous thing one could do, the Renaissance equivalent of becoming an astronaut, but the likelihood of death and disaster was far greater. These days, there are no undiscovered places on earth; in the age of the Global Positioning System, no one need get lost. But in the Age of Discovery, more than half the world was unexplored, unmapped, and misunderstood by Europeans. Mariners feared they could literally sail over the edge of world. They believed that sea monsters lurked in the briny depths, waiting to devour them. And when they crossed the equator, the ocean would boil and scald them to death.

Some of the most tenacious ideas about the world at large

derived from Pliny the Elder, who died in the eruption of Mount
Vesuvius in A.D. 79. His multivolume, encyclopedic *Natural History*,
rediscovered and widely consulted in the Renaissance, sought to
bring together everything that was known about the natural world:
mountains, continents, flora and fauna.

Pliny's chapters on humankind contained a potent mixture of
fact and fantasy. He wrote of a tribe known as the Arimaspi, "a people known for having one eye in the middle of the forehead." He
confidently cited other classical authorities, such as Herodotus, who
related tales of a "continual battle between the Arimaspi and griffins
in the vicinity of the latter's mines. The griffin is a type of wild beast
with wings, as is commonly reported, which digs gold out of tunnels.
The griffins guard the gold and the Arimaspi try to seize it, each
with remarkable greed." Pliny meant this vivid description literally,
and while it might have generated some skepticism among natural-
ists of Magellan's time, it was generally accepted as fact, as was
Pliny's curious description of "forest-dwellers who have their feet
turned back behind their leg; they run with extraordinary speed and
wander far and wide with the wild animals." India offered particularly fertile ground for extraordinary creatures. Pliny evoked "men
with dog's heads who are covered with wild beasts' skins; they bark
instead of speaking and live by hunting and fowling, for which they
use their nails." At one time, says Pliny, over 120,000 of these
hominids flourished throughout India.

Pliny assured his readers that wonders never ceased in the natural world; the result of his labors was a Ripley's Believe-It-or-Not
catalog tinged with the classics. "That women have changed into
men is not a myth," he wrote. "We find in historical records that . . .
a girl at Casinum became a boy before her parents' very eyes." To
emphasize his point, Pliny claimed to have firsthand knowledge of
the phenomenon: "In Africa, I myself saw someone who became a
man on his wedding-day." There was more; he claimed that people
in Eastern Europe had two sets of eyes, backward-facing heads, or no
heads at all. In Africa, Pliny wrote, lived people who combined both

sexes in one body, yet managed to reproduce; people who survived
without eating; people with ears large enough to blanket their entire
bodies; and people with equine feet. In India, he said, there were
people with six hands. These marvelous accounts were subsequently
retold by various respected chroniclers and widely credited up
through Magellan's time.

In the open waters of the ocean, lurked even more bizarre creatures, whales and sharks, six-foot-long lobsters and three-hundred-
foot-long eels. Sailors had no way of telling which of Pliny's
descriptions were reliable, and which were fantasies.

They were just as ignorant about major landmasses. Only three
continents were known to Europeans of the era—Europe, Asia, and
Africa—although it was suspected that more would be discovered.
The existence of an illusory island, Terra Australis, the South Land,
was accepted as fact before and long after Magellan's voyage. This
landmass was said to lurk in the Southern Hemisphere, where its
vast size supposedly counterbalanced the continents in the Northern Hemisphere. Highly schematic medieval maps depicted the
three known continents as separated by two rivers, the Nile and the
Don, as well as the Mediterranean, all of them surrounded by
the great Ocean Sea, into which other seas and rivers flowed. This
diagram resembled a *T* inside of an *O*, so medieval maps of this
genre are referred as "T in O" maps. To remain consistent with religious traditions, T in O maps located Jerusalem at dead center, with
Paradise floating vaguely at the top. To complicate matters, Asia
occupied the Northern Hemisphere of this map, with Europe and
Africa sharing the Southern. In some versions of the medieval map,
the Ocean Sea flowed out into space. One could not navigate with
such maps, or locate points of the compass on it, or plot realistic
routes; they offered a conceptual model rather than an actual representation. As such, they were utterly useless to Magellan.

In 1513, only six years before Magellan undertook his circumnavigation, Juan Ponce de León set out to find the Fountain of
Youth. Peter Martyr, another trusted authority of the Renaissance,

described the Fountain of Youth as "a spring of running water of such marvelous virtue that the water thereof being drunk, perhaps with some diet, makes old men young again." According to tradition, the fountain was located on the island of Bimini, in the Bahamas. On the strength of his reputation as a soldier, nobleman, and participant in Columbus's second voyage to the New World, Ponce de León received a commission from King Ferdinand to claim Bimini for Spain. In a fruitless search, Ponce de León explored the Bahamas and Puerto Rico, but his failure to find the Fountain of Youth did not put the myth to rest. As late as 1601, the respected Spanish historian Antonio de Herrera y Tordesillas wrote confidently about the fountain's great efficacy in restoring youth and potency to aging men.

Although his quest seems fanciful and absurd today, Ponce de León was a man of his times. Superstition governed popular impressions, and even scholarly accounts, of the world at large. A work published in 1560 contained descriptions of various sea monsters infesting the oceans. One, known as the Whirlpool, was said to have a human countenance. Another, supposedly sighted in 1531, had hideous scaly skin. There were others: the Satyr of the Sea; the Rosmarus, which rivaled an elephant in size; and the wondrous Socolopendra, with its face of flames. Voyagers across the seas, especially those attempting to circumnavigate the globe, could expect to encounter all these creatures, and more, in the course of their journey.

*E*ven educated people placed credence in fantastic realms on earth, for instance, the persistent belief in the existence of the kingdom of Prester John. It is difficult to overestimate the influence of this fabulous personage, Prester John ("Prester" is an archaic word for presbyter, or priest), on the European imagination during the late Middle Ages and early Renaissance. He was part Christian ruler and part Kublai Khan. Despite an enormous number of inconsistencies and improbabilities in the details surrounding Prester John and his realm, his existence was widely believed in for several hundred years. In an era of violent conflict between Christianity and Islam, and unsuccessful Crusades, it was vastly reassuring to the faithful to believe that a sprawling and wealthy Christian outpost existed beyond Europe.

The legend originated in 1165 when a lengthy letter began to circulate among various Christian leaders; as time passed, the letter became more elaborate as anonymous authors added beguiling, utterly fantastic details; so great was its appeal that it became one of the most widely circulated and discussed documents of the Middle Ages, translated into French, German, Russian, Hebrew, English, among other languages, and with the introduction of movable type, it was reprinted in countless editions.

Addressed to Manuel, the Constantinopolitan emperor, and to Frederick, the emperor of the Romans, the letter read, "If you should wish to come here to our kingdom, we will place you in the highest and most exalted position in our household, and you may freely partake of all that we possess. Should you desire to return, you will go laden with treasures. If indeed you wish to know wherein consists our great power, then believe without doubting that I, Prester John, who reign supreme, exceed in riches, virtue, and power all creatures who dwell under heaven. Seventy-two kings pay tribute to me. I am a devout Christian and everywhere protect the Christians of our empire, nourishing them with alms." As it continued, the letter became overtly symbolic, yet it was taken to be factual: "Our magnificence dominates the Three Indias, and extends to Farther India, where the body of St. Thomas the Apostle rests. It reaches through the desert toward the palace of the rising sun, and continues through the valley of the deserted Babylon close by the Tower of Babel." By "India," Prester John, or whoever wrote this missive, meant more than just the Indian subcontinent. During the Middle Ages, India was believed to include a good portion of northeastern Africa. It was an elastic term, and medieval geographers obeyed the convention that there were several Indias, some near, and some far.

Prester John described his kingdom as an enchanted realm, far more luxurious than European countries beaten down by war, plague, famine, and, among less memorialized miseries, the hardships inflicted by the Little Ice Age. In contrast, Prester John boasted of the wonders of his kingdom: "In our territories are found elephants, dromedaries, and camels, and almost every kind of beast that is under heaven. Honey flows in our land, and milk everywhere abounds. In one of our territories no poison can do harm and no noisy frog croaks, no scorpions are there, and no serpents creep through the grass. No venomous reptiles can exist there or use their deadly power. In one of the heathen provinces flows a river called the Physon, which, emerging from Paradise, winds and wanders through the entire province; and in it are found emeralds, sapphires, carbuncles, topazes, chrysolites, onyxes, beryls, sardonyxes, and many other precious stones."

And there was much more; this mysterious religious leader claimed his dominion reached from Eastern Europe to India and contained satyrs, griffins, a phoenix, and other wonderful creatures. He lived, or so he said, in a palace without doors or windows, built of precious stones cemented with gold.

Prester John's letter was actually written by imaginative monks toiling in anonymity, and the result begged to be read as a symbolic document, an allegory, or an expression of faith. Yet it was taken as a factual account and diplomatic initiative. Those who read the letter or heard about it wanted to know where Prester John actually lived. By 1177, the letter's renown had grown to the point where Pope Alexander III issued a reply addressed to the "illustrious and magnificent king of the Indies and a beloved son of Christ," and pilgrims went in search of the elusive Prester John.

Over time, the letter, like Pinocchio's nose, grew and grew; copyists embellished the text, adding ingredients to Prester John's domain. One important interpolation described spices in vivid detail: "In another of our provinces pepper is grown and gathered, to be exchanged for corn, grain, cloth, and leather"—which sounded plausible enough, but then the interpolation took an allegorical twist—"that district is thickly wooded and full of serpents, which are of great size and have two heads and horns like rams, and eyes which shine as brightly as lamps. When the pepper is ripe, all the people come from the surrounding countryside, bringing with them chaff, straw, and very dry wood with which they encircle the entire forest, and, when the wind blows strongly, they light fires inside and outside the forest, so that the serpents will be trapped. Thus the serpents perish in the fire, which burns very fiercely, except those which take shelter in their caves." In the Age of Faith, the serpents were representative of the devil, which invades the Edenic garden of peppers, and which could be defeated only by the fire of faith.

*G*enerous swaths of the Prester John letter found their way into the two most popular travel books of the Middle Ages: *The Travels of Marco Polo* and *The Travels of Sir John Mandeville,* lending credence to the travel accounts and to the Prester John legend.

Polo's account, the earlier of the two, was written when he was a prisoner of war in Genoa in 1298 and 1299, with the help of a writer of romances known as Rustichello of Pisa. The son of a prominent Venetian merchant, Marco Polo had spent two decades in the East, traveling throughout the Mongol empire and China, and made it as far east as Burma. His father and uncle spent years in exile at the summer court of the Grand Khan, known as Shang-tu, whose kingdom served as the inspiration for Samuel Taylor Coleridge's Xanadu, and eventually they returned to Europe as the khan's emissaries. Marco Polo had spent much of his youth in their company.

As might be expected of its co-authors, *The Travels of Marco Polo* is not strictly a travelog, and it is replete with inconsistencies. It has even been suggested that Marco Polo never made it to China, despite his apparently firsthand descriptions of that region. Why did he not mention the Great Wall, for instance, or tea? Although *Travels* included Polo's experiences closer to home, enlivened with shrewd observations, the account was embellished with various wonders of the East, notably Prester John, which added to its readability and appeal, even as they compromised its claims to authenticity. To compound the problem, the manuscript was written in a French-Italian dialect that defied easy translation. Nor was there anything like a definitive text; over one hundred manuscripts, all of them different, were in circulation.

Polo, a tireless name-dropper, says he first encountered Prester John by reputation, as the lord of the Tatars, the inhabitants of northern China, who "paid him a tribute of one beast in every ten." Polo and his collaborator merged the Prester John legend with another figure at least partly inspired by an actual person, his Tatar rival. In 1200, Polo says, Genghis Khan sent word to Prester John to announce that he wished to marry the priest's daughter. "Is not Genghis Khan ashamed to seek my daughter in marriage?" Prester John exclaimed to the messengers. "Does he not know he is my vassal and my thrall? Go back to him and tell him that I would sooner commit my daughter to flames than give her to him as his wife." Polo's collaborator displayed a fanciful touch by explaining that Genghis Khan became so distressed that "his heart swelled within him to such a pitch that it came near to bursting within his breast." When he recovered, predictably enough, he decided to go to war against Prester John.

The battle—an epic struggle, according to Marco Polo—pitted the largest armies ever assembled on a "wide and pleasant plain called Tenduc, which belonged to Prester John." This is thought to be Mongolia, but as with so much else to do with Prester John, it is impossible to know for certain. Just before taking up arms, Ghengis Khan asked his astrologers to predict the outcome, and to his delight they announced that he would carry the day. Two days later, the battle began in earnest: "This was the greatest battle that was ever seen. Heavy losses were suffered on both sides; but in the end the fight was won by Ghengis Khan. In this battle Prester John was killed. And from that day he lost his land, which Ghengis Khan continued to subdue."

Polo adds a curious postscript to the defeat of Prester John and Christianity in China. Tenduc, Polo says, became the home for descendants of both Genghis Khan and Prester John. "The province is ruled by a king of the lineage of Prester John, who is a Christian and a priest and also bears the title 'Prester John.' His personal name is George. He holds the land as a vassal of the Great Khan— not all the land that was held by Prester John, but a great part of it. I may tell you that the Great Khans have always given one of their daughters or kinswomen to reigning princes of the lineage of Prester John." Polo populates Tenduc with all sorts of marvelous creatures; even the biblical Gog and Magog can be found there. Despite these imaginative excesses, *The Travels of Marco Polo* inspired Europe to conceive of trading with the kingdoms of Asia, and of exploring the world. Many of the sailors on Magellan's voyage were familiar with it, and at least one brought a copy of Polo's account along with him.

*J*ohn Mandeville served as the other great traveler and storyteller of the era. With suave assurance, he deftly mixed accounts from ancient authors with what he claimed were his personal experiences, but Mandeville was actually a compiler rather than a traveler, and he drew much of his material directly from *Speculum Mundi*, a medieval encyclopedia, which contained extracts from Pliny and Marco Polo, among other authorities. As a finishing touch, he wove long passages from the Prester John letter into his account and passed it off as his own work.

Mandeville told jaw-dropping stories of his pilgrimages to the Holy Land, an unlikely event; he probably never got any farther than a noble's well-stocked library. He claimed he traversed India, which he said was filled with yellow and green people; visited Prester John's kingdom, without giving comprehensible directions; and even made it all the way to the borders of Paradise, but failed to enter because he considered himself unworthy. He naturally claimed to have found the Fountain of Youth in the course of his travels, and imbibed three draughts of its life-giving waters: "And evermore since that time I feel me the better and wholer."

No account of the exotic East would be complete without a discussion of spices, and when it came to this subject, Mandeville skirted close enough to the truth to lure unsuspecting readers into taking his description seriously. He sounded entirely knowledgeable about a "forest" of pepper in a Neverland he called Combar, which might or might not have been based on the Spice Islands or some other actual place. "You must know that pepper grows in the manner of wild vines beside the trees of the forest, so that it can rely on them for support. Its fruit hangs in great clusters, like bunches of grapes; they hang so thick that unless they were supported by other trees, the vines could not carry their fruit. When the fruit is ripe, it is all green like the berries of ivy. They gather the fruit and dry it in the sun, then lay it on a drying floor until it is black and wrinkled." This account was convincing enough to inspire European merchants and governments to attempt to find the mythical spice.

Sailors setting out to sea with Magellan paid special attention to Mandeville's unnerving descriptions of powerful magnetic rocks capable of destroying unwary ships, warning of "great sea rocks of the stone that is called adamant ... which draws ... iron." In consequence, "There should pass no ships that had nails of iron there away because of the foresaid stone, for he should draw them to him, therefore they dare not wend thither." If they did, the magnetic rock would draw the nails from the hulls, and the ships would leak and even sink.

Among other far-fetched tales that Mandeville tried to pass off as fact were talking birds (perhaps he was thinking of parrots); trees that sprout at dawn, bear fruit by midday, and die before dusk; sixty-foot-high cannibals; and women who rejoice at the rebirth of their deceased infants. For good measure, he dusted off the legend of Amazons, but made his account more explicit than those of antiq- uity. "These women are noble and wise warriors," he claimed, "and therefore kings of neighboring realms hire them to help them in their wars. This land of Amazons is an island, surrounded by water, except at two points where there are two ways in. Beyond the water live their lovers to whom they go when it pleases them to have bodily pleasure with them."

This was, in short, a book of marvels. Despite all its improbabil- ities, Mandeville's account was taken to be true. It was widely anthologized, and its most blatant inaccuracies excused on the basis that they must have been errors or corruptions of the original text committed by scribes and copyists over the years. His many bor- rowings from classical authors, rather than being seen as a form of plagiarism, added to his stature as a scholarly authority.

Mandeville argued that it was possible for people to circumnavi- gate the globe, but he warned, "There are so many routes, and coun- tries, where a man can go wrong, except by special grace of God." He proposed one man who had accomplished the trick. "He passed India and so many isles beyond India, where there are more than 5,000 isles, and traveled so far by land and sea, girdling the globe, that he found an isle where he heard his own language being spo- ken," Mandeville wrote. "He marveled greatly, for he did not under- stand how that could be. But I conjecture that he had traveled so far over land and sea, circumnavigating the earth, that he had come to his own border; if he had gone a bit further, he would have come to his own district. But after he heard that marvel, he could not get transport any further, so he turned back the way he had come; so he had a long journey!"

Accounts of the natural world circulating throughout Europe were so terrifying and fantastic that François Rabelais, the French friar and physician turned popular author, enthusiastically satirized them in his comic epic *Gargantua and Pantagruel*, which appeared as a series of books beginning in 1532. Rabelais mocked the unreliable accounts by the revered figures of antiquity with his own farcical version of exotic lands and the strange creatures to be found there. Among his authorities on the world was a blind old hunchback called Hearsay, who possessed seven tongues, each divided into seven parts, and maintained a school. In Rabelais's hands, this figure becomes a vicious parody of a cosmologist and his entourage of flunkies. "Around him I saw innumerable men and women listening to him attentively, and among the group I recognized several with very important looks, among them one who held a chart of the world and was explaining it to them succinctly. Thus they became clerks and scholars in no time, and spoke in choice language—having good memories—about a host of tremendous matters, which a man's whole lifetime would not be enough for him to know a hundredth part of. They spoke about the Pyramids, the Nile, Babylon, the Troglodites, the Himantopodes, the Blemmyae, the Pygmies, the Cannibals, the Hyperborean Mountains, the Aegipans, and all the devils—and all from Hearsay." Rabelais had a serious point to make; he was directing his readers back to the classical Greek concept of *autopsis,* seeing for one's self (and the origin of our word "autopsy"). *Autopsis* stressed the value of firsthand reporting; the next best thing was obtaining a reliable account from an eyewitness with firsthand knowledge.

   This was a revolutionary concept in the Age of Discovery, to go see for one's self, to study the world as it was, not as myths and sacred texts suggested that it should be. And that was exactly what Magellan proposed to do; he would see for himself if there was a water route to the Spice Islands, he would find the strait leading to them if it existed, and then he would report back to King Charles on his findings. So Magellan stood on the knife-edge dividing the ancient and medieval worlds from the modern. His voyage would be a completely practical and empirical approach to discovery. He would go and see for himself: the first-ever global *autopsis*. That ambition alone made it a daring and significant endeavor. The time was ripe for Magellan and his armada to sweep away a thousand years of accumulated cobwebs. The reign of Hearsay was coming to an end.

*F*air weather favored the Armada de Molucca, and gusts carried the black ships southwest to the Canary Islands, off the coast of the western Sahara. "We left Sanlúcar on Tuesday, September twentieth of the said year, laying a course by the southwest wind," Pigafetta noted. "And on the sixteenth of the said month, we arrived at an island of the Grand Canary named Tenerife...where we remained three and a half days to take in provisions and other things which we needed."

For centuries this group of seven volcanic islands (Grand Canary, Fuerteventura, Lanzarote, Tenerife, La Palma, Gomera, and Hierro) had served as a stopover for ships bound to and from the Iberian peninsula. They were known to Pliny, and classical historians may have been referring to the Canaries when they wrote of The Fortunate Islands. Later, a succession of Arab and European voyagers, carried by strong, favorable winds, frequently called on the Canaries to replenish their supplies, convert the islanders, or capture slaves; the islands began to appear on maps in 1341. At the time of Magellan's arrival, in late September, the Canaries glistened in the waters of the Atlantic.

While there, Pigafetta confirmed an ancient story about the Canaries: "Know that among the other islands that belong to the said Grand Canary, there is one where no drop of water coming from spring or river is found, save that once a day at the hour of noon there descends from heaven a cloud which encompasses a great tree in the said island, then all its leaves fall from it, and from the leaves is distilled a great abundance of water that it seems a living fountain. And from this water the inhabitants of the said place

are satisfied, and the animals both domestic and wild." This observation marked the first time that Pigafetta tested his firsthand experiences against the claims of ancient writers, in this case Pliny, who
wrote of a magical fountain in the Canaries with no source. It
seemed to Pigafetta that there was a natural source of water, a rain
cloud. Though hardly a revolutionary insight, the comment set
Pigafetta apart from sages such as Pliny and Marco Polo, who relied
on hearsay or the artful blending of hearsay with fact. If Pigafetta
had any idea of emulating Polo, he gave up that notion now. Instead
of embellishing timeworn legends about the world, he would present
phenomena as he observed them with his own eyes. And he would
test the legends against what he actually saw and experienced. With
this entirely factual approach, Pigafetta broke with a tradition that
reached back to antiquity.

During those brief days in the Canaries, Magellan busied himself
with the final provisioning of his fleet. He worked quickly—too
quickly, as he would later discover to his horror, for the merchants
and chandlers of the Canaries, practiced in deception, swindled Magellan by falsifying their bills of lading. They vastly overstated the
amount of supplies they sold to the fleet, and what they did sell was
in poor condition. This type of cheating was common, and very dangerous to the expeditions whose lives depended on the food acquired
in the Canaries. Although Magellan was normally meticulous in
preparing the ships, this time he was too trusting of his suppliers.

After three busy days in one of Tenerife's harbors, Pigafetta
wrote, "We departed thence and came to a port called Monterose,
where we remained two days to furnish ourselves with pitch, which
is a thing very necessary for ships." While there, Magellan heard disturbing news: The king of Portugal had dispatched not one but two
fleets of caravels to arrest him—a drastic measure, but not without
precedent. A generation earlier, Manuel's father had sent ships to
intercept Columbus. Magellan also received a secret communiqué

from his father-in-law, Diogo Barbosa, warning that the Castilian captains in the Armada de Molucca planned to mutiny at the very first chance. They might even kill Magellan to attain their goal. "Keep a good watch," Barbosa admonished. The ringleader's name came as no surprise to Magellan: Juan de Cartagena, the Castilian with blood ties to Bishop Fonseca.

In his reply to Barbosa, Magellan insisted he had accepted command of the fleet, come what may, but he promised that he would work closely with his captains for the good of the fleet and of Spain. Barbosa showed these conciliatory words to the Casa de Contratación, and Magellan won praise for gracious sentiments, at least in the short term. Despite this display of diplomacy, Magellan's concern about the safety of his fleet, and his own life, could only have increased as he contemplated the Portuguese ships in hot pursuit. Unwilling to give his rebellious captains further cause for alarm, he kept both warnings to himself.

Under the circumstances, Magellan decided that the best course of action was to leave the Canaries immediately. If the Portuguese caravels caught up with Magellan, they would return him in shackles to the Portuguese court, where he would be convicted of treason, tortured, and perhaps executed. Poorly provisioned, but afraid for his life and the welfare of the fleet, Magellan gave the order to raise anchor and set sail at midnight, October 3.

"We sailed on the course to the south," Pigafetta wrote. "Engulfing ourselves in the Ocean Sea, we passed Cape Verde and sailed for many days along the coast of Guinea or Ethiopia, where there is a mountain called Sierra Leone, which is in eight degrees of latitude."

Magellan ordered the fleet to sail both day and night, attempting to place as much distance as possible between his ships and the Portuguese caravels and to take evasive action by following an unexpected course. He led the fleet southwest, hugging the coast of Africa, rather than west across the Atlantic. From the deck of *San*

Antonio, following closely behind the flagship, Cartagena immediately challenged Magellan's orders. Why, he demanded, was Magellan following this unusual route?

Follow and do not ask questions, instructed the Captain General.

Cartagena continued to protest, insisting that Magellan should have consulted his captains and his pilots. Was he trying to get them all killed by following this dangerous course? Magellan did not attempt to explain; he simply reminded the other captains to follow, and that they did. The mutiny that he expected to break out any moment failed to materialize, and order reigned aboard the ships, at least for the time being.

*F*or the next fifteen days, the Armada de Molucca ran before the wind; the favorable conditions placated the irritable captains and gave Magellan time to strategize about the best way to avoid his Portuguese pursuers. Although he had seen no evidence of them, he continued to follow the coast of Africa rather than head west. But as they worked their way farther south, the weather turned foul, the winds confused and contrary day after day. They had no reliable nautical charts, no indications of rocks or other hazards that might have been lying in wait, and no idea when their miserable weather would change. Cooking fires were extinguished, the men went without sleep, and life on board the battered vessels became exceedingly precarious. One slip, and a sailor could plunge into the sea without hope of rescue.

The changeable winds blew the ships sideways into the troughs between waves. As the ships were tossed about, their yardarms dipped into the seething water, a prelude to a possible shipwreck. To keep from being dragged under, the captains on several occasions came close to ordering their men to chop down the masts, a desperate measure that would have disabled the fleet once the weather began to clear. Instead, they cleared nearly all their sail, offering bare masts to the relentless wind. "Thus we sailed for sixty days of rain to the equinoctial line," Pigafetta wrote. "Which was a thing very strange and uncommon, in the opinion of the old people and of those who had sailed there several times before." They were buffeted "by squalls and by wind and currents that came head on to us so that we could not advance. And in order that our ships should not perish or broach to us (as often happens when squalls come together), we struck the sails. In this manner we did wander hither and yon on the sea."

Throughout the ordeal, sharks constantly circled the ships, terrifying the crew. "They have terrible teeth," Pigafetta wrote, plainly aghast at the sight, "and eat men when they find them alive or dead in the sea. And the said fish are caught with a hook of iron, with which some were taken by our people. But they are not good to eat when large. And even the small ones are not much good."

After weeks of constant, life-threatening storms, several hissing, incandescent globes mysteriously appeared on the yardarms of Magellan's ship, *Trinidad.* Saint Elmo's fire!

Here was a natural phenomenon to rival any fanciful, supernatural apparition cataloged by Pliny or Sir John Mandeville. Saint Elmo's fire is a dramatic electrical discharge that looks like a stream of fire as it trails from the mast of a ship; it can even play about someone's head, causing an eerie tingling sensation. The superstitious sailors, always alert to omens, associated the phenomenon with Saint Peter Gonzalez, a Dominican priest who was considered the patron saint of mariners and who had acquired the name Saint Elmo; the "fire" was regarded as a sign of his protection.

This is how Saint Elmo's fire first appeared to the terrified, storm-tossed crew: It assumed "the form of a lighted torch at the height of the maintop, and remained there more than two hours and a half, to the comfort of us all. For we were in tears, expecting only the hour of death. And when this holy light was about to leave us, it was so bright to the eyes of all that we were for more than a quarter of an hour as blind as men calling for mercy. For without any doubt, no man thought he would escape from that storm." Once the apparition subsided, some crew members believed that supernatural powers had singled out the Captain General for a special destiny. But their deliverance from the perils of the sea proved brief, and their faith in Magellan's ability to save them would soon be tested again.

For the moment, Magellan's official chronicler, Antonio Pigafetta, enjoyed a rare moment of repose and pondered the mysteries of the sea. No monsters with flaming faces menaced the ships; instead, flying fish leaped from the water, and not just a few, but "so great a quantity together that it seemed an island in the sea." The wonderful sight, half real, half illusion, mesmerized Pigafetta. In the sea below, as in the heavens above, there were marvels and perils beyond comprehension. This was not the world as described by the speculative historians of antiquity and the Middle Ages; it was stranger and richer, and even more dangerous.

CHAPTER IV

"The Church of the Lawless"

There passed a weary time. Each throat
Was parched, and glazed each eye.
A weary time! a weary time!
How glazed each weary eye,
When looking westward, I beheld
A something in the sky.

Sixty days of furious storms left the ships of the Armada de
Molucca in need of repair and ruined a good part of the precious
food supply. Magellan found it necessary to reduce rations. Each
man received only four pints of drinking water a day, and half that
amount of wine. Hardtack, a staple of the sailors' diet, was also
reduced to a pound and a half a day. As with his other decisions,
Magellan did not explain why he was reducing the amount of food
and drink, and no other decision he could take was as likely to create resentment among the captains and the crew.

Once the gales abated, the battered black ships drifted into
equatorial calms. As the sails luffed lamely amid rising tempera-
tures, the ships rode helplessly in the water. The rebellious Spanish
captains, with time on their hands, resumed plotting against the
Captain General. They avoided overt violence on this occasion;
rather, they displayed a pointed lack of regard for the status of a
man they considered their social inferior.

Magellan inadvertently set the stage for their mutiny when he reminded his officers that the instructions he had received from King Charles gave him full authority over the fleet. The captain of each ship was to approach *Trinidad* at dusk to pay his respects to Magellan and to receive orders. Cartagena chose to defy Magellan in a studied manner. When *San Antonio* approached the flagship, the quartermaster rather than Cartagena spoke up and, worse, he refused to address Magellan by the correct title. Cartagena should have said, *"Dios vos salve, señor capitán-general, y maestro y buena campaña."* ("God keep you, sir Captain General, and master and good company.") Instead, the lowly quartermaster called Magellan "Captain" rather than "Captain General."

Magellan sharply reminded Cartagena of the proper form of address, but the Castilian captain took the opportunity to insult Magellan again. If he did not approve of *San Antonio's* quartermaster offering the ceremonial salute, Cartagena would select a lowly page next time. For several days after that exchange, Cartagena neglected all forms of salute. Magellan had to devise an effective way to handle Cartagena's defiant attitude or risk losing control over the entire fleet.

At this tense moment, a new crisis erupted aboard *Victoria*. Magellan learned that *Victoria's* master, a Sicilian named Antonio Salamón, had been discovered sodomizing a cabin boy, Antonio Ginovés. There was no question as to whether the incident had taken place, because the two had been caught in flagrante delicto; the question was what to do about it.

Under Spanish law, homosexuality was punishable by death. Spanish authorities and the Catholic Church condemned homosexuality in the harshest language possible, despite its prevalence. As Captain General of the fleet, Magellan had little choice but to take disciplinary action, but he found himself in an impossible predicament, caught between the cruelty of Spanish law and the reality of homosexuality at sea. In practice, homosexuality among sailors confined to ships over long periods of time was inevitable. There are few accounts of captains attempting to punish sailors for this behavior; instead, they simply looked the other way. Magellan took a harsher course of action. He held a court-martial of Salamón, serving as both judge and jury. The outcome of the proceeding was swift, and Salamón was condemned to death by strangulation. The deed was to be carried out several weeks hence, on December 20.

After the hearing, Magellan held a tense meeting with the other captains of the fleet in his cabin; there was Cartagena from *San Antonio*, Quesada from *Concepción*, Mendoza from *Victoria*, and Serrano from *Santiago*. As Magellan realized, all the captains, except Serrano, were determined to lead a mutiny. Cartagena immediately began attacking Magellan about the eccentric and dangerous course they had been following along the coast of Africa. First Magellan had led them into storms, Cartagena complained, and now he had gotten them trapped in equatorial calms. Cartagena insisted that the only explanation for this bizarre behavior was that Magellan intended to subvert the fleet, because no matter how loyal to King Charles he claimed to be, Magellan's true loyalty belonged with the king of Portugal.

In his fervor to usurp Magellan, Cartagena had been misled by appearances. In fact, the Captain General had chosen the risky, unorthodox course to avoid the Portuguese caravels pursuing him and was actually doing his best to frustrate Spain's enemies.

Another resentment fueled Cartagena's passion for mutiny. He believed that King Charles had appointed the two of them as co-admirals of the fleet. Although Cartagena carried the title inspector general, and had been appointed *persona conjunta*, King Charles had intended no such power-sharing arrangement. Cartagena had little if any experience as a navigator, certainly had nothing to recommend him as an admiral of the most ambitious ocean expedition Spain ever mounted; rather, he was to serve as a symbol of the fleet's Span-ish identity. His chief qualification, besides his relationship to Arch-bishop Fonseca, was that he was a Castilian. On that basis, the privileged Cartagena believed he was entitled to share power equally with Magellan. Had Cartagena known the truth, that Ma-gellan was fleeing the Portuguese to save the fleet rather than destroy it, the revelation might have defeated the Castilian's para-noid logic, but it would not have restrained his unbridled chauvin-ism and his sense of entitlement.

As a Castilian loyal to his sovereign, Cartagena declared he would no longer take orders from Magellan.

Fully prepared to counter Cartegena's challenge, the Captain Gen-eral gave a sign, and *Trinidad's alguacil,* or master-at-arms, Gonzalo Gómez de Espinosa, stormed the cabin. Right behind him came two loyalists, Duarte Barbosa and Magellan's illegitimate son, Cristóvão Rebêlo, all with swords drawn. Magellan leaped at Cartagena, catch-ing the Castilian by the ruff of his shirt, and shoved him into a chair. "Rebel!" Magellan shouted, "this is mutiny! You are my prisoner, in the King's name."

At that, Cartagena barked at the other traitorous captains, Que-sada and Mendoza, to stab Magellan with their daggers. From the way he spoke, it was apparent that the three of them had plotted to overthrow the Captain General, but now, at the crucial moment, lost their resolve to act.

Seizing the initiative, Espinosa, in his role as *alguacil,* picked up Cartagena and shoved him out of the captain's cabin to the main deck, where he was secured to stocks intended for common seamen who had committed minor offenses. The indignity of seeing a Castilian officer subjected to this ignominy was more than Quesada and Mendoza could bear. They pleaded with Magellan to free Cartagena or, failing that, to release him into their custody. They reminded their Captain General that they had demonstrated their loyalty by ignoring Cartagena. They persuaded Magellan that he had nothing to fear from them, and he agreed to free Cartagena on condition that Mendoza confine him aboard *Victoria*. Cartagena was immediately relieved of command.

Had he chosen, Magellan could have convened a court-martial and sentenced Cartagena to death. As Captain General, he would have been within his rights because Cartagena had plotted to kill Magellan: Nothing could be more serious. But Magellan was acutely aware of Cartagena's privileged position and concerned that executing or severely punishing him would be inflammatory, so for once he erred on the side of caution. The lack of disciplinary action made it a certainty that the irascible Castilian would continue to challenge Magellan until only one of them remained.

With the brief mutiny at an end, Magellan ordered the trumpets aboard the flagship to sound, alerting the other ships, and he announced that henceforth, *San Antonio* would be commanded by Antonio de Coca.

Stripped of his command, and having learned nothing from the experience of his failed mutiny, Cartagena grew intensely resentful of his inexperienced replacement. From that moment, he burned with desire for revenge against Magellan, no matter what the cost to the expedition, and as Fonseca's son, Cartagena had power to make great trouble. Of all the perils that Magellan faced on the journey's first leg, the greatest was Cartagena's treachery.

With Cartagena removed from power, at least temporarily, Magellan turned his attention to his long-delayed crossing of the Atlantic. For three weeks in late October and November, the fleet headed south, vainly awaiting favorable winds. At last the sails began to fill, and Magellan ordered the ships to set a southwesterly course toward Rio de Janeiro. Learning that *Concepción*'s pilot, João Lopes Carvalho, had visited Rio several years before on an earlier expedition, Magellan brought him over to *Trinidad* to serve as pilot. To supplement Carvalho's expertise, the Captain General carried with him a reliable, though not flawless, map of the Brazilian coast known as the Livro da Marinharia—the Book of the Sea. At about the same time, Francisco Albo began keeping a navigational log intended for use by those following in the wake of the Armada de Molucca.

Neither of these expert pilots knew of the South Equatorial Current, which carried the fleet west of its intended heading. Rather than Rio de Janeiro, the fleet raised Cape Saint Augustine on November 29. Here, Pigafetta relates, the fleet paused to take on fresh food and water, and quickly resumed following the Brazilian coast in search of Rio de Janeiro, as the best navigational minds aboard the ships puzzled over why they had veered off course. Albo recorded, "We arose in the morning to the right of St. Thomas, on a great mountain, and south slopes along the coast in the S.W. direction; and on this coast, at four leagues to sea, we found bottom at twenty-five fathoms, free from shoals; and the mountains are separated from one another, and have many reefs around them." Finally, two weeks later, on December 13, 1519, the fleet entered the lush and gorgeous Bay of Saint Lucy and approached the mouth of the River of January—Rio de Janeiro.

Trinidad went first, slipping past Sugar Loaf and coming quietly to anchor in the harbor. Magellan had arrived in the New World.

*I*n the final days of 1499, Vicente Yáñez Pinzón, a Spanish mariner, first saw the coast of what would later be called Brazil. Pinzón explored the easternmost shores of Brazil and ventured into the mouth of the Amazon River, but Spain failed to maintain a settlement in the newly discovered wilderness. Months later, a Portuguese explorer named Pedro Álvares Cabral claimed the entire region—whose contours were poorly mapped and poorly understood—for his king and country. For tiny Portugal, hemmed in by the Atlantic and by Spain, the newly discovered land contained great commercial and psychological promise, but it lacked quantities of gold and spices. Unsure about how to exploit their find, the Portuguese became lackadaisical in the administration of the distant realm.

For ten years, the newly discovered land went by various names; not until 1511 did "Brazil" first appear on a map, and its origins are something of a mystery. The name might have derived from the Portuguese word *brasa,* meaning glowing coal, thought to resemble the color of the dark red wood that came to be prized by the Portuguese. Or it might have derived from "bresel wood," which had been imported to Europe from India since the Middle Ages. The bright red wood was used for cabinets, violin bows, and dyeing. The newly discovered South American variety resembled the traditional Indian tree, but was easier and cheaper to obtain. No matter what its derivation, the name "Brazil" was slow to catch on. In his diary, Pigafetta called the land "Verzin," derived from the Italian word for brazilwood.

The Portuguese bestowed a valuable brazilwood monopoly lasting ten years on an influential businessman, Fernão de Noronha, in exchange for large fees, and for a while commerce flourished under his management. The coast abounded in the trees; the Portuguese cut them down, sawed the trunks and branches into a manageable size, and stored the wood in a *feitoria,* or factory, until a ship came to collect and transport the valuable cargo back to Lisbon. (This activity had first brought *Concepción's* pilot, João Lopes Carvalho, to Brazil in 1512, aboard a commercial Portuguese ship called *Bertoa.* The ship soon departed, but Carvalho remained to oversee the factory, a sojourn that lasted four years.)

The Portuguese dealings in brazilwood served as a model of how that country planned to exploit the natural resources of distant lands they claimed for their own. The most unpredictable part of the enterprise proved to be the transatlantic crossings, and even they became increasingly manageable as Portuguese navigators learned the winds and currents affecting their route. In practice, though, the brazilwood trade was too far-flung to administer with any coherence. The French were already helping themselves to brazilwood without interference. The unchallenged presence of the five ships comprising the Armada de Molucca in Brazil showed how porous and vulnerable the Portuguese "monopoly" actually was. Despite Brazil's importance, the Portuguese did not maintain a permanent settlement there. A small abandoned customshouse served as the sole evidence of the Portuguese occupation. No Portuguese ships occupied the harbor when Magellan arrived, and he felt safe enough to drop anchor.

Although this was his first visit to Brazil, Magellan was familiar with the brilliantly evocative descriptions of the land written by Amerigo Vespucci after his visit in 1502. In his words Brazil and its natural wonders were the closest approximation to Paradise that Magellan was likely to encounter during his entire voyage around the world. "This land is very delightful, and covered with an infinite number of green trees and very big ones which never lose their foliage, and through the year yield the sweetest aromatic perfumes and produce an infinite variety of fruit, gratifying to the taste and healthful to the body," Vespucci reported. "And the fields produce herbs and flowers and many sweet and good roots, which are so marvelous . . . that I fancied myself to be near the terrestrial paradise." Vespucci's descriptions, for all their charm, were not the elaborately embellished creations of Sir John Mandeville; they were generally reliable accounts looking forward to the Age of Discovery rather than backward to the Age of Faith.

Discussing the region's indigenous tribes, Vespucci wrote out of his own experience: "I tried very hard to understand their life and customs because for twenty-seven days I ate and slept with them." He assembled a disturbing if tantalizing picture of the Indians whom Magellan and his crew would encounter in Rio de Janeiro: "They have no laws or faith, and live according to nature. They do not recognize the immortality of the soul, they have among them no private property, because everything is common; they have no boundaries of kingdoms and provinces, and no king! They obey nobody, each is lord unto himself; no justice, no gratitude, which to them is unnecessary because it is not part of their code." Vespucci thrilled readers with gruesome accounts of the Indians' customs. "These men are accustomed to bore holes in their lips and cheeks, and in these holes they place bones and stones; and don't believe that they are little. Most of them have at least three holes and some seven and some nine, in which they place stones of green and white alabaster, and which are as large as a Catalan plum, which seems unnatural; they say they do this to appear more ferocious, an infi-nitely brutal thing." Even more repugnant—yet fascinating to Vespucci—were their marital and sexual customs. "Their marriages are not with one woman but with as many as they like, and without much ceremony, and we have known someone who had ten women; they are jealous of them, and if it happens that one of these women is unfaithful, he punishes her and beats her."

More troubling, the Indians practiced cannibalism and human sac-rifice in the course of their battles. "They are a warlike people, and among them is much cruelty," he warned. "Nor do they follow any tac-tics in their wars, except that they take counsel of old men; and when they fight they do so very cruelly, and that side which is lord of the battlefield bury their own, but the enemy dead they cut up and eat. Those whom they capture they take home as slaves, and if [they are] women, they sleep with them; if [they are] men, they marry them to their girls, and at certain times when diabolic fury comes over them they sacrifice the mother with all the children she has had, and with certain ceremonies kill and eat them, and they did the same to the said slaves and the children who were born of them." Vespucci con-cluded, "One of their men confessed to me that he had eaten of the flesh of more of than 200 bodies, and this I believe for certain."

Vespucci's Indians were most likely representatives of the vast network of Guaraní tribes. At the time of Magellan's arrival, there may have been as many as 400,000 Guaraní Indians, grouped by dialects. They occupied huge regions of South America extending all the way to the Andes, and lived communally in huts sheltering about a dozen families each; polygamy was not unknown to them, but it was not common. They were short—rarely more than five feet tall—and, by European standards, stout. The men wore a simple G-string and occasionally a headpiece made of feathers; the women were fully clothed. They were adept at pottery, wood carving, and skillful in their weapons of choice: the bow and arrow and the blowgun. The origin of the name Guaraní, by which they were known to the outside world, is unclear; they called themselves Abá, their word for "men."

The arrival of the Armada de Molucca in Rio de Janeiro coincided with heavy rains that ended a two-month drought in the region. "The day we arrived the rain began," Pigafetta noted, "so that the people of the place said that we came from heaven and had brought the rain with us." The sight of strange ships arriving in the harbor inspired benign rather than warlike feelings in the hearts of the Indians, as Pigafetta later learned. "They thought that the small boats of the ships were the children of the ships, and that the said ships gave birth to them when the boats were lowered to send the men hither and yon."

Yet the Guaraní Indians disturbed Pigafetta as much as they had Vespucci. Pigafetta had no doubt that the Indians practiced cannibalism, and even contributed a story about the origins of the practice, "an established custom begun by an old woman who had but one son who was killed by his enemies." Pigafetta continued, "Some days later, that old woman's friends captured one of the company who had killed her son, and brought him to the place of her abode. She, seeing him and remembering her son, ran upon him like an infuriated bitch and bit him on one shoulder. Shortly afterward, he escaped to his own people, whom he told they had tried to eat him, showing them the marks on his shoulder." The incident led to a never-ending cycle of attacks, followed by cannibalistic practices, or so Pigafetta claimed. He provided a gruesome description of how it had become part of everyday life: "They do not eat the bodies all at once, but everyone cuts off a piece, and carries it to his house, where he smokes it. Then, every week, he cuts off a small bit, which he eats thus smoked with his other food to remind him of his enemies."

*A*s Magellan's ships came to rest, a throng of women—all of them naked and eager for contact with the sojourners—swam out to greet them. Deprived of the company of women for months, the sailors believed they had found an earthly paradise. Any fear they might have had of Indian cannibals melted in the flame of carnal pleasure.

Discovering that the women of Verzin were for sale, the sailors gladly exchanged their cheap German knives for sexual favors. Night after night on the beach the sailors and the Indian women drank, danced, and exchanged partners in moonlit orgies. But there were limits: "The men gave us one or two of their young daughters as slaves for one hatchet or one large knife, but they would not give us their wives in exchange for anything at all. The women will not shame their husbands under any considerations whatever, and as was told us, refuse to consent to their husbands by day, but only by night." Even so, the sailors found it easy to take advantage of the women, and one of the women, in turn, tried to take advantage of the fleet.

"One day, a beautiful woman came to the flagship, where I was," Pigafetta wrote, "for no other purpose than to seek what chance might offer. While there and waiting, she cast her eyes upon the master's room, and saw a nail longer than one finger. Picking it up very delightedly and neatly, she thrust it between the lips of her vagina and, bending down low, immediately departed, the Captain General and I having seen the action." The reason for the astonishing behavior was the great value the Guaraní Indians placed on metal objects such as nails, hammers, hooks, and mirrors, all of which were considered to be more valuable than gold, more valuable, perhaps, than life itself.

That was not the only disturbing incident involving these women. Under the strain of temptation, one of Magellan's most trusted allies, Duarte Barbosa, who had offered critical assistance when Cartagena mutinied, all but lost his head in Rio de Janeiro. Falling under the women's spell and envisioning a life of ease as a trader on these distant shores, he decided to desert the fleet. Magellan learned of the plan and intervened at the last minute, sending sailors to arrest Barbosa onshore and drag him back to the ships. The poor man spent the rest of the layover in Rio de Janeiro confined in fetters aboard his ship, gazing on the women and the self-indulgent life that Magellan—and duty—denied him.

*W*hile the sailors pursued their casual liaisons with the Indian women, Magellan transacted business with their men. He took on fresh supplies of water and provisions, trading insignificant trinkets, such as tiny bells that he had brought with him from Seville, for precious food. "The people of this place gave for a knife or fishhook five or six fowls, and for a comb a brace of geese," Pigafetta wrote. "For a small mirror or a pair of scissors, they gave as many fish as ten men could have eaten. For a bell or a leather lace, they gave us a basketful of . . . fruit. And for a king of playing cards, of the kind used in Italy, they gave me five fowls, and even thought they cheated me."

The Captain General and the fleet's three priests intended to maintain strict religious observance throughout the voyage, both to keep their own sailors faithful and to impress the local inhabitants with the power of Christianity, and the impressionable Indians eagerly accepted Magellan's invitation to attend worship. "Mass was said twice on shore, during which those people remained on their knees with so great contrition and with clasped hands raised aloft that it was an exceedingly great pleasure to behold them," Pigafetta reported, with obvious gratification and pride. Only later did Magellan learn that the Indians regarded the fleet as harbingers of good fortune because its arrival coincided with rain. Whatever the reason, "Those people could be converted easily to the faith of Jesus Christ," Pigafetta concluded.

*T*he tranquillity of the fleet's layover in Rio de Janeiro was interrupted by a traumatic event: carrying out Antonio Salamón's death sentence on December 20. On the appointed day, Magellan summoned the officers and crew of *Trinidad* to watch the execution of the man who had committed a "crime against nature." One of the sailors, never named, his face likely hooded to preserve his anonymity, strangled Salamón in full view of the other men, as a warning. The grisly spectacle, performed with military efficiency, increased resentment among the crew against the Captain General.

There are conflicting accounts concerning Antonio Ginovés, the cabin boy whose life Magellan had spared. In one version, Ginovés suffered such extreme ridicule from other crew members that he threw himself overboard and was lost. And in another, the cabin boy, an object of scorn, was thrown overboard to his death. No matter which version was correct, the double tragedy marked the only time Magellan addressed the subject of homosexuality throughout the voyage. If homosexual relationships flourished again aboard the ships—and they likely did—Magellan decided to follow the tradition of looking the other way.

*F*ive days later, the Armada de Molucca observed its first Christmas away from Spain in the shelter of Rio's harbor, but there was little time to reflect on the holiday because the men busily prepared the ships for departure. Before weighing anchor, Magellan, together with his trusted pilots and navigators, attempted to determine the coordinates of Rio de Janeiro. Although they lacked the skills and instruments necessary to calibrate longitude with accuracy, they believed they could make useful calculations with the help of Ruy Faleiro's tables and the advice of Andrés de San Martín, the fleet's astrologer and astronomer. Not surprisingly, they arrived at an unreliable estimate, but they did make reasonably accurate calculations of the latitudes of several landmarks they had visited. Even Magellan's best measurements, good to within a degree or two, were not accurate enough to warn subsequent travelers away from hazards such as shoals and rocks; they were, at best, rough approximations.

Just before sailing, Magellan replaced Antonio de Coca, the fleet accountant who had briefly assumed command of *San Antonio* from Cartagena, with the inexperienced Álvaro de Mesquita. Both de Coca and Cartagena took the shuffle as an insult, because Mesquita had shipped out aboard the flagship from Seville as a mere supernumerary. The deposed captains cried nepotism, which was true, because Mesquita was Magellan's cousin. The lack of qualified captains in the fleet's roster would trouble Magellan throughout the voyage. Although he had a surplus of qualified pilots, most were Portuguese, and so excluded from the top ranks of this Spanish expedition. As the voyage continued, these professional, accomplished pilots served resentfully under the figurehead captains.

*A*fter two weeks of sensual indulgence, the fleet's departure from Rio de Janeiro on December 27 became an emotionally charged affair. João Lopes Carvalho, Magellan's pilot, returning to Brazil after a seven-year absence, happily reunited with his former mistress, who introduced him to their son. Carvalho took an immediate liking to the lad, whom he called Joãozito, and enlisted him as a servant aboard ship. As the fleet prepared to embark, the pilot beseeched Magellan for permission to take along the mother of his child, but Magellan allowed absolutely no women on the ships. Car-valho would sail alone.

Alarmed by the prospect of other liaisons affecting the crew, Magellan ordered an inspection of every inch of every ship for female stowaways. Several were found and swiftly returned to shore. When the fleet finally weighed anchor and sailed away, Indian women followed them in canoes, tearfully pleading with the men from distant shores to stay with them forever.

$\mathcal{R}$esuming a southerly course, the fleet, helped by favorable winds, reached Paranaguá Bay, off the coast of Brazil, by the last day of 1519. Intent on making up for time, Magellan ordered the ships to remain offshore rather than exploring the bay, one of the largest estuaries in the southwest Atlantic. Fully provisioned, the Armada de Molucca sailed on, day and night, until January 8, 1520, when Magellan spied a stretch of shoal extending as far as the eye could see. Concerned about hitting a concealed formation, he gave an order to drop anchor, but only for the night; in the morning, the fleet sailed on.

On January 10, the rolling hills and mountains of the South American coast yielded to barely discernible hummocks and the suggestion of offshore islands. Carvalho declared that they had arrived at Cape Santa María, rumored to be the gateway to the strait. If luck favored the fleet, Magellan could reach his goal ahead of winter storms. It was now summer in these subequatorial regions, and he wanted to take advantage of the relatively mild weather and traverse the strait before the weather turned cold. Just when he believed he was approaching the mouth of the strait, all his maps turned to blank wastes and speculative renderings, and the monoto-nous barrier of South American coast continued without relief.

Magellan's hope for a swift completion to the expedition would not be fulfilled.

*F*ive months from Seville, the crew and officers had become familiar with the ships as well as the rigors and deprivations of life at sea. They had learned of the violence of storms, the life-and-death necessity of sounding the bottom, and the limits of the proud vessels in which they sailed over the surface of the limitless sea. The misery of seasickness was at last behind them. There had been no escape from the ordeal; even veteran mariners were vulnerable to its pains. According to folk wisdom, sexual activity increased the likelihood of seasickness, but it was a rare sailor who could resist the opportunity for coupling before setting out on a long voyage.

At sea, sleep became the ultimate luxury, a solace nearly impossible to come by. The crew took naps whenever they could, night or day. Hammocks had yet to be introduced on board ships, so exhausted sailors appropriated a plank or, better still, a sheltered area of the deck where they could sprawl. They eased the wood's bruising hardness with a straw pallet, and shielded themselves against the cold and wet with heavy blankets. Even then, comfort eluded them. The men never became accustomed to the foul odors brewing aboard their ships. Water seeping into the hold stank despite the efforts to disinfect it with vinegar; animals such as cows and pigs added to the reek, as did the slowly rotting food supply and the sickening smell of salted fish wafting from the hold.

Pests were ubiquitous, an inescapable fact of life at sea. Teredos, or shipworms, bored through the hull, slowly compromising the seaworthiness of the entire vessel, and one ship in Magellan's fleet eventually disintegrated because of the wretched little creatures. Rats and mice infested every ship, and the sailors learned to live with them and even to play with them. Magellan's crew might have brought along a domestic creature new to Europe at the time—the cat—to hunt the rodents, following the practice of the day, although no record confirms that they did. It is recorded, however, that the men of the Armada de Molucca were plagued with all manner of lice, bedbugs, and cockroaches. When conditions turned hot and humid, the insects infested the clothing, the sails, the food supply, and even the rigging. The sailors scratched and complained, but they had no defense against the pests. Even worse, weevils invaded the hardtack, and it was further contaminated with the urine and feces of rodents. Crew members with growling stomachs forced themselves to overcome their inhibitions and swallow this disgusting, contaminated provender.

Sailors found it nearly impossible to keep clean; many brought along soap and a rag for washing, but the only available water—seawater—caused itching and irritation. The sailors washed their clothes in seawater as well, with limited results. To keep warm and dry, sailors wore baggy, loosely fitting clothes consisting of a floppy shirt, often with a hood, over which they wore a woolen pullover known as a *sayuelo*, which was cinched at the waist. Sailors were known everywhere for their floppy, pajamalike pants (*zaragüelles*), which reached below the knees. Depending on the rank of the sailor, and the money at his disposal, *zaragüelles* could be made of anything from the cheap coarse linen known as *anjeo* (after Anjou, in France) to fine wool lined with silk taffeta.

In foul weather, sailors and officers alike donned great blue capes called *capotes de la mar;* it was a common sight to see a watchman huddled within his cape, with only his head exposed, peering across a storm-tossed deck for hours on end. Sailors protected their heads (and ears) with a woolen cap called a *bonete:* more than any other article of clothing, the *bonete* was the mark of a sailor. Magellan brought along a number of caps, mostly in red, to befriend the Indians he expected to encounter along his route to the Spice Islands, but most sailors wore a *bonete* of a more dignified black or blue. Frayed from hard use and harsh conditions, the clothing demanded constant repair, and the sailors learned to become handy with a needle. Many sailors carried knives tucked into their waistbands for safekeeping.

Sailors stored their gear in large chests. In addition to clothes, the chests contained simple wooden plates (useful for hurling during impromptu fights), eating utensils, and a jug to hold the daily ration of wine. The chests frequently contained a supply of playing cards—probably the most popular pastime aboard the ships of the Armada de Molucca—and books.

The Inquisition imposed strict censorship laws, and sailors submitted all books they brought to sea for approval. The surviving records afford a glimpse of the reading habits of these men. Most volumes were devotional—the lives of saints, profiles of popes, accounts of miracles, and prayers to recite. Almost as prevalent (and probably more carefully read) were popular stories of derring-do and chivalry, of knights and damsels and vanquished villains. A few histories found their way aboard these ships, but the more literate sailors favored the most celebrated precedent for their own journey, Marco Polo's *Travels.*

*M*agellan's crew was overwhelmingly Castilian and Portuguese, but representatives of every major country in western Europe, as well as North Africa, Greece, Rhodes, and Sicily filled the ranks. Their number included alliances of natural enemies: Britons and Basques, Flemish and French, all speaking in mutually unintelligible tongues.

The common language aboard Magellan's fleet was nautical Castilian, which contained specialized terms for every line, clew, and device to be found aboard the ships. In this idiom Magellan and his captains gave orders to the crew. *"Izá el trinquete,"* they cried, to raise the foresail; *"Tirá de los escotines de gabia,"* to haul in the topsail sheets. *"Dad vuelta,"* uttered with special vehemence, meant put your back into it. And there were many other orders, enough to cover every operation a sailor could be expected to perform. *"Dejad las chafaldetas"*...well the clew lines. *Alzá aquel briol*...heave on that buntline. *Levá el papahigo*...hoist the main course. *Pon la mesana*...set the mizzen. *Tirá de los scotines de gabia*...haul in the topsail sheets. The cry of *"Suban dos á los penoles"* dispatched two sailors, scampering in tandem up the mast, trying not to look down on the heaving deck as they hauled themselves toward the sky; and the order *"Juegue el guimbalate para que la bomba achique"* sent more hands below to perform the backbreaking labor of working the pumps until the blasted thing sucked water out of the hold. The bilgewater around the pumps was also the most noxious to be found anywhere on the ship, and sailors retched from the stench. Despite the various hardships involved with operating the pumps, they were an absolute necessity at sea; without them, ships slowly took on water till they sank, and operating them exhausted teams of able-bodied seamen. It was not unheard-of for mariners to collapse and die during the ordeal of working the pump to save a ship.

The sailors had their secular chants, or *saloma,* for the arduous routine tasks they performed. The men knew them all by heart. If they were hauling the anchor, the chanteyman would shout or perhaps sing the first half of the line, and the others, gripping the rope, would complete the second half. *"O dio,"* cried the chanteyman, *"Ayuta noy,"* the men replied in unison. *"O que somo,"* he sang out; *"Servi soy"* came the reply. *"O voleamo . . . Ben servir."* And so on until the order came to make fast the line, and the men fell out to catch their breaths.

*T*he men quickly left behind the identities they had maintained on land for those imposed on them at sea. No longer did it matter if they were Castilians, Greeks, Portuguese, or Genoese; life aboard ship was lived according to a rigid social structure segregating men who nonetheless lived in extremely close quarters and who depended on each other for their survival.

A strict division of labor ruled over all. At the bottom were the pages, assigned to the ships in pairs. Many pages were mere children, as young as eight years old; none were older than fifteen. They were commonly orphans. Not all pages were created equal. Some had been virtually kidnapped from the quays of Seville and pressed into service; if they had not been on ships, they would have been roaming the streets, learning to pick pockets and getting into minor scrapes. They were treated harshly, exploited shamelessly, deprived of adequate pay, and occasionally made the victims of sexual predators among older crew members. Their chores included scrubbing the decks with saltwater hauled from the sea in buckets, serving and cleaning after meals, and performing any menial task assigned to them.

Another class of page lived a very different life, privileged and relatively free of demand, under the protection of officers. These handpicked young men generally came from good, well-connected families, and worked as apprentices for their protectors; they were expected to learn their trade and to rise through the ranks. Their duties were far lighter than those of the unfortunate boys who had been pressed into service.

The privileged pages maintained the sixteen Venetian sand clocks—or *ampolletas*—carried by Magellan's ships. Basically a large hourglass, the sand clock had been in use since Egyptian times; it was essential for both timekeeping and for navigation. The *ampolletas* consisted of a glass vessel divided into two compartments. The upper chamber contained a quantity of sand trickling into the lower over a precisely measured period of time, usually a half hour or an hour. Maintaining the *ampolletas* was simple enough—the pages turned them over every half hour, night and day—but the task was critical. Aboard a swaying ship, the *ampolletas* were the only reliable timepiece, and the captain depended on them for dead reckoning and changing the watches. A ship without a functioning *ampolleta* was effectively disabled.

Operating the *ampolletas* aboard the armada had religious over-tones, and the pages, in their presumed innocence, doubled as the ships' acolytes. When they turned over the sand clocks, they recited psalms or prayers invoking divine guidance for a safe voyage. Usu-ally, the prayers required a chorus, and they had to chant loudly enough to demonstrate that they were on the job and fulfilling their duties promptly. At the end of the day, their high voices could be heard above the ship's bawdy clamor, reciting prayers to the Virgin Mary, reminding all of their religious obligations, even here, thousands of miles from home.

*F*inished with their prayers, the boys called the new watch to their post. *"Al cuarto!"* they cried. *Al cuarto!* On deck! On deck! And the members of the day watch staggered to their accustomed places, where they could crouch comfortably against a sheltering plank or overhanging wooden ornament. They might carry a fistful of hardtack or salted fish, and they almost certainly regretted their chronic lack of sleep because night aboard ship was as noisy as day; the ocean never slept, and neither did they.

If the sailors had a moment before reporting for duty, they might relieve themselves, an unpleasant, even ridiculous chore aboard the ships. To urinate, they simply stood and faced the ocean wherever they could be sure that the wind would not send the stream back on them, or anyone else. Defecating was even more difficult, calling for a precarious balancing act as a sailor eased himself over the rail and lowered himself onto a crude seat suspended high above the waves. There were two such seats, fore and aft, known as *jardines,* a name ironically suggesting flowers. After he lowered his breeches and eased himself into the seat, the sailor had to void himself in full view of anyone who cared to watch—privacy did not exist aboard these ships—and if the sea happened to be rough, the frigid spray splattered his exposed bottom. (More than one sailor lost his life when he plunged from the *jardines* to the ocean.) When he was done, he wiped himself with a length of pitch-covered rope, and then climbed back on deck, where he no doubt breathed a sigh of relief.

Was it any wonder that the ship, with all its filth and noise and nauseating odor, was called *pájaro puerco,* a flying pig?

Once they had taken up their posts, the weary sailors studied the sea for buried shoals, examined the rigging, dried the dew from the lines, and checked the sails for damage. They scrubbed, repaired, overhauled, and polished every exposed surface of the ships. They applied pitch to fraying hemp, and repaired torn or stressed sails. They made their weapons gleam, and fought a constant, losing battle against protecting their food supply from vermin. After several months at sea, the five ships of the Armada de Molucca were in far better shape than they had been when they sailed from Seville.

*J*ust above the pages in rank came the apprentices, or *grumetes*, the most expendable and vulnerable members of all the crew. Ranging in age from seventeen to twenty, they were the ones who sprang on the rigging the moment the captain ordered them to furl or unfurl the sail; to scamper to the dangerous lookout posts atop the masts, to pull on the oars in the longboats, and to operate the complex mechanical devices aboard ship, the pulleys and cranes, the cables and anchors, the fixed and movable rigging. They teamed up to operate the capstan, rotating its drum with levers to load (or unload) heavy supplies, weapons, and ballast. They even shaved the legs and manicured the toenails of their masters, perhaps setting the stage for sexual relations between the two, even though such behavior was strictly forbidden. Apprentices were the group most likely to be disciplined, to be whipped for disobedience, or to be confined to the stocks for as long as a week.

If an apprentice survived all the ordeals and hazards of life at sea, he could apply for certification as a "sailor," receiving a document signed by the ship's pilot, boatswain, and master. He was now a professional mariner, and could look forward to a career lasting about twenty years, if he lived that long. Sailors advanced through the ranks by learning how to handle the helm, deploy the sounding line, splice cables, and, if they were mathematically inclined, marking charts and taking measurements of celestial objects to fix the ship's position.

Most sailors were in their teens or twenties. Anyone who had reached his thirties was considered a veteran scalawag; by the time

he had survived to that age, he had seen what life at sea held: brutality, loneliness, and disease; he had experienced flashes of cama-
raderie and heroism, as well as persistent dishonesty and callousness.
He knew all about the avarice of shipowners, the uncomprehending
indifference of kings under whose flags the expedition sailed, and
the tyranny of captains. Men rarely went to sea beyond the age of
forty. Magellan, nearly that age when he left Seville, was among the
oldest, if not *the* oldest person aboard the Armada de Molucca.

No matter how high an ordinary sailor rose, he was outranked by
specialists such as gunners, essential to expeditions exploring new
lands but hard to come by. Skilled in the use of cannon, in the
preparation of gunpowder, and the selection of projectiles, a gunner
tended to his weapons throughout the voyage, keeping them secure,
clean, free of rust, and ready for battle at all times. Although most
gunners were Flemings, Germans, or Italians, the Casa de Contratación kept a gunnery instructor on its staff to train Castilians. The
Casa provided the cannon, but the gunners-in-training had to pay
the instructor's fees, as well as the cost of gunpowder, which was
enough to discourage many potential students. Less glamorous but
equally necessary fields of specialization included carpenters, caulkers, and coopers. This last group repaired the hundreds of casks and
buckets aboard the ships by replacing hoops or staves and sealing
leaks. There was also a complement of divers aboard the fleet, whose
job it was to swim under the ships and, when necessary, clear seaweed from the rudder and keel, and inspect the hull for signs of
exterior damage and leaks.

The ship's barber, another specialist, was deceptively named
because trimming beards was the least of his responsibilities. He
served as the onboard dentist, doctor, and surgeon, ministering to
the crew out of his chest of nostrums, herbs, and folk remedies. The
fleet's barber was named Hernando Bustamente, who shipped out
aboard *Concepción*. Records show that his medical supplies were purchased from an apothecary named Johan Vernal on July 19, 1519,
shortly before departure. Included were distillations of various

herbs, among them fennel, thistle, and chicory; a purgative known as diacatholicon; turpentine; lard; various unguents and oils; six pounds of chamomile; honey; incense; and quicksilver—all of them carefully stored in canisters. Bustamente also carried an assortment of tools with him. Medical chests of the era contained a brass mortar and pestle to grind compounds, and a selection of surgical instruments including scissors, a lancet, a tooth extractor, an enema syringe made from copper, and a scale. This slender store of medical supplies and equipment would have to serve the needs of 260 men of the fleet in all climates and conditions for several years. In practice, Bustamente's most frequent duty at sea was extracting teeth, not treating disease.

No one answered to the description of cook aboard these ships because the job was considered too demeaning. One sailor telling another that his beard smelled of smoke was tantamount to provoking a fight. So the crew took turns cooking, or paid the apprentices to cook for them. And during foul weather, there was no cooking at all, and the sailors endured cold repasts of hardtack, salted meat, and wine.

In addition to these traditional roles, the armada's roster included phantom crew members: saints who, by custom of the sea, found their way onto the ships' rosters. Magellan's fleet included Santo Adelmo, the patron saint of Burgos; Santo António de Lisboa, the popular patron saint of Lisbon, who was reputed to rescue shipwrecked sailors and provide favorable winds to their ships; Santa Bárbara, whom Spaniards invoked as a safeguard against violent storms; and Nuestra Señora de Montserrat, to whom a famous Benedictine shrine was dedicated. Even more remarkable, each of these ghostly personages was accorded a share of the fleet's profits in return for divine protection; the arrangement was a clever way of donating a portion of the expedition's profits to the Church.

Officers ranked just above the sailors and specialists in the fleet's hierarchy. One tier consisted of the steward, charged with keeping an eye on the food supply; the boatswain, or *contramaestre*, the boatswain's mate; and the *alguacil*. The *alguacil*, for which there is no exact translation, served as the king's representative aboard the ship and served as a master-at-arms or military officer. If Magellan needed to arrest a crew member, he ordered the *alguacil* to perform the deed. This was not a job designed to endear him to the other crew members, and the *alguacil* stood apart from the rest of the crew.

At the top of the pinnacle came the pilot, who plotted the ship's route; the master, who supervised the precious cargo; and finally the captain. Each of the top three officers had his own page (as Captain General, Magellan had several, including his illegitimate son), and they lived a life as separate as possible from the rank-and-file sailors and apprentices. The officers had their own cabins, cramped, to be sure, but a mark of distinction, and they rarely ate with the crew. To most of the men aboard the fleet, even the flagship, *Trinidad*, Ferdinand Magellan seemed a remote, imperious figure, authoritarian and arbitrary, a man whose every word was law, and on whose skill, luck, and good judgment their lives depended.

*A*lthough sea captains, Magellan included, could be notoriously high-handed, the sailors' lot was governed, at least in theory, by the Consulado del Mare, the Spanish maritime code that had been in existence—and in force—for several centuries before it was formally compiled in 1494. The code described approved methods for hiring and paying sailors, and spelled out the ordinary seaman's exhausting chores ("to go to the forest and fetch wood, to saw and to make planks, to make spars and ropes, to bake, to man the boat with the boatswain, to stow goods and to unstow them; and at every hour when the mate shall order him to go and fetch spars and ropes, to carry planks, and to put on board all the victuals of the merchants, to heave the vessel over"), as well as the punishments they could expect to receive if they failed to follow orders ("A mariner ought not to undress himself if he is not in a port for wintering. And if he does so, for each time he ought to be plunged into the sea with a rope from the yard arm three times; and after three times offending, he ought to lose his salary and the goods which he has in the ship"). In addition, a sailor was bound to go wherever the captain ordered, even "to the end of the world." So, under the Consulado, Magellan had the right to take his crew wherever he wished, all the way to the Spice Islands, and even beyond.

The provisions of the Consulado afforded some protection to sailors by specifying their diet. They were entitled to meat three days a week, "That is to say on Sundays, Tuesday, and Thursdays." Other days, they were to be served "porridge, and every evening of every day accompaniment with bread, and also on the same three days in the morning he ought to give them wine, and also he ought to give them the same quantity of wine in the evening." Accounts of exactly how much wine Magellan allowed his crew members vary, but it probably came to two liters per man each day. And on Feast Days, which were frequent, the Consulado specified that the captain was to double his crew's rations. Magellan, from all accounts, followed these guidelines scrupulously, except when he had to cut back on rations to prevent starvation. As the voyage unfolded, it became apparent that he, like other captains of the day, had two obsessions: maintaining the seaworthiness of his fragile ships, and acquiring enough food for his unruly men.

*W*hy did sailors put up with it all? Why did the ordinary seamen and trained officers abandon hearth and home to live amid these grim circumstances for years on end? Why did they endure starvation rations, the indignity and agony of the lash and the stocks, torment by vermin, thirst, sunstroke, and the lack of women? They went to sea for a variety of reasons, for glory and greed, for escape, out of habit, out of desperation, and through pure chance. To Juan de Escalante de Mendoza, the veteran Spanish mariner, sailors came in two varieties. "The first sort includes all those who commence to sail as a livelihood, such as poor men. . . . Seafaring is the most suit-

able occupation they can find to sustain themselves, especially for
those born in ports and maritime areas. This sort is the most
numerous among mariners," he noted. "Although they might want
to be schooled for some other occupation, they do not have the disposition or the means to be able to do it." So they went to sea
because it was their livelihood, and in all likelihood their fathers'
before them; because they knew the sea better than they knew land;
because they could throw off the concerns of ordinary life; because,
if they stayed home, they knew the dreary routines life held in store
for them, whereas at sea anything could happen; because, if they
survived the ordeal of an ocean voyage, they would have a fund of
stories to draw on for the rest of their lives; and finally because
if they successfully smuggled even a small amount of gold or spices,
they would have a nest egg to sustain them and their families against
the vicissitudes of life.

Many of the men went to sea simply to escape. Some were fleeing jail, hanging, or torture; others were abandoning their families
and responsibilities. Others were avoiding debtors' prison; once they
obtained a berth on a ship, they would be immune from arrest, safe
for as long as they were at sea. Many sailors planned to desert their
ships once they reached the fabled Indies, with their gold and
women and luxury. For them, the Indies served, in Cervantes's
words, as "the shelter and refuge of Spain's desperadoes, the church
of the lawless, the safe haven of murderers, the native land and cover
for cardsharps, the general lure for loose women, and the common
deception of the many and the remedy of the particular few."

*I*n the late hours of January 10, 1520, a severe storm descended on
the Armada de Molucca, forcing Magellan to seek shelter. He
ordered the fleet to reverse course and head north, toward the shelter offered by Paranaguá Bay. During the journey to safety, fierce
but erratic winds blew the fleet off course, and Magellan found himself in dangerously shallow waters. Before him stretched the mouth

of the Río de la Plata, a funnel-shaped river located on the coast of what is now Argentina.

We know, though Magellan did not, that the Río de la Plata is fed by two important rivers, the Río Uruguay and the Río Paraná, whose headwaters originate in the Andes. Sailing into these shallow, sediment-rich waters, Magellan thought he might have been entering the waterway leading to Asia, but the weather frustrated his efforts at reconnaissance. The region's climate is typical of the temperate middle latitudes. Dry winds, called *zondas*, swoop down from the Andes; when they combine with cold offshore currents in the Atlantic, the result can be coastal storms called *sudestadas*, and it was probably a robust *sudestada* that caused Magellan to turn back and seek shelter.

Magellan faced difficult choices. If he lowered sail and tried to ride out the storm, the winds might blow his helpless fleet onto the shoals, or even ashore, where disaster awaited. But if he attempted to enter the harbor under short sail, he might run aground in the shallow water. He chose to proceed north with extreme caution; he made sure to sound the waters, and learned to his relief that they were deep enough for his ships to pass unharmed.

When the storm finally relented, Magellan turned south again and returned to the Río de la Plata. Although many on board the fleet argued that the river led to the strait, Magellan remained skeptical. Still, he would have to conduct a careful surveillance, just in case. And even if there was no strait, they had at least found abundant provisions. During the next two weeks, the men took on water and caught fish, or rather, learned how to catch fish.

Years before Magellan arrived at the Río de la Plata, both Spanish and Portuguese ships had searched for the strait at this very point. Antonio Galvão, who served as the Portuguese governor of the Moluccas, wrote about a "most rare and excellent map of the world, which was a great helpe to Don Henry (the Navigator) in his discouries." In 1428, Galvão said, the king of Portugal's eldest son made a journey through England, France, Germany, and Italy "from whence he brought a map of the world which had all the parts of the world and earth described. The Streight of Magelan was called in it the Dragon's taile." A dragon's tail was a fitting image for the strait, suggesting that it was dangerous, sinuous, and possibly mythological. Columbus believed in its existence, too. That mystical explorer supposedly received a vision prior to his fourth voyage in which he saw a map depicting the strait. He never found it, of course.

In 1506, Ferdinand of Aragon and Philip I of Castile commissioned two explorers, Juan de Solis and Vicente Yáñez Pinzón, to undertake an expedition to determine the position of the line of demarcation and to find a strait to the Indies. Like Magellan, Solis was a skilled and ambitious Portuguese mariner who found a receptive patron in Spain, but quite unlike Magellan, he was a fugitive from justice who had fled to Spain after murdering his wife. The Solis-Pinzón expedition, which embarked in 1508, discovered nothing, and when the expedition's two ships returned to Spain, an enraged and disappointed King Ferdinand clapped Solis in jail.

Two years later, in 1512, Solis, deftly manipulating the levers of influence, rehabilitated himself, and King Ferdinand made him pilot major; he then received an ambitious new commission to claim the Spice Islands for Spain. When King Manuel of Portugal protested, Ferdinand, shading the truth a bit, explained that Solis's task was simply to find the line of demarcation, and nothing more. Soon after, Ferdinand canceled the expedition, but he sent word to his representatives in the Caribbean to look for any sign of a strait and to arrest any Portuguese ships that might be searching for the same thing. Sure enough, the authorities in that distant outpost of the Spanish empire found a Portuguese caravel that had wandered into the Caribbean. She turned out to be a ship filled with secrets.

In 1511, Cristóbal de Haro had backed a covert Portuguese expedition to Brazil. The fleet consisted of two caravels commanded by Estêvão Froes and João de Lisboa. The Spanish knew nothing of the expedition until Froes's ship arrived in the Caribbean for repairs before heading northeast across the Atlantic to Portugal. The Spanish authorities seized the crew and threw them into jail. Meanwhile, the other ship returned to Spain, where Lisboa revealed his discoveries to an agent of his financiers, the Fuggers of Germany. After that, Lisboa's secrets gradually became public knowledge.

In 1514, a published account of Lisboa's exploits surfaced in Germany. *Newen Zeytung auss Presillg Landt,* or "News from the Land of Brazil," as the broadsheet was called, indicated that Lisboa had ventured seven hundred miles farther south than any prior expedition. According to this account, the expedition came to a strait, entered it, and sailed west until violent storms forced the ships to turn back. Lisboa might even have navigated the strait all the way to the Pacific. Although incomplete, the description of Lisboa's clandestine voyage was consistent with the strait that Magellan eventually explored. In Spain and Portugal, mariners and cosmographers alike seized on this remarkable document.

At the same time, a report circulated throughout Spain that Vasco Núñez de Balboa had glimpsed the vast ocean to the west: the Pacific. Within months of hearing the news, King Ferdinand once again sent Juan de Solis to find the strait, or, as *El Rey* put it, "to discover the back parts of Golden Castile." The strait, according to the best information of the day, ran through what is now Panama. The expedition, consisting of three ships and seventy men, embarked on October 8, 1515. Solis reached South America, sailed along its coast, and spotted a tribe that seemed friendly, at least from a distance. In good spirits, he went ashore with a landing party of seven men to greet them.

The best record of what befell the explorers comes from the pen of Peter Martyr, writing close to the time of the events. Martyr's account, in Latin, was translated into English in 1555 by Richard Eden, a Cambridge-educated scholar, in his best-known work, *The Decades of the Newe Worlde or West India, Conteyning the Navigations and Conquestes of the Spanyards, with the Particular Description of the most Rych and Large Landes and Islandes lately Founde in the West*

Ocean. In this popular work, he recorded the appalling turn of events in robust language:

> Sodenly a great multitude of the inhabitants burst forth upon them, and slue them every man with clubbes, even in the sight of their fellows, not one escaping. Their furie not thus satisfied, they cut the slayne men in peeces, even upon the shore, where their fellows might behold this horrible spectacle from the sea. But they being stricken with feare through this example, durst not come forth of their shippes, or devise how to revenge the death of their Captayne and companions. They departed therefore from these unfortunate coastes, and... returned home agayne with losse, and heavie cheare.

It became part of the lore of the expedition that the unfortunate Europeans had not merely been killed but devoured as their shipmates looked on helplessly.

*M*agellan's crew displayed considerable courage, even foolhardiness, when they confronted Indians in the region where Solis had met disaster. Magellan dispatched not one but three longboats. The men were armed, which gave them an advantage, but otherwise at the mercy of the indigenous people of the river basin. No sooner had the boats landed than the men jumped into the surf and chased the Indians observing them. Rather than standing and fighting, the Indians simply outran them. "They made such enormous strides that with all our running and jumping we could not overtake them," Pigafetta noted.

That night, a large canoe left the shore and approached the *Trinidad.* Standing upright in the middle of the vessel was an Indian covered with animal skins, apparently a chief. As the canoe drew close, the men aboard the flagship noticed that he exhibited no sign of fear. He indicated that he wished to come aboard, and Magellan agreed.

When they were face to face, Magellan offered the Indian two gifts, a shirt and a jersey. The Captain General then displayed a piece of metal, hoping to learn if the Indian was familiar with it. Recognizing the object, the Indian indicated that his tribe possessed some form of metal. Assuming the Indian would leap at the chance to obtain more, Magellan expected to barter metal objects such as bells and scissors for food and scouting assistance, but after the Indian left *Trinidad,* he never returned. The fleeting encounter with the indifferent tribal leader baffled Magellan and his officers. If they were received well, the sailors were ready for orgies, and the priests for conversions; if they were attacked, they were ready for battle. But they were not prepared to be ignored.

During the fleet's layover, Magellan constantly sounded the depths of the Río de la Plata, hoping that the water would swallow the lead, indicating that he had found the strait, but the stream remained precariously shallow. A channel or a strait would be deeper, he reasoned, and its current would run faster.

Unwilling to commit the entire fleet to the river, he dispatched *Santiago,* the smallest ship, and the one with the shallowest draught, to explore its murky and seductive reaches. *Santiago* spent two days sailing upstream, constantly sounding the river, trying to avoid run-ning aground.

Magellan meanwhile temporarily abandoned the flagship, *Trinidad,* to explore the waterway for himself aboard *Santiago.* At no point was the river deeper than three fathoms, too shallow for the ships to pass safely, and too shallow to suggest that it was a strait running all the way to Asia and the Spice Islands. Despite the many indications that they had found nothing but a large river, the other captains held fast to the belief that the Río de la Plata would lead them to the Indies, and they urged Magellan not to abandon his reconnaissance. But he had made up his mind to turn back, and once Magellan decided on a course of action, nothing could deter him. By the end of January, Magellan gave up and reversed direction, now facing directly into winds that made his return to the coast slow and erratic.

*O*n February 3, 1520, the fleet resumed its southward course in search of the real strait—if it existed—but *San Antonio* was found to be leaking badly. Within two days, the leak was repaired, and the Armada de Molucca rounded what is now known as Cape Corrientes.

Magellan adopted measures to ensure that he did not sail past the strait for which he was searching. He dropped anchor at night and resumed sailing in the morning as close to shore as he dared, always on the lookout for any formation suggesting a strait. As they ventured toward 40 degrees latitude, passing along the eastern coast of what is now Argentina, the weather steadily turned colder, a warning of the discomfort and hazards that awaited them. Their deliverance from the brisk days and frigid nights at sea would come only in the form of the strait, if such a strait existed, but it proved maddeningly elusive. Without realizing it, they were heading into latitudes notorious for sudden, frequent, and violent squalls, and on February 13, they ran into another storm, tossing the boats, damaging *Victoria's* keel, and terrifying the sailors with thunder and lightning and torrential downpours. When the storm finally blew itself out, Saint Elmo's fire once again appeared on the masts of the flagship, lighting the way, reassuring the sailors that they enjoyed divine protection.

The next day, the fleet hoisted sail, but without reliable maps to point the way, the ships were in danger of running aground on treacherous shoals—a submerged reef or outcropping of sand. The water was so shallow, and the shoals so well concealed, that *Victoria* struck bottom, not once, but several times. Colliding with a shoal was a sickening sensation dreaded by every sailor. It began with a shudder arresting the ship's progress. Those aboard the afflicted ship would cry out in dismay, fearing the worst. If the shoal was rocky, it might slice open the hull, and the ship would sink. If it was sandy, or covered with seaweed, it might hold the ship in a death grip. To clear seaweed from the rudder, those few sailors who knew how to swim would enter the water, fearing the appearance of sharks at any moment, dive beneath the ship, and with bursting lungs remove the vegetation with their bare hands. Tides were critically important to a ship stuck on a shoal; a rising tide could free her, and a low tide could leave her beached, trapped, impossible to move. By waiting for a rising tide, *Victoria* managed to free herself from the shoal's grip each time, but Magellan, in search of deeper water, eventually decided to lead the fleet away from shore and shoals even though he could no longer see land—or a strait.

The farther south he went, the more concerned Magellan became that he had accidentally passed the strait. On February 23, he retraced part of his route, and the following day, the black ships reached the expansive mouth of San Matías Gulf, on the coast of Argentina. To Magellan, the gulf appeared far more likely to lead to the strait than the Río de la Plata because the water was deep and blue and chilly. The men of the fleet might have seen whales because this was the principal breeding site of the Southern Right Whale. If they sailed close to land, they would have spotted penguins, sea lions, and even huge elephant seals lolling on the rocky shores. And if they had gone ashore, they would have encountered an animal paradise of foxes, hares, puma, peregrine falcons, owls, flamingos, hairy armadillos, and parrots. But Magellan preferred to anchor offshore, away from danger, as he continued his single-minded quest for the strait.

The fate of the expedition depended on finding it.

Book Two

The Edge of the World

CHAPTER V

The Crucible of Leadership

With sloping masts and dipping prow,
As who pursued with yell and blow
Still treads the shadow of his foe,
And forward bends his head,
The ship drove fast, loud roared the blast,
And southward aye we fled.

After six months at sea, Magellan's ability to lead the armada was still in grave doubt. Many of the most influential Castilian officers, and even the Portuguese pilots, were convinced their fierce and rigid Captain General was leading them all to their deaths in his zeal to find the Spice Islands. Few among them had confidence that Magellan could lead them to the edge of the world and beyond with a reasonable chance of survival.

A crucial evolution of Magellan's style of leadership, and perhaps his character, occurred over a period of nine trying months, from February to October 1520. He emerged from the ordeal a very different man from the one who had begun the voyage. The Magellan of February teetered on the brink of being murdered by the men he commanded. The Magellan of October was on the way to earning a place in history. In the intervening months, he passed a series of tests that forced him to confront his own limits as a leader and to change his ways, or die.

*H*ugging the coast, the fleet spent the last week of February sailing west toward Bahía Blanca, a spacious harbor worthy of investigation. Magellan led his ships in and around the islands of the bay, but
found no sign of a strait. As he familiarized himself with the coastline, he became increasingly confident of his navigational skills, and
he resumed sailing twenty-four hours a day, staying well offshore at
night to avoid rocks and reefs lurking below the dark water. On February 24, the fleet came to another possible opening. "We entered
well in," Albo recorded in his log, "but could not find the bottom
until we were entirely inside, and we found eighty fathoms, and it
has a circuit of 50 leagues." Magellan refused to consider this huge
bay as anything more than that. His surmise proved correct and
saved the fleet days of aimless investigation.

Finally, on February 27, the armada explored a promising inlet
with two islands sheltering what appeared to be numerous ducks.
Magellan named the inlet Bahía de los Patos, Duck Bay, and carefully explored it to locate an entrance to the strait. He cautiously
committed only six seamen to a landing party charged with fetching
supplies, mainly wood for fire and fresh water. Fearful of stumbling
across warlike tribes that might be prowling in the forest, the landing party confined their activities to a diminutive island lacking in
either fresh water or wood but seething with wildlife. On closer
inspection, what appeared to be ducks turned out to be something
quite different. Pigafetta identified them as "geese" and "goslings."
There were too many to count, he said, and wonderfully easy to
catch. "We loaded all the ships with them in an hour," he claimed,
and they were soon salted and consumed by the voracious sailors.
From his description, it is apparent that the "geese" were actually
penguins: "These goslings are black and have feathers over their
whole body of the same size and fashion, and they do not fly, and
they live on fish. And they were so fat that we did not pluck them
but skinned them, and they have a beak like a crow's."

Pigafetta marveled at another beast they encountered, one wor-thy of Pliny himself and all the more wonderful because it was absolutely real. "The sea wolves of these two islands are of various colors and of the size and thickness of a calf, and they have a head like that of a calf, and small round ears. They have large teeth and no legs, but they have feet attached to their body and resembling a human hand. And they have feet, nails on their feet, and skin between the toes like goslings. And if the animals could run, they would be very fierce and cruel. But they do not leave the water, where they swim and live on fish."

By "sea wolves" Pigafetta meant the sub-Antarctic sea lion or the sea elephant, usually distinguished by its inflatable snout. Although these mammals spend most of their time in the ocean, diving to depths of over four thousand feet, they occasionally spend relaxing months frolicking onshore in uncannily human family groups, lolling, stretching, yawning, scratching themselves, and peering lazily at their surroundings. Each male keeps a large harem of females, as many as fifteen, and often carries deep scars from fights with other males during the mating season. The adults weigh a thou-sand pounds, and if butchered properly, their rich meat and blubber could provide abundant food, and their thick, glossy, silvery-gray pelts a sorely needed source of warmth in these frigid latitudes.

The six enterprising seamen crept up on family groups of "sea wolves," stunned them with clubs, and lugged as many as they could into the longboat. Before the landing party could return to the fleet, a severe storm sprang up. The strong offshore winds blew Magellan's ships out to sea, stranding the six seamen on the little island. They passed a wretched night fearing that they would either be devoured by the "sea wolves" or die from exposure to the extreme cold.

In the morning, Magellan dispatched a rescue team. When they found only the abandoned longboat, they feared the worst. They carefully explored the island, calling out for their lost crewmates, but succeeded only in scaring the "sea wolves," several of which they slaughtered. Approaching the creatures, the rescue party came upon the lost men huddled beneath the lifeless "sea wolves," spattered with mud, exhausted, giving off a dreadful smell, but alive. They had settled next to the creatures to find shelter from the violent storm and enough warmth to sustain them through the night.

*A*s if these men had not suffered enough, another storm blasted the island just as they attempted to return to the waiting fleet. They managed to make it back safely to the ships, but the squall was fierce enough that *Trinidad*'s mooring cables parted, one after the other.

Helpless in the storm, pitching wildly, hurling her crew this way and that, the flagship veered dangerously close to the rocks near the shore. Only one cable held fast, and if it gave, *Trinidad* and her men—Magellan included—would all be lost. The sailors prayed to the Virgin and to all the saints they knew. In their abject fear, they promised to make religious pilgrimages on their return to Spain if only they survived this ordeal.

Their prayers were answered when not one but three glorious instances of Saint Elmo's fire danced on the ships' yardarms, casting an unearthly light of hope and inspiration. "We ran a very great risk of perishing," Pigafetta recorded. "But the three bodies of St. Anselm, St. Nicholas, and Saint Clare appeared to us, and forthwith the storm ceased." The last deity was especially apt, for Saint Clare was considered the patron saint of the blind and was often repre-sented holding a lantern; it was even believed that she could clear up fog and rain. To the religious sailors, the sudden manifestation of these signs was clear evidence that God still watched over them and protected them even in the remotest regions of the globe. As proof, the sole cable protecting them from disaster held until dawn, when the storm finally relented.

Battered by the storm, Magellan sought shelter in a cove, but the weather refused to cooperate. The wind disappeared, and the Armada de Molucca remained becalmed until midnight, when a third storm descended on them, the most destructive yet. The gale lasted three days and three nights, days and nights of freezing, of near starvation, of helplessness in the face of the elements. The fierce wind and seas tore away masts, castles, even poop decks. Through it all, the beleaguered sailors, trapped in disintegrating vessels that threatened to send them to their deaths at any moment, prayed for salvation with a fervor born of desperation.

Once again, their prayers were answered. The five ships rode out the great storm. The damage inflicted by the wind and waves, while serious, could be repaired. Incredibly, no lives were lost, despite all the hazards they had encountered on land and on sea. The Captain General gave the order, and the armada finally set sail.

*M*agellan resumed his search for a strait. Now that he had seen how quickly the offshore gales that raged in this region could maim or destroy his fleet, the need for an escape route became more urgent than ever. After several more days at sea, hope appeared in the form of another inviting cove. Magellan sailed into the protected waters, where he was disheartened not to find an inlet. This was merely a bay, but it would protect the fleet from severe storms—or so he thought. Six days later, another protracted tempest proved him wrong.

As before, the heavy weather stranded a landing party already ashore, this time with no "sea wolves" to provide shelter or warmth. Enduring bone-chilling cold, their skin and hair and beards soaked constantly with freezing rain, their fingers and toes numb, the men forced themselves to forage for shellfish in the freezing water. Their hands bleeding, they smashed the shells and survived on the raw flesh until, nearly a week later, they were able to return to the fleet.

*L*eaving the harbor, now named the Bay of Toil, the armada resumed its southerly course into even colder weather and the approaching subequatorial winter. The days grew shorter, and each unruly puff of wind darkened the sea and pummeled the sails, threatening to bloom into yet another squall. Finally, Magellan had had enough of exploration; he decided to suspend the search for the strait until the following spring. He turned his attention to finding a safe harbor where the fleet could ride out the approaching cold weather. On March 31, at a latitude of 49° 20', he found it. From his vantage point aboard *Trinidad*, it appeared to be an ideal haven: The harbor was sheltered, and abundant fish punctured the water's surface, as if in welcome. It was named Port Saint Julian.

The entrance to the port was framed by impressive gray cliffs rising one hundred feet as the harbor quickly contracted into a channel about half a mile in width. Although it offered protection, the narrow inlet experienced tides of over twenty feet and currents of up to six knots; in these conditions, the ships had to anchor themselves carefully and, when necessary, run cables to the shore to secure their positions.

Magellan considered Port Saint Julian a landmark of sufficient importance that he wanted to determine its longitude. He asked his pilots if they could make use of his friend Ruy Faleiro's techniques. Not possible, they told him. He consulted San Martín, his official astronomer, who tried to accommodate him; he took measurements, consulted with the pilots, and concluded that they might have strayed into Portuguese territory as defined by the Treaty of Tordesillas. The idea appalled Magellan, under orders from King Charles to avoid Portuguese waters and, at the same time, to demonstrate that the Spice Islands lay comfortably within the Spanish realm. Now it appeared the fleet had already sailed beyond the line of demarcation. Magellan realized he might be sailing halfway around the world only to demonstrate the opposite of what he had expected. The matter was potentially so serious, so damning to the entire enterprise, that the pilots deliberately obscured the location of Port Saint Julian on their charts.

*A*nticipating a long, grueling winter in Port Saint Julian, Magellan placed his crew on short rations, even though the ships groaned with the butchered meat of "geese" and "sea wolves," and fish abounded in the harbor. After the unbroken succession of life-threatening ordeals they had faced over the previous seven weeks, the seamen expected to be rewarded for their courage and perseverance, not punished. Outraged by the rationing, they turned insubordinate. Some insisted that they return to full rations, while others demanded that the fleet, or some part of it, sail back to Spain.

They did not believe the strait existed. They had tried again and again to find it, risking death while coming up against one dead end after another. If they kept going, they argued, they would eventually perish in one of the cataclysmic storms afflicting the region, or simply fall off the edge of the world when the coastline finally ended. Surely King Charles did not mean for them all to die in the attempt to find a water route to the Spice Islands. Surely human life had some value.

Magellan obstinately reminded them that they must obey their royal commission, and follow the coastline wherever it led. The king had ordered this voyage, and Magellan would persist until he reached land's end, or found the strait. How astonished he was to see bold Spaniards so fainthearted, or so he said. As far as their provisions were concerned, they had plenty of wood here in Port Saint Julian, abundant fish, fresh water, and fowl; their ships still had adequate stores of biscuit and wine, if they observed rationing. Consider the Portuguese navigators, he exhorted them, who had passed twelve degrees beyond the Tropic of Capricorn without any difficulty, and here they were, two full degrees above it. What kind of sailors were they? Magellan insisted he would rather die than return to Spain in shame, and he urged them to wait patiently until winter was over. The more they suffered, the greater the reward they might expect from King Charles. They should not question the king, he advised, but discover a world not yet known, filled with gold and spices to enrich them all.

This eloquent speech to the vacillating crew members bought Magellan a few days' respite, but only a few. His stern words had confirmed their worst fears about his behavior and his do-or-die fanaticism. On the most basic level, they believed he considered their lives expendable. In the following days, the men began to bicker; national prejudices suddenly flashed like well-oiled swords drawn from scabbards to cut and slash, usually at Magellan himself. Once again, the Castilians argued that Magellan's insistence that he intended to find the strait or die was proof that he intended to subvert the entire expedition and get them all killed in the process. All this talk of glorifying King Charles, they felt, was merely a stratagem to trick them into going along with Magellan's suicidal scheme. Anyone doubting Magellan's intention to subvert the expedition had only to examine the course they had been following, southward, ever southward, into the eternal cold, whereas the Spice Islands and the Indies lay to the west, where it was warm and sunny, and where luxury surely abounded.

*I*n the midst of this turmoil, the officers and crew observed the holiest day of the year, Easter Sunday, April 1. At that moment, Magellan had one paramount concern: Who was loyal to him, and who was not? With a sufficient number of loyal crew members, he would be able to withstand this latest, and most serious, challenge to his authority. Without them, he might be imprisoned, impaled on a halberd, or even hanged from a yardarm by hell-bent mutineers. To assess the extent of danger he faced, he carefully interviewed each member of the crew.

"With sweet words and big promises," Ginés de Mafra recalled, "[Magellan] told them the other captains were plotting against him, and he asked them to advise him what to do. They replied that their only advice was that they were willing to do as he commanded. Ma gellan... openly told his crew that the conspirators had resolved to kill him on Easter Day while he attended mass ashore, but that he would feign ignorance and go to mass all the same. This he did and, secretly armed, went to a small sandy islet where a small house had been built to accommodate the ceremony."

Magellan expected to see all four captains at Easter mass but only one, Luis de Mendoza, of *Victoria*, arrived. The air crackled with tension. "Both conversed," de Mafra says, concealing their emotions under blank countenances, and attended mass together. At the end of the ceremony, Magellan pointedly asked Mendoza why the other captains had defied his orders and failed to attend. Mendoza replied, lamely, that perhaps the others were ill.

Still feigning bonhomie, Magellan invited Mendoza to dine at the Captain General's table, a gesture that would force him to proclaim his loyalty to Magellan, but Mendoza coolly declined the request. Magellan appeared unfazed by Mendoza's insubordination, but the Captain General now knew that Mendoza was a conspirator.

Mendoza returned to *Victoria*, where he and the other captains resumed plotting against Magellan, sending messages by longboat from one ship to another. After mass, only Magellan's cousin, Álvaro de Mesquita, the recently appointed captain of *San Antonio*, came aboard *Trinidad* to dine at the Captain General's table. Magellan realized that the empty chairs made for an ominous sign.

At the moment, Magellan capitalized on a piece of luck. The longboat belonging to *Concepción*'s captain, Gaspar de Quesada, lost its way in the strong current while ferrying conspiratorial messages between the rebel ships and, to the dismay of the men aboard, found itself drifting helplessly toward the flagship and Magellan himself, the one individual they did not want to encounter at that moment. To their surprise, the crew of *Trinidad*, at Magellan's direction, rescued them from the runaway longboat. Even more amazing, Magellan welcomed them aboard the flagship and provided them with a lavish meal, which included plenty of wine.

At dinner, the band of would-be mutineers drank a great deal and decided that they had nothing to fear from the Captain General after all. They even revealed the existence of the plot to Magellan; they confided that if the plot succeded, he would be "captured and killed" that very night.

Hearing this, Magellan lost all interest in his visitors and busied himself readying the flagship against attack. Once again he questioned his crew to see who was loyal to him and who was not and, satisfied that *Trinidad*'s men would take his side when the mutiny inevitably erupted, awaited the inevitable assault.

Late that night, *Concepción* stirred with life. The captain, Quesada, lowered himself into a longboat and quietly made his way to *San Antonio*. He was joined there in the dark water lapping at the ship's hull by Juan de Cartagena, former captain, bishop's unacknowledged son, and frustrated mutineer; Juan Sebastián Elcano, a veteran Basque mariner who served as *Concepción*'s master; and a corps of thirty armed seamen.

Under cover of darkness, they boarded *San Antonio* and rushed to the captain's cabin, entering with a flourish of steel, rousting the hapless Mesquita out of his bunk. This had once been Cartagena's ship, and in his mind, it still was. Mesquita offered little resistance as the party of mutineers clapped him into irons and led him to the cabin of Gerónimo Guerra, where he was placed under guard.

By this time, word of the uprising had spread throughout the ship, and the crew sprang to life. Juan de Elorriaga, the ship's master, and a Basque, valiantly tried to dismiss Quesada from *San Antonio* before any blood was shed, but Quesada refused to stand down, whereupon Elorriaga turned to his boatswain, Diego Hernández, to order the crew to restrain Quesada and quash the mutiny.

"We cannot be foiled in our work by this fool," Quesada shouted, knowing that there could be no turning back. And he ran Elorriaga through with a dagger, again and again, four times in all, until Elorriaga, bleeding profusely, collapsed. Quesada assumed Elorriaga was dead, but the loyal master was still alive, though perhaps he would have been better off if he had died on the spot; instead, he lingered for three and a half months until he finally died from the wounds he received that night at Quesada's hand.

As the two struggled, Quesada's guard took Hernández hostage, and suddenly the ship was without officers. The bewildered crew, without anyone to give them orders and fearing for their lives, gave up their arms to the mutineers. One of their number, Antonio de Coca, the fleet's accountant, actually joined the insurgents, who stored the confiscated weapons in his cabin. The first phase of the mutiny had gone off as planned.

Pigafetta, normally a thorough chronicler of the voyage, offers little guidance to the mutiny. In this case, he was close, too close, to Magellan to be helpful. As a Magellan loyalist, he resisted the temptation to hear or repeat any ill concerning his beloved captain. He eloquently presented the Magellanic myth of the great and wise explorer, but at the same time he turned a blind eye to the scandals and mutinies surrounding Magellan throughout the voyage. In his one cursory mention of the drama at Port Saint Julian, Pigafetta even confuses the names of the principal actors. The chronicler, who could be extremely precise when he wished, likely got around to mentioning the mutiny only after the voyage, when he felt safe enough to discuss the bloody deeds happening all around him.

The mutineers in control of *San Antonio* swiftly converted her into a battleship. Elcano, the Basque mariner, took command and immediately ordered the imprisonment of two Portuguese who appeared loyal to Magellan, Antonio Fernandes and Gonçalo Rodrigues, as well as a Castilian, Diego Díaz. The mutineers, led by Antonio de Coca and Luis de Molino, Quesada's servant, raided the ship's stores, filling their hungry bellies with bread and wine, anything they could lay their hands on, and they endeared themselves to their followers by allowing them to partake of the forbidden food. Father

Valderrama, preoccupied with administering last rites to Elorriaga, watched all and vowed to report the evil deeds to Magellan, if he ever got the chance. Meanwhile, Elcano ordered firearms to be prepared; the arquebuses and crossbows—powerful, state-of-the-art weapons—were broken out. Anyone attempting to approach the renegade ship would face a barrage of lethal arrows and muzzles belching fire.

Within hours the mutiny spread like a contagion to two other ships, *Victoria,* whose captain, Luis de Mendoza, had resented Magellan from the day they left Sanlúcar de Barrameda, and to *Concepción.* It was only a matter of time until Cartagena, Quesada, and their supporters came after Magellan himself. Only *Santiago,* under the command of Juan Serrano, a Castilian, remained neutral. Quesada, for the moment, decided to leave *Santiago* alone; it was a decision that would later haunt the mutineers.

*T*he sun rose over Port Saint Julian on April 2 to reveal a scene of deceptive calm. The five ships of the Armada de Molucca rode quietly at anchor, their crew members sleeping off the previous night's excesses. For the moment, the Captain General remained secure in his stronghold, *Trinidad.* As a test of loyalty, he dispatched a longboat to *San Antonio,* where Quesada and Elcano held sway, to bring sailors ashore to fetch fresh water. As *Trinidad'*s longboat approached, the mutineers waved the sailors away and declared that *San Antonio* was no longer under the command of Mesquita or Magellan. She now belonged to the mutineer Gaspar Quesada.

When the longboat brought this disturbing news back to Magellan, he realized he faced a grave problem, but he remained oblivious to the full extent of the mutiny. He believed he had to contend with only one rebellious ship, not three, until he sent the longboat to poll the other ships and determine their loyalty. From his stronghold aboard *San Antonio,* Quesada replied, "For the King and for myself," and *Victoria* and *Concepción* followed suit.

To make his point, Quesada audaciously sent a list of demands by longboat to the flagship. Quesada believed, with good reason, that he had Magellan boxed in, and he tried to force the Captain General to yield to the mutineers. In writing, Quesada declared that he was now in charge of the fleet, and he intended to end the harsh treatment Magellan had inflicted on the officers and crew. He would feed them better, he would not endanger their lives needlessly, and he would return to Spain. If Magellan acceded to these demands, said Quesada, the mutineers would yield control of the armada to him.

To Magellan, these demands were outrageous. To comply meant ignominy in Spain, disgrace in Portugal, years in a prison cell, and even death. Under these circumstances, he might have been expected to launch a full-scale attack on *San Antonio,* but for once Magellan restrained his need to assert his authority. He sent back word that he would be pleased to hear them out—aboard the flagship, of course. The mutineers were hesitant to leave their base. Who knew what awaited them aboard *Trinidad?* They replied that they would meet him only aboard *San Antonio.* To their astonishment, Magellan agreed.

Having lulled Quesada and his followers into a sense of false security, Magellan quietly went on the offensive. By any objective measure, he operated at an enormous disadvantage. The mutineers controlled three out of the fleet's five ships and most of the captains and the crews. They had popular sentiment on their side and weapons to back up their demands. In his diminished position, Magellan did not attempt to meet force with force; instead, he sought to dismantle their revolt piece by piece, without placing himself in more peril than he already was.

He began his attempt to recover his fleet by claiming the longboat carrying Quesada's communiqué. With this equipment in hand, he turned his attention to recapturing at least one ship, and then he would go after the others. He decided not to attempt to reclaim *San Antonio,* where the mutineers were deeply entrenched, but *Victoria,* where support for the rebels might be softer, and where he would be most likely to summon support.

Victoria became the key to the whole plan, and to get her back, he resorted to a ruse. He filled the captured longboat with five carefully selected sailors and instructed them to appear sympathetic to the mutineers, at least at a distance. But beneath their loose clothing they carried weapons, which they intended to use. Their ranks included Gonzalo Gómez de Espinosa, the master-at-arms, which automatically lent authority to their mission. Magellan gave the men a letter addressed to Luis de Mendoza, *Victoria's* captain, ordering him to surrender immediately aboard the flagship. If Mendoza resisted, they were to kill him.

As soon as the longboat moved out of sight to begin its mission, the Captain General sent a second skiff into the water, filled with fifteen loyal members of the flagship's crew under the command of Duarte Barbosa, Magellan's brother-in-law.

When the first longboat pulled up to *Victoria,* Mendoza allowed the party to board his ship. De Mafra, the best eyewitness to the unfolding mutiny, relates, "Mendoza, a daring man when it came to evil deeds but too rash to take advice, told them to come aboard and give him the letter, which he set about reading in a careless manner, and not as befits a man involved in such a serious business." According to other witnesses, Mendoza responded to the letter with mockery and laughter, crumpled the orders into a ball, and carelessly tossed it overboard. At that, Espinosa, the military officer, grabbed Mendoza by the beard, violently shook his head, and plunged a dagger into this throat, as another soldier stabbed him in the head. Spurting blood, *Victoria's* captain slumped to the deck, lifeless.

With Mendoza dead, Magellan held the advantage in the life-and-death contest. No sooner had the captain breathed his last than the second longboat rowed into position beside *Victoria,* discharging its complement of loyalists who stormed the ship. As Magellan had calculated, his guard met with little or no opposition. Stunned by the death of their captain, the crew meekly submitted to Magellan's men.

As if the sight of the dead officer was not insult enough to the other Castilians, Magellan later paid off Espinosa and his henchmen for this bloody deed in plain view of everyone. "For this [action], the Captain General gave twelve ducats to Espinosa," recalled Sebastián Elcano, one of the mutineers, "and to the others six ducats each from Mendoza's and Quesada's savings." Was this the price of their lives, the Castilians asked themselves. A few ducats?

To signal Magellan's triumph, Barbosa flew the Captain General's colors from *Victoria's* mast, announcing to Quesada and the other rebels that the mutiny was ending. Magellan placed *Trinidad* securely between the two loyal ships, *Victoria* guarding one side and *Santiago,* now loyal to Magellan, the other. Together, the three vessels blockaded the inlet to the port; the two rebel holdouts, positioned deeper in the harbor, could not escape.

Magellan expected Quesada to recognize that resistance was futile. The mutiny had failed, and he would soon have to bargain not for better rations or a swift return to Spain, but for his very life. But Quesada refused to give up. *Concepción* and *San Antonio* remained at the other end of the harbor, offering no clue about the mutineers' intentions. To prevent them from slipping past the blockade at night, Magellan readied his flagship for combat. He doubled the watch and gave an order to "make a plentiful provision of much darts, lances, stones, and other weapons, both on deck and in the tops." To win the men's cooperation, he allowed them the pleasure of ample food. At the same time, he warned them not to let the two delinquent ships escape from the harbor into the open sea.

While the others were distracted, Magellan entrusted a seaman with a perilous assignment. Under cover of darkness, he was to sneak on board Quesada's ship, *Concepción*, where he would loosen or sever the anchor cable so that she would slip her mooring. Magellan calculated that the strong nocturnal ebb tide would draw her toward the blockade guarding the mouth of the harbor, giving him just the pretext he needed for launching a surprise attack. He was prepared to greet her with all the firepower he could muster.

Late that night, *Concepción* drifted mysteriously across the harbor. Because no one knew of Magellan's subterfuge, she appeared to be dragging her anchor. It was only a matter of time before she came within range of the flagship and touched off a battle at sea.

Aboard *Concepción*, the rebellion was beginning to falter. Ginés de Mafra, held hostage along with Mesquita, noticed that Quesada, the leader of the mutiny, was experiencing pangs of remorse, but he could not persuade his followers to end their rebellion now. "He summoned his crew and asked them that in case they could not get away with what they had begun, what should be done to avoid falling into the hands of Magellan." The other mutineers merely offered to "follow his decisions obediently."

Quesada's only hope, a faint one, was to slip past the blockade and escape. "He gave the order to weigh the anchor, but this did not turn out well for him, as the current brought his ship down the river to the flagship, something that neither Quesada nor those in the ship could help because of the fury of the waters," de Mafra recalled, without realizing what manner of subterfuge had placed the ship in jeopardy.

Quesada patrolled the quarterdeck, bearing sword and shield, hoping to regain control of the ship or, failing that, to slip past Magellan unnoticed. Instead, he sailed straight into a trap.

*A*s *Concepción* approached the flagship, Magellan shouted, "*Treason! Treason!*" and ordered his men to ready their weapons. "Once Quesada's ship passed by his," de Mafra continued, "[Magellan] ordered that it be shot, which made those who had offered to sacrifice their lives lose their ferocity and hide below."

Suddenly, *Trinidad* opened fire on the approaching vessel, hurling cannonballs onto her decks. "Quesada, armed, stayed on deck, receiving some spears that were hurled at him from the flagship's topsail, an attitude suggesting he wished to be killed. Magellan, realizing how slight the opposition offered by those in the *nao* was, boarded the skiff with several of his men." Before Quesada's men could offer resistance, *Trinidad*'s loyal seamen grappled *Concepción* to her side and rushed aboard as *Victoria* performed the same maneuver on the hapless ship's starboard side.

"Who are you for?" the attackers cried as they swarmed across *Concepción*'s cramped decks.

"For the King," came the response, "and Magellan!"

The mutineers' *volte-face* may have saved their lives, because Magellan's guard made straight for Quesada and his inner circle, who offered little resistance. The guard freed Mesquita, the deposed captain (and Magellan's cousin), along with the pilot, Ginés de Mafra. The coup was generally bloodless, and de Mafra was the only one who came close to harm when a ball fired from *Trinidad* passed between his legs as he sat in fetters below deck, shortly before he was freed.

With Quesada and his inner circle under arrest, and *Concepción* returned to Magellan's control, the midnight mutiny of Port Saint Julian came to an ignominious conclusion. Even Juan de Cartagena, aboard *San Antonio*, gave up hope of carrying out a mutiny. When the flagship drew alongside *San Antonio*, and Magellan demanded Cartagena's immediate surrender, the rebellious Castilian meekly complied and was confined in irons in *Trinidad*'s hold.

That morning, the Captain General had controlled two ships; now he ruled all five. Despite their overwhelming numbers, the mutineers had lost, and Magellan had emerged from the ordeal more powerful than before. His expedition, whose fate had been in grave doubt, would continue.

*N*ow that the Easter Mutiny was finally at an end, Magellan meted out punishment to the guilty parties. The mutineers were about to discover that defying Magellan was even more perilous than the most ferocious storm at sea. To begin, Magellan instructed one of his men to read an indictment of Mendoza as a traitor. The Captain General then ordered his men to draw and quarter Mendoza's body. This complicated and grotesque procedure usually began with hanging the victim, then cutting him down while he was only partly strangled. The executioner or an assistant would make an incision in the victim's abdomen, remove his intestines, and, incredibly, burn them in front of the half-dead victim. When he finally expired, his head and limbs were severed from his body, parboiled with herbs to preserve them and repel birds, and finally displayed to the public. In a variation, the victim's arms and legs were attached to four horses, who were made to walk in opposite directions, slowly tearing the victim's limbs from his body.

Magellan combined elements of both methods. Mendoza was secured to the flagship's deck, with ropes running from his wrists and ankles to the capstans, which consisted of a cable wound about a cylinder to hoist or move heavy objects. On cue, sailors pressed on levers to rotate the capstans' drum, which contained sockets to check its backward movement. Bit by bit, the pressure applied to the capstans ripped Mendoza's lifeless body to pieces.

Magellan directed that the quartered remains be spitted and displayed as a warning of exactly how traitors would be treated. The preserved body parts of Luis de Mendoza remained visible throughout the next several months in Port Saint Julian, an indelible lesson to the men concerning the consequences of mutiny. The practice, so barbarous by present standards, was in keeping with the customs of the time for those who would defy authority.

*M*agellan's display of barbarism did not end there; he was only beginning to exact revenge for the mutineers' insult to his authority and to the honor of King Charles. More than execution, torture was his ultimate weapon at sea. That he resorted to torture was not unusual; this was, after all, the era of the Spanish Inquisition, which had formally begun in 1478 and continued under the leadership of Tomás de Torquemada, the first Grand Inquisitor. To many Europeans, the mere mention of Spain summoned images of the Inquisition and of diabolical methods of torture, although Spain was hardly the only offender. Nor was torture confined to special cases of heresy; as Magellan's behavior demonstrated, it was also applied to other criminal behavior such as usury, sodomy, polygamy, and especially treason, considered the most serious crime against the state.

An inquisition was not a trial in the modern sense. The accused were presumed guilty; their reluctance to confess their crimes only added to the sum of their crimes. Torture was designed to elicit withheld confessions, and the sooner the accused confessed, the sooner the agony ended. Indeed, confessions elicited by torture were considered the best evidence of all.

One eyewitness account of a typical Spanish inquisition evokes the fear and despondency Magellan's victims likely experienced. "The place of torture in the Spanish Inquisition is generally an underground and very dark room, to which one enters through several doors. There is a tribunal erected in it, in which the Inquisitor, Inspector, and Secretary sit. When the candles are lighted, and the person to be tortured brought in, the Executioner... makes an astonishing and dreadful appearance. He is covered all over with a black linen garment down to his feet, and tied close to his body. His head and face are all hid with a long black cowl, only two little holes being left in it for him to see through. All this is intended to strike the miserable wretch with greater terror in mind and body when he sees himself going to be tortured by the hands of one who looks like the very devil." In Magellan's time, torture was a vivid, dreaded presence in daily life, and it belonged in every captain's arsenal of techniques to keep sailors in line. With its legal and religious trappings, it was far more systematic, cruel, and psychologically damaging than the traditional remedy of the lash. Even when physical pain ended, psychological wounds continued to fester deep in the victims' souls.

Magellan's use of torture inflamed early Spanish historians, who professed to be shocked by his brutality, but what upset them was not that he resorted to torture, a fact of life in Torquemada's Spain, but that he tortured *Spaniards*. Among those who denounced Magellan's conduct regarding the mutiny was Maximilian of Transylvania, the scholar who interviewed the survivors of the expedition on their return to Seville; based on their recollections, he stated flatly that Magellan's actions had been illegal. "No one, aside from Charles and his Council, can pronounce capital punishment against these dignitaries." (Pigafetta, who was at the scene of the torture, simply ignored the inconvenient display of barbarism, as he did with everything else concerning the mutiny; it would not do to portray his beloved Captain General as inflicting grievous hardships on his beleaguered men.) Early historians stress that some of the victims of Magellan's torture were Spanish officers in order to emphasize the insult to King Charles and Castile—evidence of Magellan's disloyalty toward Spain—but many victims were actually Portuguese.

Torture, no less than the skill he displayed in recapturing the mutinous ships, played an important part in Magellan's preventing further mutinies. Through his use of torture, his crew came to understand that the only thing worse than obeying Magellan's dictates, and possibly losing their lives in the process, was suffering the consequences of defying him. One of the outstanding reasons that his crew had the courage and determination to circumnavigate the globe, even if it meant sailing over the edge of the world, was that he compelled them to do so. Fear was his most important means of motivating his men; they became more afraid of Magellan than the hazards of the sea.

*T*o punish the other offenders, Magellan conducted a secular inquisition at Port Saint Julian. He appointed his cousin, Álvaro de Mesquita, as judge presiding over an exhaustive trial. First Magellan had promoted him to captain of *San Antonio* over the heads of more qualified pilots and master seamen, both Spanish and Portuguese. Now Mesquita functioned as Magellan's agent of agony, deciding who was guilty of treason and who would suffer the consequences. No wonder the men hated him.

Mesquita spent two weeks assessing the "evidence" of guilt before passing judgment. At the end of the trial, Mesquita, no doubt under orders from Magellan, let one of the accused off with a slap on the wrist. The hapless accountant Antonio de Coca was merely deprived of his rank. But Mesquita found Andrés de San Martín, the esteemed astronomer-astrologer; Hernando Morales, a pilot; and a priest all guilty of treasonous behavior.

This judgment was unquestionably excessive. Their behavior was that of frightened men rather than of conspirators. For example, when searched, San Martín was found to possess an itinerary of the expedition, as would be expected of the fleet's chief astronomer. In a panic, he threw the chart into the water. And what had the priest done to deserve the same treatment? According to the charges, he had been heard to say that the "ships did have enough provisions"— which was only the truth—"and for not having consented to com-municate to the Captain General the secrets of what the crew had told him in confession." Magellan probably expected that the priest had been privy to the plot, which sailors would have confessed, but it is unlikely they considered their deeds sinful; rather, they were justified by their desperate circumstances.

The tenuous connection of these deeds to the actual mutiny suggests that Mesquita and Magellan, for all their patient investigation, turned up little additional evidence of disloyalty and simply resorted to San Martín and the priest as scapegoats for their wrath. San Martín had been exercising his navigational skills with distinction at least since 1512, when King Ferdinand had appointed him as a royal pilot. He later tried twice to win a commission of pilot major, or chief of all pilots. Even though King Charles passed him over, San Martín replaced Ruy Faleiro as the astronomer-astrologer for the Armada de Molucca. San Martín's skills, his royal charter, the lavish pay he received, his prominence, and his long record of loyalty all made him an unlikely candidate for the role of mutineer. Unlike Quesada, Cartagena, and the other co-conspirators, he did not hunger to become a captain and harbored no resentment against Magellan. His worst offense consisted only of a moment of panic. Nevertheless, this lapse condemned him to suffer what many considered a fate worse than death.

Mesquita ordered San Martín to undergo the most common punishment of the Inquisition, the ghastly *strappado*. The *strappado* was administered in five stages of increasing agony. In the first degree, the victim was stripped, his wrists were bound behind his back, and he was threatened until he confessed. If he refused, he was subjected to the second degree. In it, the victim's arms were raised behind his back by a rope attached to a pulley secured overhead, and he was lifted off his feet for a brief period of time, and given another chance to confess. If he still refused, he faced the third degree of the *strappado*, in which he was suspended for a longer period of time, which dislocated his shoulders and broke his arms. Once again, he was given another chance to confess. If he still failed to make a satisfactory confession, he was subjected to the fourth degree: The victim was suspended and violently jerked, which

inflicted excruciating pain. Few victims of a methodically administered *strappado* lasted beyond this point without confessing. In certain cases, there was a fifth degree, as well. In the final phase of the
strappado, weights were attached to the victim's feet, and they were
often heavy enough to tear the limbs from his tormented body.

Andrés de San Martín suffered the full five stages of the *strappado*.

In the last, most horrific stage of Magellan's inquisition, several
cannonballs were attached to San Martín's feet, and the additional
weight inflicted excruciating pain when he was suspended. Another
early account of this inquisition describes the final stage of *strappado*, as it might have been experienced by San Martín: "The prisoner hath his hands bound behind his back, and weights tied to his
feet, and then he is drawn up on high, till his head reaches the very
pulley. He is kept hanging in this manner for some time, that by the
greatness of the weight hanging at his feet, all his joints and limbs
may be dreadfully stretched, and of a sudden he is let down with a
jerk, by slacking the rope, but kept from coming quite to the
ground."

After enduring these torments, San Martín may have begged to
be executed rather than be made to endure any more of the *strappado*, he may have fainted from the pain, but he survived the ordeal.
In fact, he recovered sufficiently to return to his former position as
astronomer-astrologer, but from then on, he remained wary of Magellan and the entire enterprise of the armada.

The punishment Mesquita and Magellan inflicted on Hernando
Morales was even more severe than San Martín's. Accounts of the
proceedings say only that Morales's limbs were "disjointed," but the
procedure to which he was subjected was so severe that the poor pilot later died from the wounds he received; the agonies he suffered
at the hands of Mesquita and Magellan can only be imagined. He
might have undergone a variation of another common torture of the
Inquisition era, the fiendish Wooden Horse, in which the victim was

secured with metal bars to a hollowed-out bench, his feet higher than his head. "As he is lying in this posture," runs an early account, "his arms, thighs, and shins are tied round with small cords or strings, which being drawn with screws at proper distances from each other, cut into his very bones, so as to no longer be discerned. Besides this, the torturer throws over his mouth and nostrils a thin cloth, so that he is scarce able to breathe through them, and in the meanwhile a small stream of water like a thread, not drop by drop, falls from on high upon the mouth of the person lying in this miserable condition, and so easily sinks down the thin cloth to the bottom of his throat so that there is no possibility of breathing, his mouth being stopped with water, and his nostrils with cloth, so that the poor wretch is in the same agony as persons ready to die, and breathing their last. When this cloth is drawn out of his throat, as it often is, that he may answer to the questions, it is all wet with water and blood, and is like pulling his bowels through this mouth." After enduring this torture, what victim, no matter how innocent, would not willingly confess to spare himself more agony?

Both the *strappado* and the water ordeal were well-known "official" methods of torture used in the Inquisition, but there were also illegal methods, which were nearly as common, to which San Martín, Morales, and the priest might have been subjected. They might have been starved. They might have been subjected to sleeplessness. Or they might have had their feet bound and covered with the abundant natural salt found in the Port Saint Julian harbor. A goat licking the soles of the feet for a prolonged period of time was said to inflict excruciating agony, yet it left no damage to the victim's body.

*O*nce the horror of the inquisitional catharsis subsided, Mesquita (with Magellan's blessing) sentenced the other accused—in all, forty men—to death. A mass execution appeared to be in the making, but the expedition could not continue without the help of the convicted men. It was unlikely that Magellan, even in his cold wrath, would execute forty men, many of whom abandoned the mutiny soon after it began. His victory was, in this sense, all too complete, and he had to find a way out of the grim situation he had helped to create.

Magellan had succeeded in terrorizing all the men under his command, captains and commoners alike. In his letter of March 22, 1518, King Charles gave Magellan complete authority over every-one in the armada; this was the "power of rope and knife." He had demonstrated that he had, as his orders indicated, the power of life and death over all those who served under him. As brutal as his con-duct sounds, the Captain General was well within the rights granted to him by King Charles. But Magellan took his authoritarianism to an extreme, refusing to share power or even give the illusion of power to his captains, and they communicated their dissatisfaction down the chain of command to the ordinary seamen, making rebel-lion and its hideous aftermath—torture—inevitable. With his insis-tence on controlling every aspect of the expedition himself, and scorning any suggestions that threatened his master plan, Magellan made the captains who served under him feel impotent, and they directed their rage at him. Magellan insisted, but rarely troubled to persuade, and his continuous invocation of King Charles when they were thousands of miles from Spain and in great peril sounded hol-low, especially coming from the lips of a Portuguese.

Believing that he had finally demonstrated his absolute authority, Magellan commuted all forty of the death sentences to hard labor. Among the forgiven was Elcano, the Basque shipmaster who would later have his revenge on the Captain General. Those who had been freed looked on the man who controlled their fate with decidedly mixed emotions. They were overjoyed, in the short run, to be spared a gruesome death by drawing and quartering or another form of tor-ture, but as the prospect of a long winter in Port Saint Julian loomed, they realized they faced a life of daily hardship and danger. On shore, cannibals, only slightly more ruthless than the Captain

General, might attack them and devour them; on the high seas, a storm might send their ships to the bottom. Desertion was impossible; no one could survive the harsh climate unaided. The only choice left to them was a slavish adherence to Magellan's authority, even if it led over the edge of the world.

There were two important exceptions to the general clemency: Gaspar de Quesada, the leader and murderer of *San Antonio*'s master, and his servant, Luis de Molino. Magellan insisted that Quesada be executed. And he gave Molino a brutally simple choice: He could either be executed along with his master or spare his own life by beheading his master. If Molino did so, he would violate some of the most central tenets governing Spanish conduct and morality, codes of behavior going back to feudal times. As Magellan expected, Molino accepted the deal, as cruel as it was.

In full view of the crew, Quesada knelt on the deck of *Trinidad*, and Molino stood over him, sword in hand. He asked his master for forgiveness, but received none. And then with one powerful blow, he severed Quesada's head from his neck. As if that were not enough carnage for one day, Magellan ordered a detail to draw and quarter Quesada's body. His remains were displayed as a grisly warning to the others, just as Mendoza's body had been displayed several weeks before.

Days later, Magellan discovered that Cartagena, the sole surviving Spanish captain, was conspiring with a priest, Pero Sánchez de la Reina, to mount yet another mutiny. Under his real name, Bernard de Calmette, the priest, who came from the south of France, served as chaplain aboard *San Antonio*; he adopted a Spanish name so that the crew would feel more comfortable with him. It was astonishing that Magellan's nemesis would risk his life again, after all the carnage, this time with little hope that any of the seamen would follow, but Cartagena was almost as stubborn as Magellan.

The Captain General subjected the two conspirators to a fresh court-martial. His first instinct was to have both men executed; this was, after all, Cartagena's third attempt at mutiny, but Magellan found himself in a difficult position. He could not bring himself to condemn a priest—even a disloyal priest—to death. And as for Cartagena, his blood ties to Archbishop Fonseca prevented Magellan from taking severe disciplinary action such as execution or torture. Instead, Magellan devised a much worse fate for Cartagena and the priest. He decided to leave them behind to fend for themselves in the wilderness of Port Saint Julian after the fleet's departure.

In all, Magellan's conduct during the mutiny and its aftermath was worthy of Machiavelli—subtle and calculating when possible, but brutal when necessary. He had survived the testing, and emerged victorious.

Always a perfectionist about outfitting his ships, Magellan turned his attention to his neglected fleet. The ships were in a state of disrepair, their sails and rigging in disarray, their holds fetid, their hulls leaky. He ordered his men to empty the ships and give them a thorough cleaning. This exhausting chore meant removing all the provisions, even the stone ballast, which was cleansed by seawater. The forty mutineers, bound in chains, performed the most grueling labor; they operated the pumps, essential for keeping the ships afloat until the armada's carpenters made them seaworthy again. Once they had emptied the ships, the seamen scoured the holds, washed down the wooden surfaces with vinegar to eradicate the ubiquitous stench, and returned the ballast.

So the wretched winter passed, day by day, hour by hour, the men working constantly and trying to keep themselves warm as best they could, enduring life in a prison so remote it needed no walls.

Overseeing these projects, Magellan intended to keep his prisoners in chains until they left Port Saint Julian in the spring.

When the time came to load the provisions, they discovered more evidence that the dishonest chandlers in Seville and the Canary Islands had robbed them blind, and endangered their lives. Although their bills of lading showed enough supplies on board to last a year and a half, long enough to reach the Spice Islands, the ships' holds actually carried only a third of that amount. This grim discovery cast the rest of the expedition in a different light because, as Magellan realized, they would likely run out of food well before they reached their goal. The men resumed hunting to make up the difference, but they were eating their way through their supplies almost as fast as they replenished them. The only way out of their predicament was to resume the voyage as soon as possible, storms or no storms.

CHAPTER VI

Castaways

And now there came both mist and snow,
And it grew wondrous cold:
And ice, mast-high, came floating by,
As green as emerald.

*F*inding the strait leading to the Spice Islands, always a priority for Magellan, reached the level of an obsession in late April. When the oppressive weather briefly lifted, he rashly sent out a reconnaissance mission to search for the elusive waterway. He selected *Santiago,* the soundest of the vessels, for the task, with Juan Rodríguez Serrano as her Castilian captain. "An industrious man, he never rested," said one of the crew members of him. He was about to meet the ultimate challenge of his career.

Even if Serrano succeeded in finding the mouth of the strait, he would have to embark on an equally dangerous return journey to Port Saint Julian. Violent storms at sea or cannibals on land could spell disaster. And the temptation to mutiny and sail away—either east toward Spain or west through the strait—might be irresistible to *Santiago's* crew. Magellan stifled thoughts of escape by keeping provisions on board to a minimum and offering Serrano a reward of one hundred ducats if the expedition located the strait; of course, he could collect only on his return.

Favored by calm weather, the mission began auspiciously enough. On May 3, about sixty miles south of Port Saint Julian, Serrano discovered a promising inlet, which on closer inspection revealed itself as the mouth of a river, which he named Santa Cruz. More than three hundred years later, in 1834, the youthful Charles Darwin visited the Santa Cruz River aboard HMS *Beagle* on her voyage of discovery, and found the same inviting prospect. The river, he wrote, "was generally from 300 to 400 yards broad, and in the middle about 17 feet deep. . . . The water is of a fine blue colour, but with a slight milky tinge, and not so transparent as at first sight might have been expected."

Santiago's crew soon discovered that food was even more plentiful around the Santa Cruz River than at Port Saint Julian, and Serrano decided to linger for six days to fish and hunt for sea elephants. Given the urgency of finding the strait, his decision to tarry is peculiar. Perhaps neither he nor his men wished to return to Port Saint Julian and its grim reminders of the mutiny sooner than necessary; or perhaps they had no desire to risk their lives on the open water.

After the tranquil respite, *Santiago* set sail and proceeded south in search of the strait. On May 22, the wind picked up and the seas began to churn, tossing the ship as if she were nothing more than an oversized piece of flotsam. The armada had encountered many violent squalls, but little *Santiago* had stumbled into the most powerful storm her crew had ever experienced, and they would have to face it alone.

Serrano had no time to reef the sails. Fierce seas pounded the ship mercilessly, terrifying her crew. Serrano attempted to head into the wind and ride out the storm, but overpowering gusts tore the sails, and the seas battered the rudder until the device failed to respond. *Santiago* was now out of control, caught in the middle of a storm that was still building in power, her men beyond the hope of rescue. The situation was desperate.

At that moment, the storm gathered force, and the winds pushed the helpless ship toward the rocky coast and the prospect of certain death for her crew. Serrano faced every captain's nightmare as razor-sharp rocks sawed into her hull, and she began taking on water. Luck was with her crew, since *Santiago* washed ashore before breaking up. One by one, her crew of thirty-seven crawled to the end of the jib boom and jumped to a rocky beach. As soon as they had abandoned ship, *Santiago* broke up, and the storm carried away all her life-sustaining provisions—wine, hardtack, and water, to say nothing of the freshly caught sea elephants. Incredibly, all the men aboard ship survived, but once they had given thanks to the Lord for sparing their lives, they grasped the desperate situation they now faced.

The storm had stranded the castaways about seventy miles from the rest of the fleet, without food or wood or fresh water, in freezing weather. They were cold and exhausted; soon they would be starving. There was no way to get word of their plight to the Captain General. Their land route back to Port Saint Julian presented seemingly overwhelming obstacles: snow-covered mountains and the Santa Cruz River, three miles wide.

*T*he castaways spent eight days in more or less the same area, disoriented, dispirited, waiting for pieces of the wreck, possibly even food, to drift onto the pebbly beach, but the sea yielded only a few planks broken off from *Santiago's* hull. Subsisting on a diet of local vegetation and whatever shellfish they could catch, the castaways evolved a plan. They would drag the planks over the mountains until they reached the river and there, on its banks, build a raft to cross it. The river lay many miles to the north, and the task proved daunting to the crew. They left most of the planks behind, and after four wretched days of marching overland, the exhausted crew finally reached the broad expanse of the river. The weather had relented,

and fish, as they knew from their first visit to the river, were plentiful. It seemed they would not starve, after all.

Lacking planks to build a raft large enough to carry all the
men, the castaways split into two groups. The larger group—
thirty-five men—set up camp at the river's edge, while two strong
men, whose names were not recorded, set out on the tiny raft.
They intended to cross the river and walk the rest of the way
back to Port Saint Julian to seek help. It was an exceedingly risky
undertaking. Successfully crossing the three-mile expanse of river
required a combination of daring and luck, and when they reached
the other side, they faced an arduous march in freezing weather,
living off the land.

The two crew members in the vanguard succeeded in mastering
the river's breadth in their rudimentary raft, and once they had
landed on the far side, they set out in the direction of Port Saint
Julian. At first, they followed the coast, where they could be reasonably certain to find shellfish, but vast swamps barred their progress,
and they had to walk inland, over hills and mountains, eating only
ferns and roots, and suffering greatly in the freezing weather. The
trek lasted eleven harrowing days, and when they reached Port Saint
Julian, ravaged and gaunt from their ordeal, even those who knew
the survivors barely recognized them.

Once the castaways revived, they described the desperate situation of their shipmates on the far side of the Santa Cruz River.

*M*agellan had no choice but to attempt to rescue the other
thirty-five crew members of *Santiago*. Afraid to risk the loss of
another ship to a storm, he sent a rescue squad of twenty-four men,
carrying wine and hardtack, along the overland trail that the two
survivors had blazed through the harsh wilderness. "The way there
was long, twenty-four leagues, and the path was very rough and full
of thorns," said Pigafetta of their grim progress. "The men were four
days on the road, sleeping at night in the bushes. They found no

drinking water, but only ice, which caused them great hardship."
Failing to find a river or spring, they resorted to melting snow.
Finally, in a drastically weakened condition from their days in the
wild, they reached the desperate castaways, who had been camping
out along the banks of the Santa Cruz River. A pathetic reunion
ensued: exhausted men at the end of the world, suffering intensely,
expecting to die at any time, united only in the cause of survival, as
unlikely as the prospect seemed.

Relying on the small raft cobbled together from the wreck of
Santiago, the rescue party ferried the survivors back across the river
in groups of two or at most three; each trip consumed hours and was
fraught with hazard, but miraculously everyone made it to the
northern shore. Even so, they were still far from safety because they
had to make the rugged overland trek back to Port Saint Julian. As
Magellan anxiously awaited the outcome of the rescue mission, the
thirty-five castaways and twenty-four rescuers picked their way
through the snows of the Patagonian winter, fortified mainly with
wine and hardtack. About a week later, they emerged one by one
from the forest surrounding Port Saint Julian. Driven by an unshakable will to survive, everyone made it back safely.

Magellan greeted the dazed, exhausted men with ample food
and wine, and treated them all as heroes.

*T*he wreck of *Santiago* and the hardship endured by her crew troubled Magellan more deeply than the violence and torture of the
Easter Mutiny. "The loss of the ship was much regretted by Magellan," de Mafra recalled, "although it was not the pilot's fault,
because along this coast the sea rises and ebbs eight fathoms, and
this was the cause of the calamity, so that the ship found itself high
and dry."

As serious as the loss of *Santiago* might be, Magellan had more to
fear from the emotional consequences of the wreck. The disaster
confirmed his crew's fear that the Captain General was leading them

on an expedition so dangerous that they would all get killed long before they reached the Spice Islands. To ensure his control of the remaining four ships in the fleet, he saw to it that only diehard loyalists commanded them. While Álvaro de Mesquita, his first cousin, remained in command of *San Antonio,* Magellan appointed Duarte Barbosa, his brother-in-law, as captain of *Victoria,* and Juan Serrano, the unlucky skipper of *Santiago,* as the new captain of *Concepción,* the ship once commanded by Gaspar de Quesada, the mutineer whose severed head rotted on a pike. Magellan himself still ruled over all from *Trinidad.* Finally, he scattered *Santiago*'s long-suffering crew among the four remaining ships to prevent them from secretly conspiring.

In fact, Magellan's appointment of his relatives as captains served to fuel the silent resentment of many crew members, even those from Portugal. When they finally returned to Spain, if they ever did, they could tell vivid tales of Magellan's insolence toward the Spanish captains; his shameless nepotism; his reckless seamanship, culminating in the needless loss of *Santiago;* and, most blatant of all, the drawing and quartering of Gaspar de Quesada. All of these grievances remained urgent in the minds of many seamen as they awaited a time and place to act on them.

*W*inter relentlessly advanced on Port Saint Julian; the days contracted to less than four hours of light, and the snow line reached down the mountains, across the fields and swamps, eventually extending to the water's edge. If the crew members and officers ever spent an idle hour at Port Saint Julian, if they fished for the sport of it, or played cards, or indulged in practical jokes, or read the books of exploration and discovery that they carried with them, books such as *The Travels of Marco Polo,* or *The Travels of Sir John Mandeville,* or if they participated in any other pastime during the Patagonian winter, there is no record of it. Magellan kept his men too busy and afraid for their lives for such activities. Only survival mattered.

Magellan ordered his men to hunt and fish, which they did. They found mussels, as well as foxes, sparrows, and "rabbits much smaller than ours," in Pigafetta's words. They preserved their catch with salt derived from flats surrounding the bay.

When it became too cold to fish, Magellan sent a party of four armed men—all that he was willing to risk—to explore the interior. They had two goals, to plant a cross on the highest mountain they could climb and to befriend Indians, if they found any. The landscape proved more rugged than they had anticipated, and they were unable to make much progress or to ascend any of the distant mountains. Instead, they selected a lower mountain close to the harbor, named it "Mount of Christ," fixed their cross on the summit, and returned to the waiting ships, where they confidently reported that there was no sign of human life around Port Saint Julian.

*D*espite the empty winter months stretching before them, Magellan was determined to await the coming of spring before he ventured into the treacherous ocean and resumed searching for the strait. To keep his men occupied, he ordered a detachment ashore to construct a small stone enclosure for a forge to be used to repair the ships' metal fittings, but even this modest project ended in frustration because the weather became so bitter that several sailors suffered crippling frostbite on their fingers.

Amid the intense suffering and hardship, discontent spread among the crew members. As the prospect of another mutiny loomed, Magellan, along with everyone else in the fleet, was distracted by an unexpected sight: a distant plume of smoke wafting over the landscape. Perhaps they were not alone, after all.

"We remained two whole months without ever seeing anyone," wrote Pigafetta of their stay in Port Saint Julian. "But one day (without anyone expecting it) we saw a giant who was on the shore, quite naked, and who danced, leaped, and sang, while he threw sand and dust on his head. . . . Our captain sent one of his men toward him, charging him to leap and sing like the other to reassure him and show him friendship. Which he did."

The strange rite recommended by Magellan worked; after watching the European seaman imitate his gestures, the giant appeared peaceful and eager to socialize, as the dancing continued.

On seeing the giant, every sailor in Port Saint Julian thought immediately of the ghastly fate that had befallen the landing party of Juan de Solis five years before. "In time past these tall men called Canibali, in this river, ate a Spanish captain named Juan de Solis and sixty men who had gone, as we did, to discover land, trusting too much in them," Pigafetta wrote, inflating the number of victims in Solis's party, but otherwise invoking that horrible event with clarity.

Eager to make contact but wary of falling into a trap, Magellan took the precaution of inviting the giant to meet him in an isolated, protected setting, rather than allowing himself or his men to be lured to an unfamiliar spot where they might be ambushed. "Immediately the man of the ship, dancing, led this giant to a small island where the captain awaited him. And when he was before us, he began to marvel and to be afraid, and he raised one finger upward, believing that we came from heaven."

The Europeans marveled at the giant's stature. Some crew members reached only the waist of the giant, who was said to be twelve or thirteen *palmos* tall—a *palmo* being the equivalent of a hand span. By this measurement, he and giants like him stood over eight feet tall. Pigafetta, who had been silent during the mutiny, recovered his powers of description the moment the giant appeared. "He was so tall that the tallest of us only came up to his waist," the official chronicler observed. "He had a very large face, painted round with red, and his eyes also were painted round with yellow, and in the

middle of his cheeks he had two hearts painted. He had hardly any
hairs on his head, and was painted white."

The giant was a member of the tribe known as the Tehuelche
Indians, who were numerous throughout the region. In reality, the
Tehuelche measured about six feet tall; the impression of the
Indian's great stature derived in part from his costume and especially the elaborate boots he wore, which added to his height.
"When he was brought to the captain, he was clad in the skin of a
certain animal"—the guanaco, similar to the llama—"which skin
was very skillfully sewn together. And this animal has the head and
ears as large as a mule's, and a neck and body like those of a camel,
a stag's legs, and a tail like that of a horse. . . . This giant had his feet
covered with the skin of the said animal in the manner of shoes, and
he carried in his hand a thick bow, with a thick bowstring, made
from the intestines of the said animal, with a bundle of cane arrows
which were not very long and were feathered like ours, but no iron
point, but, at the tip, small black and white stones."

Emboldened after the first encounter, Magellan invited the giant
on board the flagship, where he offered his guest plentiful "food and
drink." In the midst of their feasting, the Captain General summoned his men to produce a large "steel mirror."

The reaction was swift and stunning. "The giant, seeing himself,
was greatly terrified, leaping back so that he threw four of our men
to the ground." When the mirror was removed, the giant regained
his composure. Magellan then tried to win back his guest's confidence with a gift of trinkets: "two bells, a mirror"—presumably
smaller and less alarming—"a comb, and a chaplet of paternosters."
This last item was by far the most significant of all the gifts and no
doubt included what is now known as the Lord's Prayer—in Latin,
of course. Magellan probably knew that Jesus taught this prayer to
the disciples, and perhaps he expected that the giant would take this

prayer back to the Indians. When the feast ended, a guard of four armed men escorted him to shore.

During the feast, another giant had watched the proceedings from land, and as soon as his tribesman returned safely, he tore off in the direction of their concealed huts to convey the news. Slowly, the other giants emerged from the trees to reveal themselves to the crew members, who were astounded by the sight and transfixed by the unexpected appearance of giant women:

"They placed themselves one after the other quite naked and began to leap and sing, raising one finger to the sky, and showing our people a certain white powder made from roots of herbs, which they kept in earthenware pots, and made signs that they lived on that, and that they had nothing to eat but this powder," Pigafetta wrote. "Whereupon our men made signs to them that they should come to the ships, and that they would help them carry their provisions. Then these men came, bearing only their bows in their hands. But their wives came after them loaded like asses and carrying their goods. And the women are not so tall as the men, but somewhat fat-ter. When we saw them, we were all amazed and astonished. For they had teats half a cubit long, and they were painted on the face and clad like the men. But they wore a small skin in front to cover their private parts. They brought with them four of those little animals of which they make their clothing"—guanacos—"and led them on a leash with a cord."

The guanacos intrigued the sailors almost as much as the giants did. The guanacos had adapted to life in this harsh region over thou-sands of years. Their stomachs contain three compartments to extract protein efficiently from the food they chew, and they have long legs for scrambling up and down steep mountainsides. Magel-lan's crew, eager for guanacos of their own, learned from the giants how to capture the animals. The procedure proved surprisingly sim-ple. "They tie one of the young ones to a bush, and thereupon the large ones come to play with the little ones, and the giants hidden behind some ledge kill with their arrows the large ones. Our men took eighteen of these giants, both men and women, whom they divided into parties, half on one side of the port, and the other half on the other, to catch the said animals."

Within days, the sailors were delighted to have their own guana-cos. Their tough, stringy meat provided a welcome alternative to the diet of salted sea elephant, and the guanaco wool, similar in color and texture to a sheep's, helped the men to endure the rigors of Port Saint Julian.

*I*n addition, the giant Indians provided a welcome source of com-panionship and distraction for the crew members from the bleak-ness of the empty landscape. Everything about their customs and appearance fascinated Pigafetta. "Verily those giants stand straighter than a horse, and are very jealous of their wives," he observed. "They wear a cotton cord around their head, to which they fasten their arrows when they go hunting, and bind their member close to their body by reason of the very great cold."

There was more: "When these giants have a pain in the stomach, instead of taking medicine, they put down their throat an arrow two feet or thereabout in length, then they vomit [matter] of a green color mingled with blood. And the reason they bring up this green matter is that they eat thistles."

And still more: "When they have a headache, they make a cut across their forehead, and the same on the arms and legs, to draw blood from several parts of their body."

Most of all, their belief system fascinated Pigafetta, who recorded a vivid glimpse into the inner lives of the Tehuelche Indi-ans. "When one of them dies, ten or twelve devils appear," he came to understand, "and dance around the dead man. And it seems they are painted. And one of these devils is taller than the others. And from this the giants took the fashion of painting themselves on the face and body." Pigafetta learned a few words of their language to better understand these concepts. The big devil, they told him, was called Setebos, a name to be reckoned with, and the little ones Cheleule.

Eventually, Magellan gave the Indians a name—*Pathagoni*, a neologism suggesting the Spanish word *patacones*, or dogs with great paws, by which he meant to call attention to their big feet, made even larger by the rough-hewn boots they wore. So these were the Bigfeet Indians, according to Magellan, who later gave the name to the whole region, known ever since as Patagonia.

*N*ow that they had a name, the giants came to seem even more human to Magellan and his crew, who, as Pigafetta describes, made friends with one in particular. "This giant," he wrote, "was of better disposition than the others, and was very graceful and amiable, loving to dance and leap. And when dancing he depressed the earth to a palm's depth in the spot where his feet touched. He was with us for a long time, and in the end, we baptized him, naming him John."

In this instance, the solemn rite was undertaken as a sign of kinship rather than of conquest. "The said giant pronounced the name Jesus, the paternoster, Ave Maria, and his own name as clearly as we," Pigafetta related. "He had a terribly loud and strong voice." What these incantations meant to John the Giant can only be imagined, but he no doubt associated them with the lavish gifts he received from Magellan. "The captain gave him a shirt and a cloth jerkin, and seamen's breeches, a cap, a mirror, a comb, bells, and other things, and sent him away whence he had come; and he went off very joyous and happy." The new convert returned the next day, bearing precious guanacos, and received still more gifts in trade, but then he was not seen or heard from again. "It is to be supposed that the other giants killed him because he had come to us."

This was pure conjecture. The crew found no evidence that John had been killed or ostracized for fraternizing, but their fear that he had been harmed for this reason betrayed how wary they were of the Indians, friendly or not. Relations with the Patagonian giants deteriorated when a European scouting party unearthed a cache of Indian weapons. The discovery suggested that an ambush might be in the making. All thought of further baptisms were forgotten as the crew members feared for their lives and sought the safety of the ships. For two weeks, no giants were seen, and Magellan decided it was time to change his tactics.

*O*n July 28, four Patagonian giants, two men and two boys, appeared at the water's edge, signaling to the fleet that they wished to come aboard. This was just the opportunity Magellan had been awaiting. A longboat was dispatched to bring the four unsuspecting Indians aboard *Trinidad.* Magellan bestowed presents on his guests—"knives, scissors, mirrors, bells, and glass"—and while the four held them and marveled at them, "the captain sent for large iron fetters, such as are put on the feet of malefactors." Two of the giants were shackled. Even Pigafetta recoiled at the sight, disdainfully remarking that Magellan had resorted to a "cunning trick." For once, the expedition's official chronicler found it painful to watch Magellan's scheme unfold.

Instead of resisting, "The giants took great pleasure in seeing these fetters, and did not know where they had to be put, and they were grieved that they could not take them in their hands." The two giants who were still free tried to undo the fetters, but Magellan refused to allow them to interfere. Still naïvely trusting the cunning stranger in their midst, the giants "made a sign with their heads that they were content with this." Magellan saw to it that their bellies were full, offering them "a large boxful of biscuit, and unskinned rats, and... half a pailful of water at a time." At this point in the journey, the rats were nothing more than a nuisance to the sailors. Whenever they caught one of the little beasts, they tossed it into the sea. Seeing this waste of perfectly good food, the giants begged to have them, and devoured them whole.

Magellan then proceeded with his plan. "Forthwith the captain had the fetters put on the feet of both of them. And when they saw the bolt across the fetters being struck with a hammer to rivet it and prevent them from being opened, these giants were afraid. But the captain made signs to them that they should suspect nothing. Nevertheless, perceiving the trick that had been played on them, they began to blow and foam at the mouth like bulls, loudly calling on Setebos (that is, the great devil) to help them."

Confusion engulfed both parties. Magellan underwent a sudden change of heart and decided against imprisoning the giants. He ordered a detachment of nine guards under the command of Carvalho, *Concepción*'s pilot, to escort two of the giants ashore, and to reunite one of them with the woman assumed to be his wife "because he was greatly lamenting her." As soon as his feet touched dry land, this giant managed to escape, "running with so much nimbleness that our men lost sight of him." Carvalho's detachment feared he would tell others about the trick that had been played on them, and the tribe would return, seeking revenge.

The downward spiral of brutality continued. "The other giant who had his hands bound made the utmost effort to free himself, so that to prevent him one of our men struck him and wounded his head, at which he was violently angry." In an effort to calm him, the crew members led him to the huts where the women had taken refuge, but this gesture failed to bring about peace between the Indians and the sailors.

That night, Carvalho, still in charge of the detachment, decided they would sleep ashore. In the morning, the giants' huts were deserted; all the Indian men and women had fled into the interior, perhaps for good, perhaps to regroup for a surprise attack.

Their intentions became clear when arrows began whizzing from the dark forest. Suddenly nine Indian warriors appeared. Each carried three quivers of arrows held in leather girdles. With fluid movements, they fired off one deadly arrow after another. "Fighting thus, one of these giants pierced one of our men in the thigh, who died immediately. Whereupon seeing him dead they all fled." The fallen crew member was *Trinidad*'s Diego Barrasa. The suddenness of his death suggests that the arrow carried a poisoned tip.

Enraged, the other crew members retaliated with all their might. "Our men had crossbows and muskets, but they could never hit any of those people because they never stood in one place, but leapt hither and thither." The weapons' deafening roar scattered the giants, and when quiet returned to Port Saint Julian, the detachment sorrowfully buried their fallen colleague.

Magellan still held two Indians hostage, one aboard *Trinidad,* the other assigned to *San Antonio,* and despite his prohibition against passengers, slaves, or stowaways, he intended to present these two giants to King Charles.

The encounter between the European visitors and the Tehuelche Indians deteriorated drastically from its spirited and happy beginning. Conditioned by the fate of Solis and his crew, Magellan expected the Indians to bare their fangs eventually, no matter how sociable and benign their behavior seemed at first.

Yet his response to the Indians wavered. Had he considered them nothing more than cannibals, he would not have troubled to convert John to Christianity, nor would he have offered the new convert sacred texts, gifts far more valuable than any number of shiny mirrors, tinkling bells, and other trinkets. The paternoster was not given as a trick or as bribery, but as an attempt to forge a bond between the Indians and Europeans. The conversion strongly implied a measure of trust and respect between the two parties, as well as the expectation that John would abide by Christian standards of morality. Then, inexplicably, Magellan turned away from John and the other Indians. Perhaps it was the cache of weapons gnawing at Magellan's sense of security. Perhaps he had difficulty accepting that his faith could embrace both Europeans and Indians. That startling possibility had existed ever since Columbus, who saw no contradiction between converting Indians and enslaving them; both methods caused them to submit to the will of Spain. Although he was a devout Catholic, Magellan had no policy concerning conversions, and nothing in his royal orders offered explicit guidance in this crucial matter. He was an admiral, not a missionary. His conversion of John seemed motivated more by personal conviction than by a preconceived plan.

*O*n August 11, 1520, Magellan carried out the sentence he had proclaimed for his nemesis, Juan de Cartagena, and the priest, Pero Sánchez de la Reina, who had conspired with the Castilian captain. At Magellan's order, both were marooned on a small island in sight of the ships. They had no longboat, no firewood, and scant clothing. Their supplies consisted mainly of bread and wine, enough to last them the summer, perhaps, but they would have to face the next winter in Port Saint Julian alone.

Magellan finally gave the command to weigh anchor on August 24. After the harrowing five-month layover in Port Saint Julian, the Armada de Molucca put to sea. During that interval, Magellan had endured violent storms, single-handedly defeated a seemingly insurmountable mutiny, lost *Santiago,* befriended and then antagonized the local population of Indians, and at the cost of several lives strengthened his command of the fleet. He had demonstrated that he could be as much of a trickster as Odysseus. Most important of all, he had survived, kept most of his fleet intact, and his men under his command.

Just before leaving, Magellan sent everyone ashore to attend a final religious observance. They confessed their sins, received the sacrament, and returned to the ships, humbled before their Maker, to whom they prayed to preserve their lives during the next phase of the journey. As they sailed away from the ill-starred Port Saint Julian, the crew believed the worst of their difficulties were behind them. They had been hardened by adversity, and if nothing else, they were determined to survive to the end of the voyage. The strait still eluded them but, God willing, they would find it, and reach the Spice Islands, and eventually return to Spain, where they would be rich enough to spend the rest of their lives in retirement. It was a fantastic dream, and their only hope of deliverance.

As the four remaining ships of the armada sailed into the open waters of the Atlantic, the abandoned conspirators, Cartagena and the priest, watched the spectacle from their island prison. The two condemned men kneeled at the water's edge, crying and pleading for mercy as the ships grew smaller and finally vanished over the horizon.

CHAPTER VII

Dragon's Tail

The ice was here, the ice was there,
The ice was all around:
It cracked and growled, and roared and howled,
Like noises in a swound!

*T*he earth's crust can be compared to a cracked eggshell consisting of tectonic plates that bump and grind against each other, forming our oceans and continents, creating earthquakes, moving mountains. Millions of years ago, two tectonic plates merged and created a unique landscape at the southernmost tip of South America, not far from the South Pole. Over time, the plate from the east smashed into the plate from the west, which slid underneath. As a result, the eastern sea is about fifteen hundred feet deep, but the western sea reaches depths of over fifteen thousand feet. The awkward juxtaposition of two plates formed distinctive features in the landscape; the western portion contains the southernmost extensions of the Andes Mountains, which attracts moisture, while the eastern part tends to be smoother and drier. This was the landscape awaiting the Armada de Molucca.

From the moment of their departure from Port Saint Julian, their journey southward was fraught with more difficulty. After two days at sea, as they approached the mouth of the Santa Cruz River, another storm engulfed them and threatened to drive them all ashore, where they would likely meet the same fate as unlucky *Santiago*. Magellan gave the order to enter the broad river, and there, sheltered from the worst of the winds, the fleet rode out the squall.

After the storm passed, Magellan, with every fiber of his being, wanted to put to sea and resume the search for the strait. He gambled that if he could only survive the open water long enough to reach the strait, the channel would shelter the fleet from the storms that had plagued them for months. Yet the hazards of exploring the coast in August, as winter relented, remained overwhelming, even to Magellan, who was normally fearless. With the greatest reluctance, he decided to remain here until well into the subequatorial spring; then, and only then, would his ships have any chance of surviving at sea.

Magellan made the most of this enforced layover. For the next six weeks, the seamen busied themselves catching fish, drying and salting them, and stocking the ships. They ventured on land only to chop wood and haul it back to the ships. Occasionally, they made brief excursions to the southern shore of the Santa Cruz, where *Santiago* had broken up, and salvaged whatever items the sea had thrown up on shore, mainly chests and barrels.

On October 11, a celestial event of singular importance occurred, as noted in all likelihood by San Martín, the fleet's astronomer and astrologer: "An eclipse of the sun was awaited, which in this meridian should have occurred at eight minutes past ten in the morning. When the sun reached an altitude of 42½ degrees, it appeared to alter in brilliancy, and to change to a somber color, as if inflamed of a dull crimson, and this without any cloud intervening between ourselves and the solar body.... Its clearness appeared as it might in Castile in the months of July and August when they are burning the straw in the surrounding country."

A week later, on October 18, 1520, Magellan decided to risk the open sea again. He supposed, correctly, that the weather was as calm as it ever gets in this region. If the fleet faced more storms, his best hope was to seek shelter in a safe harbor, but he would pause no longer than necessary. He was months behind schedule—he had expected to approach the Spice Islands at about this time—and he yearned to make up for lost time. The fleet departed from the Santa Cruz River, tracing the undulating eastern coast of South America, with the Captain General in the grip of his obsession to find the strait.

Once again, foul weather bedeviled the ships, but it was not quite severe enough to drive them back. After two difficult days without any progress, the direction of the wind changed; now it came from the north, and the four ships plunged before the wind, leaving sharp, bubbling wakes and making rapid progress along a south-by-southwest course. Increasingly desperate to find the strait, Magellan scrutinized every inlet, hoping it might contain a hidden channel leading inland, but in each instance he was disappointed and continued his southerly course. Finally, he noticed a significant spit of land extending into open waters: a cape. As he approached, he made out a broad sandbank strewn with the skeletons of whales—a suggestion that he had come across a migration route, perhaps leading from the Atlantic to the Pacific. The gray water churned angrily where competing tides vied with one another, but the opening was wide, a league or more.

Vasquito Gallego, an apprentice Portuguese seaman aboard *Victoria* and the son of her pilot, recalled the gradual realization that the gaping break in the land might be more than a mere bay. "As the way became narrower, they thought it was a river," he recalled, and then, with mounting excitement, recorded that the wide mouth turned into a narrows farther ahead. "Continuing that way, they found deep salt water and strong currents, appearing to be a strait and the mouth of a big gulf that might be discharging into it." Magellan ordered his ships to sail into the gulf, and when they were well within its embrace, he saw it: the outlet leading west, just as he prayed it would.

Magellan had finally found his strait.

*O*n October 21, Albo, the pilot, recorded the great event in his log: "We saw an opening like a bay, and it has at the entrance, on the right hand, a very long spit of sand, and the cape which we discovered before this spit is called the Cape of the [Eleven Thousand] Virgins, and the spit of sand is in 52 degrees latitude, 52½ longitude, and from the spit of sand to the other part there may be a matter of five leagues," he observed. This is what he saw: a series of mounds, covered with tufts of grass, rising approximately 130 feet from the water. A later explorer described the cape as "three great mountains of sand that look like islands but are not." There was no mistaking the strait for a bay or an inlet; a broad waterway cut deep into the impenetrable landmass along which the fleet had been sailing for months.

Pigafetta exulted at the sighting of the waterway. "After going and setting course to the fifty-second degree toward the said Antarctic Pole," he wrote, "on the Festival of the Eleven Thousand Virgins, we found by miracle a strait which we called the Cape of the Eleven Thousand Virgins." After all the ordeals suffered by the armada, the discovery of the strait did lay claim to being a miracle.

*T*he Cape of the Eleven Thousand Virgins marks the entrance to the strait that Magellan had sought for more than two years. Precisely how he divined its existence has been the subject of debate ever since. He might have been aware of Lisboa's expedition, which claimed to have found the strait, and he was certainly aware of maps depicting the mythical strait. According to Pigafetta, Magellan, while still in Portugal, had seen a map depicting or suggesting a strait cutting through South America, but what map had Magellan seen? "He knew where to sail to find a well-hidden strait," Pigafetta declares, "which he saw depicted on a map in the treasury of Portugal, made by that excellent man, Martin de Boemia"—who was of course, Martin Behaim, who had created a state-of-the-art globe in 1492. (The earliest Pigafetta manuscripts employ the word *carta,* which could mean either a globe, a map, or a chart.)

It is often assumed that Behaim's "well-painted globe," which Magellan had displayed to King Charles and his advisers to persuade them to back his voyage, showed the strait; in fact, Behaim's globe, or map, did no such thing. Instead, it showed a waterway cutting through eastern Asia and the island of "Seilan." To add to the cartographic confusion, it positioned other Asian islands to the east of the strait. It is unlikely that Magellan would have employed this fanciful, wildly inaccurate representation to persuade King Charles of the existence of a strait cutting through the American continent; indeed, it is unlikely that Magellan ever saw the Behaim globe, despite the linkage of their names.

Pigafetta was inadvertently responsible for the case of mistaken identity; in all likelihood, he confused Behaim's rendition with that of another Nuremberg mapmaker, Johannes Schöner, a professor of mathematics who produced two maps, one in 1515 and the other in 1520, close to the time Magellan was displaying a map to King Charles. To the nonspecialist, Schöner's maps closely resembled Behaim's, and Pigafetta could easily have mistaken one for the other, especially because Schöner did not sign his productions.

Schöner's globe depicted a strait cutting through the American continent in the approximate location of the Isthmus of Panama— several thousand miles north of the actual strait. There is no conclusive evidence that Magellan saw this map, either, but it does demonstrate that cartographers were starting to include some sort of strait in South America, however poorly it was understood. If this

was the map Magellan had in mind, it would have been nearly useless in trying to find the strait. Even the daring Schöner hesitated to
depict the western coast of South America; it was, as he termed it,
terra ulterior incog.—in other words, "the land that has been hitherto
unknown."

Everything to the west was also unknown. Schöner, like other
cartographers of his era, shrank the immense Pacific into an entic-
ingly small and apparently navigable gulf, a misunderstanding that
resulted in Magellan's conviction that he could reach the Spice
Islands within weeks, if not days, after exiting the strait. And like
other maps of the era, it placed China in close proximity to the
American continent. Finally, Schöner's globe placed the Spice
Islands well within Spanish territory as defined by the line of demarcation, and this feature—again, wildly inaccurate—might have
accounted for Magellan's conviction that the Spice Islands legitimately belonged to Spain rather than Portugal.

Magellan knew better than to take maps at face value, but he
was deeply susceptible to their influence. They were idealized pro-
jections of what the world might be like. Instead of the dragons and
magnetic islands of older maps, these contained a new marvel, one
that was possibly just as mythical: a strait. They were calls to adventure rather than a set of directions, hypotheses rather than conclu-
sions, provocative geographical cartoons that fed the fantasy of
empire.

*N*ow that Magellan had finally found the strait, he faced three
hundred miles of nautical nightmare. Navigating the waterway
would prove as daunting a challenge as simply finding it had been.
Tides in the strait run as much as twenty-four feet, making it difficult to anchor ships securely, and currents are strong. Beds of kelp
lurking below the water's surface threatened to foul lines, keels, and
rudders. But if Magellan could overcome the obstacles presented by
the strait, and keep his mutinous crew intact, he would pioneer a

new route to the Indies, to a new understanding of the continents and of the globe itself.

The ships turned west, braved the swirling tides, and entered the inland waterway. The first thing the pilots noticed was the extreme depth of the strait. "In this place it was not possible to anchor," Pigafetta observed, "because no bottom was found. Wherefore it was necessary to put cables ashore of twenty-five or thirty cubits in length."

Curious to learn exactly where they were, Magellan sent Carvalho ashore with orders to climb to the highest point to look for an opening. On his return, Carvalho reported that he failed to discern the Pacific to the west; nevertheless, Magellan was gripped with the conviction that he had found the waterway to Spice Islands. He ordered Albo to record the strait's twists and turns as accurately as possible. "Within this bay we found a strait which may be a league in width," he wrote, "and from this mouth to the spit you look east and west, and on the left hand side of the bay there is a great elbow within which are many shoals, but when you are in the strait, take care of some shallows less than three leagues from the entrance of the straits, and after them you will find two islets of sand, and then you will find the channel open. Proceed in it at your pleasure without hesitation," he recommended. "Passing this strait we found another small bay, and then we found another strait of the same kind as the first, and from one mouth to the other runs east and west, and the narrow part runs N.E. and S.W., and after we had come out of the two straits or narrows, we found a very large bay, and we found some islands, and we anchored at one of them." No doubt Albo had specific landmarks in mind as he wrote, but the strait defied even this precise chronicler, and his directions proved difficult for subsequent visitors to interpret.

Within days, the strait's gloomy enchantment impressed itself on the crew. As they negotiated its frigid waters, they observed thickly vegetated, forbidding shores sliding past, cloaked in eerie shadows. Late one night, during the few hours of darkness at that time of year, they caught glimpses of what they believed were signs of human settlements; distant fires with an indistinct source burst forth, their ruby flames glimmering like spectral apparitions in front of the dark green cypresses, vines, and ferns. The fires sent plumes of smoke into the hazy sky, and fouled the air with an acrid odor.

Magellan and his crew believed these fires had been set by Indians who lurked in the shadows, waiting to pounce—one more reason for the sailors to stay aboard ship, especially at night, even though their provisions were running low. This was a reasonable precaution, but the fires were most likely of natural origin, the result of lightning. In any event, Magellan called this region Tierra del Fuego, Land of Fire. Today, we know that Tierra del Fuego is actually an enormous triangular island buffeted by winds from both the Atlantic and Pacific oceans, and constantly beset by storms and rapidly changing weather. The Land of Fire is actually the land of storms. Tierra del Fuego covers more than 28,000 square miles of glaciers, lakes, and moraines. Magellan's crew looked on the low-lying areas, where the hills rarely reach six hundred feet; to the south and west, the southern extension of the Andes mountain range pierces the clouds, reaching heights of over seven thousand feet.

Now that they were in the strait, the pilots found that the sky was rarely clear by day or night, which made it nearly impossible to take accurate measurements either by the stars or by the sun. Gloomy, ragged, low-hanging clouds scudded over the mountains hugging the fjords through which the ships expectantly glided. Occasionally, the leaden mists parted to allow sunlight, gleaming with painful brilliance, to stream down on the impenetrable land and the surging water.

The sunlight, when it managed to break through, could be pitiless at this low latitude and appeared to illuminate the landscape with a gray, polarized radiance. Striations of light played over the stony beaches and the glaciers frosting the mountaintops. Although Magellan traversed the strait at the warmest time of year, when the wind, for all its bite, was at its lightest, and snows had receded, the enormous glaciers were plentiful and awe-inspiring. Snow nearly always fell atop the glaciers—they were endlessly renewing themselves—and at lower altitudes the ice melted into narrow waterfalls cascading over the granite outcroppings into the fjords. Invisible to the sailors, the glaciers extended across the landscape, running through thirty miles of mountains before sheering off at the water's edge.

As they continued to sail through the strait, Magellan's crew observed a solid wall of ice rising majestically before them—two hundred feet, five hundred feet, and more. They were ancient edifices, these glaciers, some of them ten thousand years old, and they looked it, with their grimy faces deeply pockmarked and weathered.

Consisting of packed snow and ice, the glaciers never rested; they cracked, they groaned, they roared, and they threatened to decompose and tumble onto the beaches and water below. Their crystalline towers leaned out over the water in irregular columns, like rotting teeth in a decaying jaw. They inclined ever more precariously over the placid water until one column after another, warmed by the sun and buffeted by the wind, calved and collapsed amid a cloud of icy dust with a shattering report followed by a drumlike roll of thunder, low and resonant, announcing destruction.

To everyone's surprise, the glaciers were neither white nor gray, but a light, almost iridescent blue that in the crevasses and seams darkened to a deep azure. The countless chunks of ice broken off from the glaciers, some as large as a whale, others as small as a penguin, had the same enigmatic bluish cast as they bobbed past the ships: an armada of sculptured ice drifting toward a mysterious location.

Groping for a plausible explanation for the glaciers' appearance, Magellan theorized that the glaciers' distinctive color had to do with their extreme age. In fact, the bluish cast was determined by the distinctive properties of snow and ice. The surface of snow and ice reflects all light, without preference for any particular color of the spectrum, but the interior handles light differently. Snow acts as a light filter, and treats the spectrum preferentially, scattering red light more strongly than blue. Photons emerging from snow and ice generally have more blue rays than red. The deeper the snow and ice, the farther the light must travel, and the darker blue it becomes, just as water appears a deeper blue as it increases in depth. For this reason, the deep crevasses in the glaciers possess an unearthly azure hue.

Every visitor to the strait has been awed by the majestic, moody spectacle it presents. It is reminiscent of Norway, or Scotland, or Nova Scotia, but ultimately it is unlike any other place on earth. In 1578, Sir Francis Drake, the English explorer and pirate, led the first expedition since Magellan's through the strait. One of his officers, Francis Pretty, was amazed by the spectacle passing before his eyes. "The land on both sides is very huge and mountainous; the lower mountains whereof, although they be monstrous and wonderful to look upon for their height, yet there are others which in height exceed them in a strange manner, reaching themselves above their fellows so high, that between them did appear three regions of clouds," Pretty marveled. "This Strait is extreme cold, with frost and snow continually; the trees seem to stoop with the burden of the weather, and yet are green continually, and many good and sweet herbs do very plentifully grow and increase under them." And when the naval historian Samuel Eliot Morison visited the strait in February 1972, he, too, fell under its spell. "One seems to be entering a completely new and strange world, a veritable Never-Never Land," he remarked. "The Strait never freezes except along its edges, and the evergreen Antarctic beech, with its tiny matted leaves, grows thickly along the lower mountain slopes. The middle slopes support a coarse grass which turns bronze in the setting sun; and above, the high peaks are snow-covered the year round; when it rains in the Strait, it snows at 6000 feet."

Although the sky was generally overcast, especially at night, it cleared at brief intervals to reveal a dazzling array of constellations competing for attention, with an unnaturally brilliant Milky Way. The familiar—Orion's belt, the Big Dipper—mingled with the unfamiliar constellations of the Southern Hemisphere, especially the Southern Cross, whose presence reinforced Magellan's conviction that the Almighty was looking over the entire venture, even here, at the end of the world.

*O*nce the armada had negotiated the first two narrows within the strait, Magellan became increasingly cautious about the hazards ahead and decided to scout the strait's uncharted waters. "The Captain General sent his cousin Álvaro de Mesquita to go in his vessel *San Antonio* through that mouth in order to find out what was inside while he and the other ships remained anchored in the wide part of the entrance until they knew what was what," Vasquito Gallego noted. Actually, Magellan dispatched two ships (the other was *Concepción*), but *San Antonio* took most of the risks. "Álvaro de Mesquita went for fifty leagues up the strait, and in some parts he found it so narrow that between one shore and the other there was no more distance than one Lombard shot, and the strait turned toward the west whence the sea currents came in full force, so strong that they could not go on, except with difficulty," Gallego remembered. "Mesquita turned back, saying that he thought that the great water came out of a big gulf and his advice was to go in search of its end and see the mystery, because not without reason came that water with such force from that direction."

All the while, *Victoria* and *Trinidad* remained tied up in Lomas Bay, on the southern shore of the strait. Here the water was shallow enough to permit the ships to drop anchor, and they seemed to be safe, but at night a "great storm," as Pigafetta called it, blew up and lasted well into the next day, battering the ships. Magellan was forced to raise anchor and let the two ships ride out the storm in the protected reaches of the bay.

The gales in this region were especially violent and seemingly appeared out of nowhere. The "great storm" of which Pigafetta spoke is called a "williwaw," and it is peculiar to the strait. A williwaw occurs when air, chilled by the glaciers surrounding the strait, becomes unstable and suddenly races down the mountains with ever-increasing velocity. By the time it reaches the fjords, it creates a squall so powerful that it never fails to terrify and disorient any sailor unlucky enough to be caught in its grip.

San Antonio and *Concepción* had an even more difficult time riding out the williwaw than the ships that stayed behind. The sailors aboard those ships had experienced terrifying storms, but nothing equal to this. The fierce winds prevented them from rounding the cape, and when they tried to rejoin the fleet, they nearly ran aground. In the darkness, the two ships became disoriented, and their pilots, without maps and unable to see the stars, feared they were lost. They hunted for a way out for the next day, and the next after that, until they finally approached a narrow channel leading to a continuation of the strait. Once they noted the exact location of the strait's extension, they sailed back through relatively calm waters to find their Captain General.

A dramatic reunion occurred, as Pigafetta explained: "We thought that they had been wrecked, first, by reason of the violent storm, and second, because two days had passed and they had not appeared, and also because of certain smoke [signals] made by two of their men who had been sent ashore to advise us. And so, while in suspense, we saw the two ships with sails full and banners flying to the wind, coming toward us. When they neared us in this manner, they suddenly discharged a number of mortars, and burst into cheers. Then all together thanking God and the Virgin Mary, we went to seek [the strait] further on." The rejoicing, the triumph over weather and geography, and the feeling of being blessed by divine authority were new to Magellan's men. For the better part of two years, they had been deeply mistrustful of their Captain General, divided from one another by language and culture, and prone to mutiny. After passing through these ordeals, they had become united and saw in each other not subversion or menace but the possibility of ultimate triumph.

Despite the euphoria Magellan felt on discovering the strait, he still faced serious obstacles. Influenced by the maps he had seen in Portugal, Magellan mistakenly conceived of the strait as a single channel running through the huge landmass blocking the route to the Indies, when in fact there was no single strait; instead, he faced a complex array of tidal estuaries snaking through the mountains at the southern limit of the Andes. Instead of a simple shortcut to the Pacific, Magellan had led his fleet into a uncharted maze that would put his navigational abilities to the ultimate test.

The waterways he explored were wide enough—never less than six hundred feet across, and generally more than several miles in width—but still treacherous. The strait largely consisted of a network of fjords, geologic evidence of deep glaciers that still held the surroundings in their icy embrace. At low levels, the glaciers melted into narrow, glistening waterfalls that cascaded across the granite face of the mountains until they emptied into the frigid water. If any of Magellan's men fell overboard, they would survive in these conditions for ten minutes at the most.

Here and there, along stony gray beaches, lolled families of sea elephants, easily distinguished by their length, about ten feet, their two flippers close to their torpedolike heads, and a broad stabilizing tail lazily patting the sand. Sea elephants could barely get around on land, so they lay at the water's edge, yawning and stretching. Other indigenous wildlife in the strait included arctic foxes and penguins crowding beaches of their own. Giant black-and-white condors wheeled overhead, their wingspan extending to ten feet. They kept close to the mountain ridges, where they circled in the rising currents of warm air known as thermals. Occasionally, they nested in pairs, patrolling their eyries, appearing at rest more like the vultures they actually were.

Despite the snow cover lasting for eight months a year, the waterfall-fed vegetation in the strait was suffocatingly lush. Within several feet of the shoreline lurked a dense forest with dozens of types of ferns; windblown, stunted trees; silky moss; and a layer of spongy tundra. There were also brightly colored clumps of tiny, hardy berries; they were bitter on the outside, sweet on the inside, their delicate fruit covered with miniature air cushions to protect them from snow. (The crew had to be careful about eating them; although the berries were not toxic, they had a severe laxative effect.) There were even small white orchids blooming in the mud. Little light penetrated the thick canopy of leaves to dispel the fertile, peaceful shade within. "So thick was the wood, that it was necessary to have constant recourse to the compass," wrote the young Charles Darwin when he visited the strait aboard HMS *Beagle* in 1834. "In the deep ravines, the death-like scene of desolation exceeded all description; outside it was blowing a gale, but in these hollows, not even a breath of wind stirred the leaves of the tallest trees. So gloomy, cold and wet was every part, that not even the fungi, mosses, or ferns, could flourish." When he at last worked his way out of the enchanted forest to a summit, Darwin described a view familiar to Magellan's crew: "irregular chains of hills, mottled with patches of snow, deep yellowish-green valleys, and arms of the sea intersecting the land in many directions. The strong wind was piercingly cold, and the atmosphere rather hazy, so that we did not stay long on top of the mountain."

The strait's thick vegetation gave the air an intoxicating fragrance and buoyancy. The breezes were scented with a damp mossy odor lightened by the scent of wildflowers, freshened by the cool glaciers, and faintly tangy with the salt from the sea. Like everything else in this region, the very air was alive with mystery and promise. The strait seemed to be a giant natural monastery in which the crew sought refuge, a place of quiet contemplation of the paradoxes of nature on a scale capable of inducing profound humility.

Since leaving Port Saint Julian, Magellan had seen no indigenous peoples, but his men remained alert, both for self-protection and the opportunity to barter for provisions. He dispatched a skiff crowded with ten men under orders to comb the landscape for signs of human life, but they found only a primitive structure sheltering two hundred gravesites. Apparently, a tribe of Fuegian Indians had used the place to bury their dead in warm weather, and then vanished into the perfumed interior. It is believed that these Indians came from Asia thousands of years earlier, and had been on the losing side of battles for land ever since, displaced again and again until they were nearly at the end of the continent, occupying territory no other tribes wanted.

Though disappointed, Magellan's scouting party was probably better off avoiding the locals. Three hundred years later, Charles Darwin encountered a canoe bearing tribal members, whose lot had scarcely changed over the intervening centuries. Indeed, Darwin felt that he was peering through eons to the dawn of human society. He judged them "the most abject and miserable creatures I have any where beheld. . . . These Fuegians in the canoe were quite naked, and even one full-grown woman was absolutely so. It was raining heavily, and the fresh water, together with the spray, trickled down her body. In another harbour not far distant, a woman, who was suckling a recently-born child, came one day alongside the vessel, and remained there whilst the sleet fell and thawed on her naked bosom, and on the skin of her naked child. These poor wretches were stunted in their growth, their hideous faces bedaubed with white paint, their skins filthy and greasy, their hair

entangled, their voices discordant, their gestures violent and without dignity. Viewing such men, one can hardly make oneself believe
they are fellow-creatures, and inhabitants of the same world."
Clinching his disgust for the Fuegians, Darwin observed, "Nor are
they exempt from famine, and, as a consequence, cannibalism
accompanied by parricide." They were, he judged, "the miserable
lord of this miserable land."

As the ships of the fleet glided along the fjords, they experienced only three hours of night, and the extended days allowed
them to make up for the time lost in Port Saint Julian. The prospect
of successfully negotiating the strait appeared increasingly likely, at
least to Magellan. But the conviction was not unanimous, as he discovered when he summoned an officers' council to discuss the fleet's
future course. He was delighted to hear that they had sufficient provisions to last three months, more than enough, he calculated, to
carry them through the strait and to the Moluccas. Encouraged, the
captains and pilots indicated they were strongly in favor of pushing
on—all but one, that is.

Estêvão Gomes, reassigned as the pilot of *San Antonio*, strongly
dissented. Now that they had found the strait, he argued, they
should sail back to Spain to assemble a better-equipped fleet. He
reminded Magellan that they still had to cross the Pacific, and while
no one knew how large it was, Gomes assumed it was a large gulf in
which they might encounter disastrous storms. The Captain General insisted they would continue at all costs, even if they were
reduced to eating the leather wrapping their masts, but not everyone shared Magellan's passionate determination. With his widely
acknowledged piloting skills, Gomes had his own supporters among
the crew, a situation that infuriated the Captain General. The council was not intended as an exercise in collective decision-making;
rather, it was a forum for Magellan to rally his men behind and to

prepare them for the challenges that lay ahead—challenges that God alone could help them meet.

Gomes's opposition set the stage for another mutiny, but unlike the previous uprisings, this was not a violent confrontation with flashing swords. It began more insidiously, as a grim debate at the end of the world between two respected rivals.

Gomes was Portuguese, so this time the dispute was not a matter of Spanish-Portuguese rivalry. In fact, Gomes had defected from Portugal with Magellan in 1517. He seemed to be an integral member of Magellan's tightly knit group of trusted mariners, but Gomes harbored ambitions of his own, and he skillfully leveraged his relationship with Magellan to further his ends. Within a few months of arriving in Seville, he received a pilot's commission, and immediately afterward began promoting his own Armada de Molucca. He nearly achieved his goal, but then Magellan, with his superior experience and connections, including his advantageous marriage to Beatriz Barbosa, appeared before the king, who promptly forgot all about Gomes. On April 19, 1519, Gomes settled for a commission as Magellan's pilot major; the appointment only served to whet his appetite for still more power and to encourage his bitterness toward the Captain General under whom he served. The enmity between the two was no secret; even Pigafetta, normally circumspect, was aware of the bitter history: "Gomes . . . hated the Captain General exceedingly, because before the fleet was fitted out, the emperor [King Charles] had ordered that he [Gomes] be given some caravels with which to discover lands, but his Majesty did not give them to him because of the coming of the Captain General."

Gomes received another blow when Magellan, perhaps sensitive to Gomes's ultimate goal to supplant him, refused to appoint him captain of *San Antonio* after the mutiny in Port Saint Julian. Instead, Gomes had to suffer the ignominy of serving as a pilot under the inexperienced but well-placed Álvaro de Mesquita; this was, if anything, a lesser position than that of pilot major of the flagship. More experienced and better qualified, Gomes seethed with resentment at having been passed over, and he transmitted his sense of outrage to *San Antonio*'s sympathetic crew.

Every time Magellan dispatched *San Antonio* on a reconnaissance mission, Gomes, her pilot, became more alarmed by the hazards of the journey. Mesquita was so inexperienced that Gomes shouldered the responsibility for exploring these unknown waters. As a result, he knew the strait better than anyone else in the expedition, the Captain General included, and he was thoroughly unnerved by what he had seen. Gomes and his crew were, in Gallego's assessment, "disgusted with that long and doubtful navigation."

The dispute between Gomes and Magellan pitted two competing visions of the expedition against each other. Magellan saw it as a divinely sanctioned quest for new worlds, undertaken in the name of the king of Spain, to whom he was, if anything, even more devoted than he had been to the king of his native Portugal. If Magellan succeeded, it would be because God meant him to. This was discovery as revelation, as prophecy, as a high-risk collaboration between God and His favored nation, Spain. Magellan, in this scheme of things, was little more than God's servant, doing His will. To Gomes, the rebellious rationalist, Magellan's exhortations sounded like the words of a fanatic who would lead them all to certain death in the name of king and country. The only sane course, in his analysis, would be to return to Spain.

Gomes did not let the matter rest there.

*U*nder Magellan's command, *Trinidad* steadfastly continued the westward exploration of the strait. According to Albo's log, on October 28, little more than a week after discovering the strait, they tied up at an island guarding the entrance to another bay; this was either Elizabeth Island or Dawson Island. Here the strait extended in two directions, Froward Reach and Magdalen Sound. To choose a course, Magellan dispatched two ships to reconnoiter. *Concepción*, under the direction of Serrano, sailed westward into Froward Reach to Sardine River. Given the paucity of navigational detail supplied by Pigafetta's diary and Albo's log, it is difficult to say for certain what the expedition meant by Sardine River; it might have been what is now called Andrews Bay.

Meanwhile, *San Antonio* entered Magdalen Sound. Magellan gave his ships four days to return with their reports, but even after six, *San Antonio* failed to reappear. "We came upon a river which we called the River of the Sardine because there were so many sardines near it," said Pigafetta of this moment of doubt and confusion, "so we stayed there for days in order to await the two ships"—*Concepción* and *San Antonio*. "During that period we sent a well-equipped boat [*Victoria*] to explore the cape of the other sea. The men returned within three days and reported that they had seen the cape and the open sea." Sighting the Pacific was itself a momentous event, but the excitement of this discovery was overshadowed by the mysterious failure of *San Antonio* to reappear at the appointed time and place. Magellan had no idea what had become of her. Perhaps she had foundered and lay at the bottom of one of the yawning fjords. Or perhaps she had deserted just when the expedition was on the verge of its great accomplishment.

At this critical moment, Magellan conferred with Andrés de San Martín, now aboard *Trinidad*. After consulting the position of the stars and planets, he concluded that *San Antonio* had indeed sailed for Spain, and worse, her captain, Mesquita, a Magellan loyalist, had been taken prisoner. His vision proved to be remarkably accurate. "The ship *San Antonio* would not await *Concepción* because she intended to flee and to return to Spain—which she did," Pigafetta tersely reported. The long-frustrated mutiny had finally succeeded; even worse, it had taken place when Magellan least expected it. *San Antonio*, and all her crew, had vanished.

Aboard the renegade *San Antonio*, the situation was more even complicated than Magellan or his astrologer realized. Mesquita, the captain, had attempted to rendezvous with the rest of the fleet, but he failed to locate the other ships in the strait's confusing network of estuaries. Gomes naturally offered little help in the endeavor. During a formal inquiry after the voyage, another usurper, Gerónimo Guerra, insisted that he had deposited papers for Magellan at the precise point where the ships were supposed to meet. These papers would serve as proof of that effort, but they were never found.

Guerra's words sound self-serving, and perhaps they were. He had worked for Cristóbal de Haro, and was rumored to be related to the financier as well. He had shipped out on *San Antonio* as a mere clerk, but his remarkably high salary, 30,000 *maravedís,* twenty times greater than an ordinary seaman's, signaled a much larger role. Guerra's real mission was to look out for Haro's interests; in other words, he was a spy. Had Magellan agreed to return to Spain, Gomes's alliance with Guerra suggests that the Haro family would have supported the decision; after all, they would have gotten their ships back safe and sound. But King Charles was another matter. At the very least, he would have sent Magellan to jail.

Exactly when *San Antonio* tried to rejoin the rest of the fleet—if she ever did—is open to question. The ships' officers later testified at the inquiry that they returned well before they were expected. If so, why had Magellan failed to locate the missing ship? There were two possibilities. Either she had gotten lost in the strait's endless estuaries, or the mutineers had seized the ship, sought refuge in a concealed bay or fjord, and slipped out of the strait under cover of darkness for Spain.

*N*o matter what the intentions of Gomes and Guerra actually were, discontent aboard *San Antonio* increased. Mesquita sent smoke signals and fired cannon to try to raise the rest of the fleet, but these signs went unseen and unheard. Mesquita stubbornly insisted on continuing his search for Magellan, but the growing uncertainty convinced Guerra, Gomes and a few like-minded sailors that the time had come to seize the wayward ship. They swiftly overpowered Mesquita, a deed for which they could pay with their lives. Once the mutiny was in progress, there was no stopping it; the mutineers had to succeed or, as they well knew, they would be drawn and quartered and displayed as so many pieces of freshly butchered meat.

Desperate, Gomes flourished a dagger and stabbed Mesquita in the leg. Battling the wound's throbbing pain, Mesquita snatched the dagger from Gomes and stabbed the attacker in the hand. Gomes howled as the iron entered his flesh, and his cries attracted reinforcements. They overwhelmed and shackled Mesquita, who was held prisoner in Guerra's cabin. Now Mesquita would receive his bitter payback for the court-martial and suffering he had overseen in Port Saint Julian. As *San Antonio* set a course for Spain, the mutineers planned to torture him into signing a confession that Magellan had tortured Spanish officers.

The thought of *San Antonio* slipping away from the rest of the fleet filled Magellan with dread. The Captain General feared that the would-be mutineers had finally found the perfect occasion for their revenge on Mesquita. Even without the prompting of his astrologer, Magellan suspected that Gomes would sail for Spain, and, once there, attempt to tarnish Magellan's name with a biased account of the tragic events at Port Saint Julian. Gomes could twist the truth to claim that his mutiny had actually been an act of heroic resistance in the face of Magellan's disloyalty. In this scenario, none other than Estêvão Gomes would be Captain General for the next expedition to the Moluccas, while Magellan would hear about it from the obscurity of a Spanish prison.

San Antonio was the largest ship in the fleet, and she carried many of the fleet's provisions in her hold, so the loss instantly put the other sailors' food supplies—indeed their very lives—in jeopardy. The rebels also carried off another prize, an affable Patagonian giant whom they had captured several months before. Magellan had to decide whether to pursue the mutineers or hope that his cousin would regain control of the ship. He elected to resume searching for the missing *San Antonio.* "We turned back to look for the two ships, but we found only *Concepción*," Pigafetta wrote. "Upon asking them where the other was, Juan Serrano, who was captain and pilot of the former ship (and also of that ship that had been wrecked), replied that he did not know, and that he had never seen it after it had entered the opening." Magellan launched a search mission to recapture the missing ship, a virtual impossibility in this watery labyrinth. "We sought it in all parts of the Strait," Pigafetta recorded, "as far as that opening whence it had fled, and the Captain General sent the ship *Victoria* back to the entrance of the Strait to ascertain whether the ship was there."

In his actions, Magellan strictly followed his royal instructions of May 8, 1519, governing ships that had gone astray, to establish prominent indicators. Pigafetta described the lengths to which Magellan went: "Orders were given, if they did not find it, to plant a banner on the summit of some small hill with a letter in an earthen pot buried... near the banner, so that if the banner were seen the letter might be found, and the ship might learn the course we were sailing. For this was the arrangement made between us in case we went astray one from the other. Two banners were planted with their letters—one on a little eminence in the first bay, and the other in an islet in the third bay, where there were many sea wolves and large birds."

Although Pigafetta provides scant clues, this was likely Santa Magdalena Island, a massive, windswept dune rising from the frigid waters. At that time of year, it was overrun with thousands of penguins, the "large birds" mentioned by Pigafetta, mating, burrowing, and most of all fouling the entire islet with their droppings, whose penetrating stench not even the brisk, salty air could mask. Denuded of vegetation, and located in open water, the islet made an excellent place for a marker to remain visible to a passing vessel.

Magellan waited for the errant *San Antonio* to return. "He had a cross set up in an islet"—in all likelihood, one of the Charles Islands—"near that river which flowed between high mountains covered with snow and emptied into the sea near the River of Sardines." The precautions were taken in vain, lonely signals at the end of the world for a phantom ship. *San Antonio* never reappeared.

*O*nce Magellan became resigned to the loss of the ship, the three remaining vessels of the Armada de Molucca pressed on. After the hardships they had endured at bleak Port Saint Julian, the crew came to welcome the variety and natural majesty the strait afforded them. As they plied its fjords, they marveled at the dolphins that swam beside their ships and jumped in agile arcs. Sailors' lore had it that when dolphins jumped straight ahead, good weather was approaching, and when they jumped to one side or the other, the weather would turn foul.

The marvelous but hazardous strait still lacked a name. At first, the men called it simply *the* strait. Pigafetta took to referring to the waterway as the Patagonian Strait, while San Martín, the astrologer-pilot, preferred the name Strait of All Saints. Still others referred to it as Victoria Strait, after the first ship to enter its waters. By 1527, six years after the expedition's conclusion, the waterway had earned the name by which it is now known, the Strait of Magellan. For all his pride, Magellan never dared to name the strait after himself; the names he did confer during his journey were either descriptive (Patagonia) or religiously inspired (Cape of the Eleven Thousand Virgins).

As one mountainous prospect gave way to another, Pigafetta wrote glowingly of the strait's natural splendor and sustaining food. "One finds the safest of ports every half league in it, water, the finest of wood (but not of cedar), fish, sardines, and *missiglioni,* while smallage, a sweet herb (although there is also some that is bitter) grows around the springs. We ate of it for many days as we had nothing else." Although the men did not realize it, their diet replenished their depleted bodies. The wild herbs they consumed contained vitamin C, which protected them against the depredations of scurvy, at least for a while.

All things considered, Pigafetta judged, "I believe there is not a more beautiful or better strait than this one."

While Pigafetta took satisfaction in the armada's accomplishment, Magellan succumbed to a rare moment of self-doubt, and sought the advice of his officers about whether to proceed with the expedition or return to Spain, just as Gomes had urged him to do. His uncharacteristic wavering suggests he dreaded the rumors that the rebellious crew of *San Antonio* would spread about his conduct if they ever reached Spain.

Magellan dictated a lengthy missive to Duarte Barbosa, *Victoria's* captain, an indication that relations had become so strained that the Captain General feared that simply bringing them together would lead to yet another mutiny. The document reveals his urgent need to build a consensus: "I, Ferdinand Magellan, Knight of the Order of Santiago, and Captain General of this Armada which His Majesty sends to the discovery of the Isles of the Spices, etc., hereby inform you, Duarte Barbosa, captain of *Victoria*, and its pilots and boatswains, that I am aware of your deeming it a very grievous thing that I shall be determined to continue onwards, because you think that time is short to accomplish our journey," he said.

> And since I am a man who never despised the advice and opinion from others, on the contrary, all of my decisions are taken jointly with everyone and notified to one and all, without my offending anyone; and because of what happened in San Julian with the deaths of Luis de Mendoza and Gaspar de Quesada, and the banishment of

Luis de Cartagena and Pero Sánchez de la Reina, priest,
you out of fear refrain from telling me and advising me
on everything you believe to be useful to His Majesty
and the Armada's well-being, but if you do not tell me so,
you are going against the service of the Emperor-King,
our lord, and against the oath and homage you took with
me; therefore I ask you on behalf of the said lord, and I
myself beg you and order you to write down your opinions, each one individually, stating the reasons why we
should continue onwards or else turn back, and all this
showing no respect for anything that may prevent you
from telling the truth.... Being aware of those reasons
and opinions, I will then say mine and my willingness to
conclude what should be done.

—Written in the Canal de Todos los Santos, opposite
   the Río de la Isleta, on the 21st of November,
   Thursday, at fifty-three degrees, of 1520.
   Ordered by Captain General Ferdinand Magellan.

This remarkable document—Magellan's longest statement to
have survived—reveals the suspicion and mistrust running rampant
at what should have been one of their most harmonious and triumphant intervals. The normally resolute Magellan sounds as
though he is about to apologize for the protracted trial and cruel
executions he ordered at Port Saint Julian, and he clearly realizes
that as a result of his severe (though legally sanctioned) disciplinary
measures, he has alienated his officers, even those closest to him.
Afraid of losing still more of his ships to mutiny, Magellan's isolation at this moment was nearly complete.

*T*hrust into an unaccustomed position of authority, Andrés de San
Martín, the fleet's astronomer, urged that they continue the expedi-

tion at least through mid-January, although he remained skeptical
that the strait would ultimately prove to be the miraculous passage
to the Spice Islands. After January, he warned, the days would grow
short, and the williwaws, whose destructive power they had already
experienced, would become even more ferocious; furthermore, they
must not sail by night because the men would be exhausted after a
long, strenuous day battling high winds and rough waters. "Most
magnificent Lord," he began,

> Having seen your lordship's command, of which I was
> notified on Friday 22nd of November of 1520 by Martín
> Méndez, clerk of the ship of His Majesty named *Victoria*,
> and which orders me to give my view as regards what I
> believe to be better for this journey, either to continue,
> or to turn back, with the reasons behind either choice, I
> say: That, aside from doubting that neither through this
> Canal de Todos los Santos, in which we now are, nor
> through the other two straits lying to the East and East-
> Northeast, there might be found any passage to the
> Moluccas, this is irrelevant to the question of what could
> be eventually found, weather permitting, insofar as we
> are in the prime of summer. And it seems that your lord-
> ship must continue ahead in search of it, and depending
> on what shall be found or discovered until the middle of
> this coming January of 1520, you may consider the possi-
> bility of returning to Spain, because from then on the
> days suddenly dwindle and the weather shall worsen. And
> since now, even though the days last seventeen hours,
> added to the dawn and dusk, we still suffer stormy and
> shifting weather, much more so can be expected when
> the days decrease from fifteen to twelve hours and much
> more in winter, as we already know. So your lordship may
> want to leave these straits and spend the month of Janu-
> ary in reaching the outside and then, after collecting

enough water and fuel, head towards Cádiz and the port
of Sanlúcar de Barrameda, whence we departed.

San Martín's position was reasonable and well argued, but cautious.

Continuing nearer the Austral Pole than we presently
are, as you instructed the captains at the river of Santa
Cruz, I do not think it feasible, due to the terrible and
stormy weather, because if at this latitude sailing proves
so hazardous and painful, what shall it be like when we
find ourselves at sixty or seventy-five degrees or more, as
your lordship said he must go in search of the Moluccas
by way of the Eastern and East-Northeastern routes,
rounding the Cape of Good Hope? By the time we
should arrive there it would already be winter, as your
lordship well knows, and also the crew is thin and lacking
in strength; moreover, if there are now sufficient provisions, they are not many nor enough to regain energies
and enable too much working without the crew's health
suffering it, and I also have noticed how it takes the ill
ones long to recover.

On the positive side of the ledger, San Martín reminded Magellan that the three remaining ships of the fleet were still seaworthy,
but, he warned, their reduced provisions would not be sufficient to
last them all the way to the Moluccas. "Even though your lordship's
ships are good and well equipped (praise be God), some ropes are
missing, especially in *Victoria*, and besides, the crew is thin and
weak, and the provisions are not enough to reach the Moluccas by
the aforesaid route, and then return to Spain."

And he had a final word of advice for the Captain General:

I also believe that your lordship should not sail along
these coasts at night, both because of the ships' safety and

the crew's need to rest a little; since there are seventeen
hours of daylight, let your lordship have the ships lie at
anchor for the four or five nightly hours so that, as I said,
the people can rest instead of having to bustle about the
ships with the rigging; and, most importantly, in order to
spare ourselves the blows that an untoward fate could
inflict on us, may Heaven forbid it. For, if such blows
befall us when things can be seen and observed, it should
not be unfitting to fear them when nothing can be seen or
known or well watched, so let your lordship have the ships
anchor one hour before sunset rather than continue forward at night to cover two leagues. I have said as I feel and
understand in order to serve both God and your lordship
with what I believe is best for the Armada and your lordship; your lordship shall do as your lordship sees fit and as
God shall guide your lordship. Please He that your lordship's life and condition be successful, as it is my wish.

San Martín dared to express what nearly everyone on the voyage
whispered: There was great danger ahead, and chances were they
would not make it to the Spice Islands, wherever they were; their
maps had long since proved to be useless. Give it until January, he
advised, and if they had not reached their goal by then, return to
Spain, and try again.

Magellan considered these carefully thought-out admonitions, but
he was nevertheless inclined to proceed, no matter how long it took to
reach the Spice Islands. They had at least three months' provisions, by
his reckoning. More important, he believed that God would assist
them in achieving their goal; after all, He had permitted them to discover the strait, and He would guide them to their final goal.

The next day, Magellan gave the order to weigh anchor. The ships
fired a salvo of cannon that reverberated among the splendid dark

green mountains, gray ravines, and azure glaciers of the strait, and the armada set sail once again, heading west, always west.

At last, the churning, metallic waters of the Pacific came into view, and they realized they had reached the end of the strait. Magellan had done it; he had found the waterway, just as he had promised King Charles. Now that the armada had accomplished this feat, all the arguments for turning back by mid-January were never again discussed. "Everyone thought himself fortunate to be where none had been before," Ginés de Mafra exulted.

Magellan was overwhelmed to have completed his navigation of the strait, at last. Pigafetta records that the Captain General "wept for joy." When he recovered, he named the just-discovered Pacific cape "Cape Desire, for we had been desiring it for a long time."

As the armada approached the Pacific, the seas turned gray and rough. It was late in the day, and the dull skies were fading to darkness as the three ships put the western mouth of the strait to stern. "Wednesday, November 28, 1520, we debouched from that strait, engulfing ourselves in the Pacific sea," noted Pigafetta with quiet satisfaction. Even with the mutiny of the *San Antonio,* and the time spent trying to recover the ship, not to mention the ubiquitous dead ends the strait presented and at least one fierce williwaw, Magellan needed only thirty-eight days and nights to travel from the Atlantic Ocean to the Pacific.

For Magellan and his crew, it had been a remarkable rite of passage. As they sailed beyond the strait into the open water, how could they doubt that their expedition was indeed blessed by the Almighty? Although Magellan and his crew appeared vulnerable to the elements, to starvation, to the local tribes they encountered, and most of all to each other, this was not how they saw themselves. They all believed that a supernatural power looked after them and conferred on them the unique status of global travelers.

But how much of this accomplishment of navigating the strait derived from Magellan's skill, and how much could be attributed to plain good luck? Magellan was fortunate that the weather was relatively mild; after the intense williwaw that had menaced his ships, no other squalls surprised them, no glaciers collapsed on them, and the temperature, fluctuating as it does at that time of year between 35 degrees and 50 degrees Fahrenheit, remained within normal bounds, so the men were spared the intense cold they had suffered at Port Saint Julian. Their scouting excursions, as well as the addition of fresh vegetables to their diet, boosted both their spirits and their health. The passage through the strait, while strenuous, was far healthier than being at sea for long stretches, within the unsanitary confines of the ships, subsisting on a diet of salty, spoiled food and wine.

Although the armada enjoyed reasonably good fortune, Magellan's extraordinary skill as a strategist proved to be the decisive factor in negotiating the entire length of the Dragon's Tail. He ordered lookouts scrambling to the highest perch on the ships, where they could see the waterways and obstacles that lay ahead. In addition, he regularly sent small scouting parties in the longboats. "They would go on and come back with news of the findings, and then the rest of the armada would follow. This is the way the armada operated for the whole passage of the strait," Ginés de Mafra recalled. The information they brought back helped Magellan plot his next move; they warned him against rocky shoals, bays that deceptively resembled a continuation of the strait, and other dead ends that would have delayed his passage. Magellan even relied on the taste of seawater to guide the fleet. As the water became fresher, he knew he was traveling inland, and once it turned salty, he realized he was approaching the Pacific on the western side of the strait.

This array of tactics saved tedious days of wandering up and down dead-end channels and harbors. If one approach failed, he always had others on which to fall back. Not even the loss of his best pilot, Estêvão Gomes, and his biggest ship, *San Antonio,* defeated him; the more the fleet shrank, the more nimble it became. His sophisticated approach to navigating uncharted waters went far beyond technical ability in boat handling and direction finding; it revealed an ability to deploy novel tactics to overcome one of the great challenges of the Age of Discovery: namely, how to guide a fleet of ships through hundreds of miles of unmapped archipelagos in rough weather.

Magellan's skill in negotiating the entire length of the strait is acknowledged as the single greatest feat in the history of maritime exploration. It was, perhaps, an even greater accomplishment than Columbus's discovery of the New World, because the Genoan, thinking he had arrived in China, remained befuddled to the end of his days about where he was, and what he had accomplished, and as a result he misled others. Magellan, in contrast, realized exactly what he had done; he had, at long last, begun to correct Columbus's great navigational error.

When the fog receded and the sun broke through the low clouds, the Western Sea, as the Pacific was then called, turned from lifeless gray to seductive cobalt, its surface mottled with frothy whitecaps that melted into the frigid air. The water boiled menacingly and surged over the rocks and cliffs emerging from its inscrutable depths. Fearing shoals, Magellan adjusted his navigational technique; instead of gliding through deep fjords, he steered a course in rough water between two rocks later named, with a bitter irony best appreciated by wary sailors, The Evangelists and Good Hope. A cold miasma descended, blinding the pilots. "The western exit of the strait is very narrow and foggy, and there is no sign of it," de Mafra wrote. "Having exited it and sailed three leagues into the sea, its mouth cannot be descried."

Magellan set a northerly course along the coast of Chile. The strait they had just left seemed an enchanted refuge by comparison to the ocean they now faced. Darwin, on his journey, found the vista so horrifying that he was moved to comment: "One sight of such a coast is enough to make a landsman dream for a week of shipwreck, peril, and death."

The men of the Armada de Molucca looked on the scene with

the same foreboding. They knew the voyage was far from over; in a sense, it had only just begun. No matter how great the feat of navigating the strait from one ocean to another, it would have little value unless the armada reached the Spice Islands, wherever they were. No one aboard the fleet's three remaining ships suspected they were about to traverse the largest body of water in the world to get there.

CHAPTER VIII

A Race Against Death

The fair breeze blew, the white foam flew,
The furrow followed free;
We were the first that ever burst
Into that silent sea.

The scale of the Pacific Ocean was past imagining to Magellan. It encompasses one-third of the earth's surface, covers twice the area of the Atlantic Ocean, and contains more than twice as much water volume. It extends over a greater area than all the dry land on the planet, more than sixty-three million square miles. Lost in this immensity are twenty-five thousand islands, and concealed beneath its waters lurks the lowest point on earth, the Mariana Trench, buried in inky blackness thirty-six thousand feet beneath the shimmering surface. The Pacific had had the same appearance and character for tens of millions of years before Magellan and his men sailed across its surface, yet they knew nothing of these geological wonders. The men of the Armada de Molucca might as well have been sailing across the dark side of the moon.

Even today, the Pacific remains mysterious and alluring to scientists and oceanographers. Until recently, more was known about the surfaces of Mars or Venus than about the depths of the Pacific.

Nor does the scientific community agree about the origin of the oceans. One hypothesis maintains that in the first billion or so years after the earth was formed, comets—space ice—continually crashed to the surface, and they melted to form our oceans. Another suggests that the most ancient building blocks of earth—asteroidal material in the solar nebula and space dust—began to accrete and to heat. The heavier material sank to the center of the planet, and the lighter material remained nearer the surface. When the earth's crust was formed, water may have been released and formed oceans. As Magellan's men journeyed across the Pacific, they slowly and painfully came to realize what everyone knows now: Oceans cover 70 percent of the earth's surface. Our planet has been misnamed; it is the *ocean* planet.

Magellan anticipated a short cruise to the Spice Islands, followed by a longer but untroubled voyage home through familiar waters. He believed that his men had learned from their ordeals. The mutinies had weeded out the faint of heart and the uncooperative. The crew, once numbering 260 men and boys in five ships, was now less than 200 in three vessels: *Trinidad,* still the flagship of the fleet; *Concepción,* where Juan Serrano ruled; and *Victoria,* under Duarte Barbosa's command. Still, he had no idea of the real challenge that lay ahead, not one of shoals or climate but of distance.

The fleet's progress was rapid, but just how rapid is open to question. In his log, Albo noted, "On the morning of December 1, [1520,] we saw bits of land like hillocks." The usually scrupulous pilot gives his latitude as 48 degrees south, but his calculations may have been off by as much as one degree south; thus the fleet might have traveled even farther and faster than he supposed. In a cryptic entry, Pigafetta noted in his diary: "Daily we made runs of 50, 60, or 70 leagues *a la catena ho apopa"*—a phrase generally taken to mean "at the stern." Pigafetta might have been referring to Magellan's method of dead reckoning—the time it took for a log or other object to pass from one end of the ship to the other—but he did not furnish enough details to explain the fleet's exact speed or distance. In any event, the days at sea went by in a trance throughout December and most of January 1521.

To while away the idle hours, Pigafetta turned his attention to birds that occasionally flew overhead. He was of the opinion that they were undiscovered species. Swooping and diving into the waters of the Pacific, the birds hunted for flying fish, which occasionally lifted themselves out of the sea and landed on the deck of the ships with a distinctive thud. Pigafetta called the flying fish *colondrini,* by which he probably meant the flying gurnard, also known as the Oriental helmet gurnard, whose fins can expand into an impressive display of fanlike wings tipped with bluish spines. An exotic, forbidding-looking creature, the gurnard served as a reliable supply of food for the crew.

"In the Ocean Sea one sees a very amusing fish hunt," Pigafetta wrote. "The fish are of three sorts, and are a cubit or more in length, and are called dorado, albacore, and bonito. They follow and hunt another kind of fish that flies and is called *colondrini,* a foot or more in length and very good to eat. When the above three find any of those flying fish, the latter immediately leap from the water and fly as long as their wings are wet—more than a crossbow's flight," Pigafetta marveled. "While they are flying, the others run along back of them under the water following the shadow of the flying fish. The latter have no sooner fallen into the water than the others immediately seize and eat them. It is a very fine and amusing thing to watch."

Life at sea—so uncertain during the Port Saint Julian mutiny and the intricate maneuvering through the strait—became routine. From the first light of dawn, the crew kept time with an hourglass; when it was turned over, the pages sang their familiar incantations. Each day at noon the pilot, Albo, shot the sun and determined latitude, generally with considerable accuracy. Every evening, the other two captains went on deck, drew close to *Trinidad,* and saluted Magellan: *¡Dios vos salve, señor capitán-general, y señor maestro y buena compaña!*

Magellan and his captains held morning and evening prayers each day. The nights brought respite from the heat, and the sailors remained on deck to escape their cramped, stinking, and suffocating sleeping quarters. At rest, they observed the diamond-bright stars etched on the canopy of the heavens. Pigafetta turned his ever curious mind to making astronomical observations: "The Antarctic Pole is not so starry as the Arctic. Many small stars clustered together are seen, which have the appearance of two clouds of mist."

Without realizing it, Pigafetta had just recorded an observation of great consequence. These "clouds" are in fact two irregular dwarf galaxies orbiting our own galaxy and containing billions of stars enveloped in a gaseous blanket; they are known today as the Magellanic Clouds. The larger one, Nubecula Major, is about 150,000 light-years away, the smaller, Nubecula Minor, even farther, about 200,000 light-years. To the naked eye, they resemble pieces of the Milky Way torn off and flung across the heavens. Until 1994, they were considered the galaxies nearest to ours. The larger of the two covers an area in the night sky about two hundred times greater than that covered by the moon, while the smaller covers an area fifty times larger.

Pigafetta's observations continued: "In the midst of them are two large and not very luminous stars, which move only slightly. Those two stars are the Antarctic Pole." He may have been referring to the constellation Hydra, which is near the southern celestial pole. And as the fleet moved away from land into the open expanses of the Pacific, he noted, "We saw a cross with five extremely bright stars being exactly placed with regard to one another." This has usually been taken to be the Southern Cross, the most familiar constellation in the Southern Hemisphere, but that constellation would have been very low in the night sky, and Pigafetta might have confused it with Orion's Belt or another constellation. Although the Southern Cross is small, the mere sight of it was so compelling to Magellan's sailors that it became an important marker for both faith and navigation.

The absence of visible landforms meant the fleet's pilots relied on celestial navigation, using the Southern Cross and other constellations as their guide. Magellan, ever vigilant, constantly double-checked their course, lest they change direction under cover of night, as Pigafetta relates. "The Captain General asked all the pilots, always keeping our course, what sailing track we should prick [that is, mark] on the charts. They replied, 'By his course exactly as laid down.' And he replied that they pricked it wrong (and it was so) and that the needle of navigation should be adjusted."

*O*n December 18, 1520, Magellan finally changed course. At this point, they were between the mainland and the Juan Fernández Islands, which lie roughly east of what is now Santiago, Chile. Their new course took them west, away from South America into the Pacific. Soon the mainland, hardly more than a smudge on the horizon, disappeared from view, increasing the crew's sense of isolation and anxiety. If there was ever a time for a monster to appear on the horizon, for the ocean to boil, or for a magnetic island to pull the nails from the hulls of their ships, this was it.

Nothing quite that supernatural occurred. Instead, the armada encountered a different kind of miracle: the steady trade winds at its back. The wind still lacked a name, and the crew did not realize how extraordinary this current of air was until they had experienced it for some weeks. As they reached higher and higher latitudes, the Pacific, so forbidding when they first encountered it in the south, gradually metamorphosed into an undulating silken sheet. The mysterious change was brought about by solar heating— the effect of the sun warming the atmosphere. Solar heating is

greatest at the equator, where heated air rises high into the atmosphere and then divides into two streams, one flowing to the north
and the other to the south. As the streams move toward the poles,
they cool down, and the air, now relatively heavy, descends at about
30 degrees of latitude north and south. Eventually, the streams
encounter what is known as the Coriolis force; the earth's easterly
rotation causes the wind to veer in a westerly direction; in the
Southern Hemisphere, the location of the Armada de Molucca,
the winds come from the southeast. These are the trade winds,
named for the crucial role they played in facilitating transoceanic
trade routes. Even better, from Magellan's perspective, the Coriolis
force increases toward the equator. As the fleet worked its way
north, it was getting the benefit of some of the steadiest winds on
the planet.

A succession of placid, soporific days ensued. For hours on end,
the waves slapped rhythmically against the hulls, the sails sighed and
swelled contentedly in their fittings, and seamen spent their idle
hours playing card games or sleeping. Pigafetta, short on patience,
diverted himself by attempting to converse with their captive, cooperative Patagonian giant. In the process, he became the first European to learn and to transcribe the Tehuelche language of Patagonia.
He was undoubtedly influenced by earlier explorers, such as Columbus, who had attempted to record South American languages with
simple phonetic notations, but Pigafetta faced a complex language
that defied reduction. Linguists have identified about one thousand
languages in South America, and Tehuelche, or some variant of it,
was the principal tongue of Patagonia. Exactly what dialect the
Patagonian giant spoke is unknown. Despite their limitations,
Pigafetta's vocabulary lists rank among the expedition's most significant discoveries. They lacked the commercial value of spices or gold,
or the prestige of conquered territories, but they marked the begin-

ning of the modern study of linguistics, and to later generations of scholars they offered clues to the migrations of various tribes across the South American continent.

Pigafetta described the modus operandi that evolved between the two of them: "When he, asking me for *capac,* this is to say, bread (for so they call the root which they use for bread) and *oli,* that is to say, water, saw me write down these names, and afterward, when I asked him for others, pen in hand, he understood me." Their collaboration resulted in a phrase book called "Words of the Patagonian Giants." "All these words are pronounced in the throat," he advised, for they pronounce them thus." Pigafetta began with the Tehuelche word for "head," which he transcribed as *her.* "Eyes" sounded to him like *other.* Nose: *or.* Ears: *sane.* Mouth: *xiam.* And so on through subjects of interest to him.

Armpits: *salischen.* Breast: *ochii.* Thumb: *ochon.* Body: *gechel.* Penis: *scachet.* Testicles: *sacaneos.* Vagina: *isse.* Intercourse with women: *iohói.* The thighs: *chiaue.*

Hour after hour, the tall, bronzed, clean-shaven, nearly naked Patagonian huddled in earnest conversation with the much shorter and paler European in his breeches and loose-fitting shirt, scratching eagerly with his pen, gesticulating, querying with his hands and fingers, the two of them engaged in a game of mutual comprehension, surrounded by an ocean of incomprehensible dimensions.

Pigafetta was plainly delighted by the captive's range of vocabulary and his willingness to follow directions; at the same time, he was pleased with his own ability to capture the Tehuelche language on paper. Displaying the transcriptions to the giant, Pigafetta introduced the Patagonian to writing, and the power of the written word to speak silently across widely separated cultures, and, ultimately, across time. Imagine the captive's sense of wonder at the use of magic symbols to capture and transmit his language and thoughts. The use of linguistic symbols became the best way—in fact, the only way—for these two men to understand one another. Of all the weapons the Europeans brought to the Pacific, guns included, none was more powerful and more capable of effecting lasting change than written language.

As their intellectual labors continued, Pigafetta's queries moved from the concrete to the conceptual. What was the Patagonian word for the sun? he asked. *Calex cheni.* The stars? *Settere.* The sea: *aro.* Wind: *oni.* Storm: *ohone.* How does one say "Come here"? *Haisi,* replied the giant. To look? *Conne.* And to fight? *Oamaghce.*

*P*igafetta also introduced his cooperative prisoner to Catholicism. "I made the sign of the cross," Pigafetta recalled, "and kissed the cross, showing it to him. But at once he cried out *Setebos,* and he made signs to me that, if I made the sign of the cross again, it would enter my stomach and cause me to burst." *Setebos,* Pigafetta learned, meant "the great devil," the opposite of everything the cross represented in Christendom. The giant intuited that the cross represented a spiritual power, and eventually Pigafetta persuaded him that it symbolized a source of strength rather than danger.

At about that time, the Patagonian began to weaken and fall ill. No one could say what afflicted him; perhaps it was the change of diet, or a virus he caught from the Europeans. The sicker he became, the more he relied on the cross. Pigafetta gave him a real cross to hold, and, as instructed, the giant brought it to his lips, seeking its strength and healing power. But the illness worsened— Pigafetta does not supply any symptoms—and it became apparent that the giant was dying. Their conversations turned to religion, and Pigafetta persuaded the prisoner to convert to Christianity. He was baptized, and the giant, whose original name Pigafetta never mentioned, became known as Paul. He died shortly thereafter, a Patagonian Christian who met a unique and tragic fate. Pigafetta did not record what kind of funeral rites Father Valderrama accorded Paul, but presumably he was given a proper burial at sea.

*A*bout ninety years later, Pigafetta's affecting account of the curtailed education and conversion of the Patagonian giant drew the attention of William Shakespeare, who read an English translation by Richard Eden of Pigafetta's diary. Distinctive fingerprints in Shakespeare's resulting play, *The Tempest,* first performed in 1611, could only have come from Pigafetta's account.

In Shakespeare's imagination, the humble details of Pigafetta's encounter with the Patagonian giant are woven into an immense cosmological tapestry. The playwright sets the scene on an en-chanted magical island ruled by Prospero, the duke of Milan, who, with his daughter Miranda, had been set adrift by his brother Antonio, a usurper. Shipwrecked, Prospero learns magic and man-ages to remain on good terms with spirits inhabiting the island, especially Ariel, a sprite whom Prospero had freed from an evil sorceress known as Sycorax. But Sycorax also has a son, Caliban, one of the most compelling yet enigmatic characters in the Shake-spearean canon, and a character inspired in part by the Patagonian giant.

The clash between Prospero and Caliban offers a vivid image of the impact of European discovery and conquest on indigenous peo-ples throughout the world, and Shakespeare dramatizes the encounter with wit and a frisson of horror.

> *You taught me language; and my profit on't*
> *Is, I know how to curse: the red plague rid you,*
> *For learning me your language!*

Later, Caliban quotes Pigafetta's account of the Patagonian giant:

> *I must obey: his art is of such power,*
> *It would control my dam's god, Setebos,*
> *And make a vassal of him.*

Although Shakespeare keeps the setting vague, this mystical play demonstrates, if nothing else, that the New World, with its splendor and barbarism, had taken up residence in the European consciousness.

*A*lthough the weather remained perfect, the winds strong and constant, the armada failed to encounter islands with the food and water needed to sustain life. The ships had passed east of the Juan Fernández Islands, then north of the Marshall Islands: Bikar, Bikini, and Eniwetok. Had their course varied by only a few degrees south, they would been able to explore Easter Island or, farther west, the Society Islands and Tahiti. Had their course varied by only a few degrees north, they might have eventually encountered the Marquesas or Christmas Island. At the same time, the ship also narrowly avoided marine hazards such as razor-sharp reefs that could have sliced their hulls. A roaring surf concealed subsurface coral towers. Magellan's ships passed within a hundred miles of such hazards, and emerged unscathed.

To look at Magellan's course through the Pacific, it may seem as though he deliberately avoided the islands, and the chance to seek supplies, but he had no such plan in mind. None of these islands appeared on maps in his day, and if Magellan or anyone else spied telltale signs of landmass—a faint soaring plume, or the water turning light green—no one left any record of it. The two most reliable accounts, Pigafetta's diary and Albo's log, are silent on the subject of avoiding islands. Even if Magellan had known of their existence, he would not have felt any special urgency to land on their shores, for he expected to reach the Spice Islands or some other point in Asia within days. Headed toward an illusory destination, the fleet remained isolated in the Pacific, three little ships suspended in an infinite cerulean sea.

Thirst and hunger tormented the crew. The seals they had

butchered and salted in Patagonia turned putrid and became infested with maggots, which devoured the sails, rigging, and even the sailors' clothing, rendering them all useless. Pigafetta chronicled the appalling deterioration in their food supply. "We were three months and twenty days without getting any kind of fresh food. We ate biscuit which was no longer biscuit, but powder of biscuits swarming with worms, for they had eaten the good. It stank strongly of the urine of rats. We drank yellow water that had been putrid for many days. We also ate some ox hides that covered the top of the mainyard to prevent the yard from chafing the shrouds, and which had become exceedingly hard because of the sun, rain, and wind. We left them in the sea for four or five days, and then placed them for a few moments on top of the embers, and so ate them; and often we ate sawdust from boards. Rats were sold for one-half ducado apiece, and even then we could not get them."

The rats commanded a premium because sailors believed that eating them might offer protection against the disease they all feared: scurvy.

*S*curvy posed the single greatest danger to the health of the men during the entire voyage. There was no known cure, and if unchecked, it could claim the lives of them all. Magellan's only defense was against scurvy was an assortment of folk remedies. Once scurvy struck the crew, the voyage became a race against death itself.

One by one, the men began to suffer from the disease. In his diary, Pigafetta described its dreaded symptoms. "The gums of both the lower and upper teeth of some of our men swelled, so that they could not eat under any circumstances." A sense of exhaustion gradually overtook the men, and their gums began to feel sore and spongy. When they pushed with their tongues, even gently, their teeth wobbled. As the disease progressed, their teeth began to fall

out, and their gums bled uncontrollably and festered with exqui-
sitely painful boils.

Even though they suffered terribly from scurvy, sailors were still
expected to work. If they failed to appear on deck, the boatswain
whipped them with the end of a rope and then dragged them up on
deck, where the sunlight pitilessly revealed their deteriorated condition. Their skin seemed to be falling from their bones, and old scars
and sores, long healed, reopened. Their bodies were literally falling
apart.

As scurvy claimed one life after another, burials at sea became
commonplace. Sailors, many of them suffering from the early stages
of scurvy themselves and seeing their own deaths foretold, wrapped
the body in a remnant of an old, tattered sail, secured it with rope,
and tied cannonballs to the feet. A priest, and on occasion the captain, uttered a brief prayer; two sailors lifted the corpse onto a
plank, tilted it, and committed their crewmate's mortal remains to
the hungry sea.

Pigafetta put the grim tally of those who died from scurvy at
twenty-nine, in addition to the sole remaining Indian passenger they
had captured. Many others suffered grievously. "Besides those who
died, twenty-five or thirty fell sick of divers maladies, whether of the
arms or of the legs or other parts of the body, so there remained
very few healthy men."

*I*n Magellan's day, scurvy was a disease new to Europe, a terrible
by-product of the Age of Discovery. In 1498, Vasco da Gama's crew,
exploring the African coast for Portugal, suffered the first widely
noted outbreak. Da Gama observed that his men developed the telltale swelling of hands, feet, and gums. He also wrote of Arab traders
offering oranges to the afflicted sailors, and the men making a
miraculous recovery thereafter; the clear implication is that the
Arabs, more accustomed to long ocean voyages than their European
counterparts, knew the affliction and its cure. During a three-

month-long passage across the Indian Ocean, Vasco da Gama's crew again fell victim to scurvy, and this time thirty men died. "In another two weeks there would have been no men at all to navigate the ships," da Gama wrote. Deliverance came when they reached land and again feasted on life-giving oranges. Despite abundant evidence to the contrary, Vasco da Gama and other early European explorers believed that unhealthy air—not dietary deficiencies—caused scurvy.

The intense suffering experienced by da Gama's men, and later by Magellan's, could have been prevented by a daily dose of one spoonful of lemon juice, for that is the amount of vitamin C necessary to prevent scurvy. In the body, vitamin C, or ascorbic acid, helps to manufacture the enzyme prolyl hydroxylase, which in turn synthesizes a protein collagen used for connective tissues such as skin, ligaments, tendons, and bones, all of which give our bodies tensile strength. A vitamin C deficiency leads to the melting of the collagen fibers and a breakdown in the connective tissues, especially in bones and in dentin, the building block of teeth. Collagen acts as a glue binding connective tissues together, and when it disintegrates, the tissues separate and capillaries hemorrhage, creating black-and-blue patches on the skin. (Curiously, Magellan's men's desperate hope that eating rats would avert scurvy had a basis in fact; unlike humans, rats synthesize and store vitamin C.)

Scurvy continued to afflict explorers for more than two hundred years. Often, the difficulty of obtaining oranges during voyages was to blame, but even the most dedicated investigators remained befuddled, while thousands died at sea. Finally, in 1746, James Lind, a Scottish naval surgeon, turned his attention to the problem of scurvy, then afflicting sailors in the Royal Navy. To determine the cause, he conducted the first modern clinical trials on record. He isolated a dozen sailors suffering from scurvy and fed them the same diet. Then he subjected each to different treatments, administered daily. Some received seawater, some nutmeg and other spices, some vinegar, and others two oranges and one lemon. "The consequence was that the most sudden and visible good effects were perceived from the use of the oranges and lemons," Lind observed, "one of those who had taken them being at the end of six days fit for duty."

Despite the overwhelming evidence, Lind's findings were not widely accepted. He persisted. After leaving the navy, Lind was elected a Fellow of Edinburgh's Royal College of Physicians and subsequently published an exhaustive study entitled *A Treatise of the Scurvy Containing an Inquiry into the Nature, Causes and Cure of That Disease*. In the four-hundred-page treatise, Lind offered his own bizarre theory of the origins of scurvy; he claimed that a cold and wet climate clogged the pores and set the stage for the disease. This was nothing more than an updating of theories prevalent in Magellan's era.

Not until 1795 did the British Royal Navy finally insist that sailors receive a daily ration of the juice of lemons or limes to combat scurvy, a practice leading to the term "limeys" to refer to British sailors. (At the time, a "lime" meant both lemons and limes.) This was an act of faith more than science because it was still not known why lemons, limes, oranges, and other fruits and vegetables prevented scurvy. Finally, in 1932, three medical researchers, W. A. Waugh, C. G. King, and Albert Szent-Gyorgyi, managed to isolate and synthesize ascorbic acid; they offered a scientific explanation of vitamin C's effect on the body, and showed how a vitamin C deficiency leads to scurvy.

*W*hile their men suffered and died around them, Magellan, Pigafetta, and several other officers remained mysteriously healthy. "By the grace of our Lord I had no illness," Pigafetta marveled. Neither he nor anyone else knew why, but there was an outstanding reason why they had escaped scurvy. Throughout the ordeal, the officers regularly dipped into their supply of preserved quince, an applelike fruit, without realizing it was actually a potent anti-

scorbutic. Saved by this fluke, the good fortune seemingly conferred on Magellan by Saint Elmo appeared to hold, at least for
the present.

Nothing in Pigafetta's diary suggests that the officers conspired
to keep their supply of quince to themselves at the cost of their
men's lives. Magellan and the others remained oblivious to its life-
sustaining properties, and they continued to believe that their men
suffered from a variety of afflictions, most of them caused by "bad
air." Since Magellan was known for personally ministering to his
men when they became ill, he would likely have insisted they take
daily rations of quince had he known of its benefits.

During these three months and twenty days," wrote Pigafetta,
"we made a good four thousand leagues across the Pacific Sea, which
was rightly so named. For during this time we had no storm, and we
saw no land except two small uninhabited islands, where we found
only birds and trees." Their first landfall occurred on January 24,
and very disappointing it proved to be: a simple atoll rising enig-
matically from the ocean. Magellan named it San Pablo because the
sighting occurred on the Feast of the Conversion of Saint Paul. (This
tiny atoll was also the explorer Thor Heyerdahl's first sighting of
land during his transpacific crossing in 1947 aboard the balsa raft
Kon Tiki.) The atoll proved useless to Magellan's vessels; he saw
neither evidence of human habitation nor a safe place to drop
anchor. After sailing completely around the island, he signaled the
fleet to proceed on its course. San Pablo could not come to their aid.

Eleven days later, on February 4, 1521, Magellan spotted another
islet—most likely Caroline Island, in Micronesia. The fleet
approached, and once more tried to find an anchorage, but did not
succeed. The water, complained Pigafetta, was so deep that "there is
no place for anchoring because no bottom can be found." De Mafra,
writing long after the event, recalled an impenetrable reef that
repelled the ships: "It seemed as if Nature had armed it against the

sea." And Albo's log notes: "In this latitude we found an uninhabited island, where we caught many sharks, and therefore we gave it the name of Isla de los Tiburones"—Shark Island. Stunned from monotony and debilitated by illness, the crew watched the large, menacing creatures circle, apparitions in a scene of despair. Even Magellan, normally possessed of superhuman determination and indifference to hardship, became depressed and unstable as the transpacific crossing wore on. In a rage, he flung his useless maps overboard, crying, "With the pardon of the cartographers, the Moluccas are not to be found in their appointed place!"

Unable to make a landfall, they were carried by substantial trade winds over astounding distances. "We made each day fifty or sixty leagues or more," Pigafetta wrote of their seemingly miraculous progress westward. "And if our Lord and the Virgin Mother had not aided us by giving good weather to refresh ourselves with provisions and other things we had died in this very great sea. And I believe that nevermore will any man undertake to make such a voyage."

Their progress north was so rapid that they crossed the equator on February 13. Magellan was utterly confounded by this time. He had expected to reach the Spice Islands long before this point in the voyage; according to the maps he had studied, he had already covered the entire Pacific and should have been in Asia by now. Worse, he had entered Portuguese waters, as defined by the Treaty of Tordesillas; if he discovered that the Spice Islands lay squarely within Portuguese territory, the finding would defeat the purpose of the expedition, and he could not claim them for Spain. To add to the pressure, he was running out of water, and his men were dying of scurvy. He needed to find safe harbor soon, if he was to survive the ordeal of crossing the Pacific.

*D*eliverance finally came ninety-eight days after the fleet left the strait. At about 6:00 A.M. on March 6, 1521, two landmasses slowly rose from the sea; they appeared to be about twenty-five miles away.

Eventually, a third mass came into view. From his perch in a crow's nest sixty feet above deck, Lope Navarro, *Victoria's* lookout, peered into the indistinct glow, trying to distinguish between these promising outlines and mere cloud formations. Throughout that anxious morning, the ships made directly for the shapes at a rate of about six knots.

When Navarro was convinced of their true nature, he announced from on high, *"¡Tierra!"*

Tierra! Tierra!

The shrill cry tore through the silent morning. *Tierra!*

"These sudden words made everyone happy," de Mafra recalled of that miraculous sighting of land, "so much so that he who showed less signs of joy was taken for a mad man, as anyone who has found himself in such conditions would understand."

CHAPTER IX

A Vanished Empire

The upper air burst into life!
And a hundred fire-flags sheen,
To and fro they were hurried about!
And to and fro, and in and out,
The wan stars danced between.

*P*hysically and emotionally exhausted, Magellan climbed partway to the crow's nest to see the prospect for himself. His men, many of them about to succumb to scurvy, starvation, and dehydration, their tongues swollen, breath foul, and eyes glassy, raised their shaggy heads to glimpse their salvation. As the islands grew more distinct in the morning light, the lookout shouted again, *"¡Tierra!"* and gestured to the south, where cliffs rose from the sea. Overjoyed, Magellan awarded the fortunate lookout a bonus of one hundred ducats.

The first landmass Navarro had spotted was likely mountainous Rota. Thanks to the earth's curvature and the angle from which the armada approached, it initially appeared to be two islands. Rota's deceptive appearance confused Pigafetta, and has led to centuries of debate concerning which landmasses the lookout actually spotted. The other island, the one where the armada would eventually land, is now Guam, an unincorporated territory of the United States. About thirty miles long, covering 209 square miles, Guam is the largest of an archipelago of volcanic islands known as the Marianas, which lie about three thousand miles west of the Hawaiian Islands.

For Magellan, the landfall on Guam came as a mixed blessing. Although the island provided shelter from the misery he and his men had endured during their ninety-eight-day Pacific cruise, nothing about it suggested they were anywhere near their goal, the Spice Islands. Nevertheless, it was land. Since leaving the western mouth of the strait, Magellan had traveled more than seven thousand miles without interruption: the longest ocean voyage recorded until that time.

On Wednesday, the 6th of March, we discovered a small island in the north-west direction, and two others lying to the south-west," Pigafetta wrote of the momentous event. "One of these islands was larger and higher than the other two. The Captain General wished to touch at the largest of these islands to get refreshments of provisions." Pigafetta even sketched this sight for his diary, but the illustration, depicting three irregular blobs floating in a shimmering sea, is so crude that it has no value for navigation. Even more confusing, Pigafetta followed the practice of his time and placed north at the bottom of his maps and south at the top. The completed drawings suggest that after the journey's conclusion he furnished a rough description to an illustrator, who turned Pigafetta's sketch into a charming and colorful cartoon in which the ocean's azure blue was accented with flecks of gold and the islands seemed to float on the surface like giant potatoes. Nevertheless, his images are the only surviving cartographic record of the voyage.

Albo's log for the same day includes a slightly different and more scrupulous account of their discovery. "On this day we saw land and went to it, and there were two islands, which were not very large, and when we came between them we headed to the southwest, and we left one to the northwest." And he adds, ominously, "We saw many small sailboats approaching us, and they were going so fast they seemed to fly." The secret of their astonishing velocity was the unusual design of their sails, which caught Albo's attention. "They had mat sails of a triangular shape, and they went both ways, for they made of the poop the prow, and of the prow the poop, as they wished, and they came many times to us."

Albo was getting his first good look at the highly maneuverable outrigger canoe known as a *proa,* and often called a "flying *proa,*" because it was able to attain speeds of up to twenty knots and seemed to fly over the water's surface, exactly as Albo recorded. The *proa's* secret of speed derived from its unusual design. Unlike European sailing vessels, its prow and stern were identical, but its sides were different: the windward side was rounded for maximum aerodynamic efficiency, and the leeward side was flat. The interchangeability of the stern and prow, combined with a maneuverable lateen sail, meant it could head into the wind without strain, and coast from one island to another without having to come about.

The *proas* approaching Magellan's fleet were manned by a Polynesian tribe now known as the Chamorros, although this was not the name by which it was known in Magellan's day. Initially, Magellan's crew referred to all the tribes they encountered in the Pacific as *Indios,* Indians, in the mistaken belief that the Indies must be nearby. Succeeding generations of Spanish visitors gave the indigenous people of Guam the name "Chamurres," which derived from the local name for the upper caste; later, they were called Chamorros, the old Spanish word for "bald," or in Portuguese, "clean-shaven," possibly in reference to the Chamorran men's habit of shaving their heads.

How Guam and thousands of other isolated islands came to be inhabited has puzzled ethnologists to this day. Migrations from Southeast Asia gradually fanned out across the Pacific, into what we today call Melanesia and Polynesia, beginning about three or four thousand years ago, perhaps in light craft reminiscent of the outrigger canoes that advanced on Magellan's fleet. Today's Chamorros are a mixture of Malaysian, Indonesian, Filipino, Mexican, and Spanish, and speak a distinct language, also called Chamorro. Whether the tribe whom Magellan first encountered that morning in 1521 was the direct ancestor of the region's current inhabitants remains an open question.

Four hours after sighting land, the Armada de Molucca, sur-rounded by a welcoming party of outrigger canoes, entered a deep turquoise lagoon of exceptionally warm, clear water. As they approached, the sailors could see beaches, rocky cliffs, and steep, thickly forested slopes. The verdant landscape contained a paradise of springs, streams, and waterfalls: everything a sailor who has been too long at sea could want. The possibility of deliverance put the entire crew on edge. Jubilation alternated with watchfulness. The moment of contact between two societies, until now wholly ignorant of one other's existence, had finally arrived.

At first, the Chamorros—hundreds of them in their small, maneuverable canoes—encircled the fleet. "Fearing nothing, they got aboard, and there were so many of them, especially in the flag-ship, that some of our men asked the captain to have them thrown out," de Mafra related. The Chamorros, taller and stronger than the Europeans, boarded the flagship and stole everything they could—rigging, crockery, weapons, and anything made of iron—as the crew, in their weakened condition, pleaded with Magellan to force them to leave.

Eventually, one sailor summoned the strength to retaliate. "The boatswain of the flagship slapped one of those Indians for a small reason, and the Indian slapped him back. Insulted, the boatswain stabbed him in the back with a machete that he carried on his waist." At that, the Chamorros—a "mob of barbarians," de Mafra now called them—hurled themselves overboard. "Once they were aboard their shoddy boats they began fighting with their sticks, for they had nothing else. Some arrows were cast at them from the ships, but, being so many, the Indians managed to wound some of our men."

In the middle of the fracas, a second wave of Chamorros skimmed across the azure water in their *proas*, and, to the Europeans' astonishment, distributed food to the starving sailors. Once they had fed the Europeans, the Chamorros took up their sticks and began fighting again, this time more viciously.

De Mafra described how the Captain General narrowly averted disaster. "Magellan, seeing that the number of people was increasing, ordered those in the ship to stop throwing arrows; with this, the Indians stopped, the fighting subsided, and they resumed selling food as before, the kind of food in those islands being coconuts and fish aplenty, which were purchased in exchange for some glass beads brought from Castile." Magellan's show of restraint turned out to be just the right gesture. The modest yet historic encounter between the Armada de Molucca and a few dozen outrigger canoes contained in microcosm the conflicting impulses of the European colonialist adventure—from initial innocence and curiosity through confusion, fear, and bloodshed, all of it resolving in commercial activity.

If only matters had ended on this harmonious note. Unaccustomed to European concepts of trade and property, the Chamorros, while happy to feed the sailors, failed to comprehend that some things aboard ship were simply off limits. "Whilst we were striking and lowering the sails to go ashore, they stole away ... with the small boat called the skiff, which was made fast to the poop of the captain's ship," wrote Pigafetta. To judge from the chronicler's description, the Chamorros had made off with Magellan's personal dinghy. The robbery could only be interpreted as an insult to the Captain General himself.

The next day, Magellan, "much irritated," according to Pigafetta, retaliated. He was not about to let thieves make off with his personal vessel. He ordered forty men into the two remaining longboats. Rowing mightily, the crews pushed past the reef's spume and reached the shore—the first landing by Europeans on an inhabited Pacific island. Then they went on a rampage. "The Captain General in wrath went ashore with forty armed men, who burned some forty or fifty houses together with many boats, and killed seven men," Pigafetta related without comment. The sailors who remained on board the ships, many close to death from the effects of scurvy, implored the landing party to return with the internal organs of the slain Chamorros, which they thought would cure their scurvy. Their willingness to turn to cannibalism shows how desperate they had become.

During the rampage, the stunned Chamorros offered no resis-tance, and the Europeans held their fire. But their crossbows were brutally effective. "When we wounded many of this kind of people with our arrows, which entered inside their bodies," Pigafetta wrote, "they looked at the arrow, and then drew it forth with much aston-ishment, and immediately afterwards they died. Others who were wounded in the breast did the same, which moved us to great com-passion." Amid the carnage, Magellan "recovered the small boat, and we departed immediately, pursuing the same course."

Although Pigafetta mentions only a frenzied raiding party, he spent considerable time ashore over the course of the next few days, and recorded his carefully considered impressions of Chamorro soci-ety. "These people live in liberty and according to their will," he remarked, clearly disturbed by the lack of a well-defined social order. Like Magellan, he felt at home in a hierarchical, authoritarian society in which loyalty to the king and the Church mattered most. During the mutinies he had faced, Magellan had always struggled to defend the social order and maintain his primacy over rebellious captains and crew members. But here, on the open waters of the Pacific, was a tribe that lived by different rules, or no rules at all. Chamorro society appeared to be assembled horizontally rather than vertically. If there was a leader, Magellan could not determine who it was. Subsequent Spanish visitors to the island learned that the structure of Chamorro society was actually intricate and subtle; it was matrilineal and heavily committed to ancestor worship. Chamorran women performed the central roles in family life, and although Chamorro men had appeared hostile to Magellan, their bellicose gestures were essentially ritualistic; they merely played at war.

Captivated by the Chamorros' habits, Pigafetta recorded simple ethnographic details. "Some of them wear beards, and have hair down to their waist. They wear small hats ... made of palm leaves. The people are as tall as us, and well made.... When they are born they are white, later they become brown, and have their teeth black and red." Their teeth were stained from constantly chewing the betel nut, called *pugua* or *mama'on* by the locals. It grew on the areca tree, which resembles a coconut palm. They frequently chewed the nuts along with the betel leaf, *pupulu,* which tasted fresh and peppery. The islanders preferred the hard reddish nut variety called *ugam,* with its granular texture. It was their chewing gum, their tobacco, their coveted tradition.

This was the first time the men in the crew had laid eyes on women since their departure from the strait three months earlier, and they were a source of fascination. "The women also go naked," Pigafetta was pleased to observe, "except when they cover their nature with a thin bark, pliable like paper, which grows between the tree and the bark of the palm. They are beautiful and delicate, and whiter than the men, and have hair loose and flowing, very black and long, down to the earth. They do not go to work in their fields, nor stir from their houses, making cloth and baskets of palm leaves."

The ever-curious Pigafetta also describes the interior of the Chamorros' huts, so it is likely that the Europeans and the Chamorros enjoyed other, more enjoyable encounters than their violent first meeting. Members of the crew probably stayed overnight because Pigafetta was able to offer vignettes of their domestic life. "Their houses are constructed of wood, covered with planks, with fig leaves, which are two ells in length: they have only one floor; their rooms and beds are furnished with mats, which we call matting, which are made of palm leaves, and are very beautiful, and they lie down on palm straw, which is soft and fine."

During his visit, Pigafetta examined the Chamorros' most advanced piece of technology, their highly maneuverable *proas*, paying special attention to their ingenious counterweight. "Some are black and white, and others red. And on the other side of the sail they have a large spar pointed at the top. Their sails are of palm leaf sewn together like a lateen sail to the right of the tiller. And they have for steering oars certain blades like a shovel. And there is no difference between the stern and the bow in the said boats, which resemble dolphins jumping from wave to wave." He even included a crude sketch showing a small vessel with two oarsmen facing each other; in the middle of the craft, a single mast holds a lateen sail, and most strikingly, the counterweight balancing the hull, projecting straight toward the viewer. Curiously, Pigafetta (or whoever made these sketches for him) depicted the Chamorros as water-borne warriors in hoods and tunics, giving them a decidedly European appearance; in reality, they were naked, or nearly so.

The European visitors were surprised to find that the Chamorros possessed very few arms; their most dangerous weapon consisted of a stick with a fishbone attached to one end, and it was used not for combat but to catch flying fish. It now appeared that the armada's initial encounter with the Chamorros might have been a tragic misunderstanding, because Pigafetta, trying as usual to communicate with the local populace, determined that they had been startled more than anything else. "According to the signs they made," he wrote, the Chamorros thought that "there were no other men in the world besides them." If this was the case, and the armada had disturbed an isolated island society, the Chamorros' hostile response becomes understandable, as does their fascination with *Trinidad*'s skiff, the one piece of equipment in the armada that bore resemblance to their own canoes. In addition, the Chamorros had no concept of private property, and so they believed the newcomers' possessions belonged to one and all. On this basis, they had been equally pleased to share their food and supplies with the starving intruders. Nevertheless, Pigafetta and Magellan decided that the Chamarros' worst offense was their thievery, and the Captain General christened the island, as well as two others nearby, the Islas de los Ladrones—the Islands of the Thieves.

A more accurate name might have been the Islands of the Sharers.

*O*n March 9, 1521, as the armada left the island, the Chamorros reacted with anger, perhaps feeling insulted or betrayed by the unexpected departure. Over a hundred *proas* took to the water. "They approached our ships, showing us fish, and feigning to give it to us. But they threw stones at us, and then ran away, and in their flight they passed with their little boats between the boat which is tied at the poop and the ship going at full sail; but they did this so quickly, and with such skill, that it is a wonder."

As Magellan led his enfeebled crew out of the harbor, they observed the effects of the violence they had visited on the Chamorros. "We saw some of these women, who cried out and tore their hair, and I believe that it was for the love of those whom we had killed," Pigafetta recorded.

Although the fruit and vegetables they had acquired would soon begin to restore the scurvy-ridden crew to health, one was too sick to recover. Master Andrew of Bristol, as he was listed in the fleet's roster, died, and his earthly remains joined those of his other deceased shipmates in watery repose. The only British crew member, he had served as the fleet's master gunner; the post was immediately filled by Hans Bergen, a Norwegian.

*O*nce again, the fleet plunged blindly into the expanses of the western Pacific, with no clear idea how to reach the Spice Islands.

Had Magellan tarried among the Chamorros, he might have

learned valuable lessons about navigating across the Pacific. Like
other island tribes, the Chamorros had techniques for identifying
distant landmasses. They were adept at reading the ocean swells to
maintain a course; they could distinguish between distracting swells
raised by winds in the area and the widely spaced, regular swells useful for orienting a ship. The swells contained other clues to the
whereabouts of remote islands, because they tended to bounce off
islands or even to curve around them. By studying the patterns of
the swells, an experienced navigator could make educated guesses
about the distances and locations of various islands.

Island tribes also studied birds for signs of land. By simply following a bird's trail at the end of the day, when it flew to its nest after
a day's fishing on the open ocean, island navigators could reach land.

They studied clouds. The higher islands in the Pacific interfere
with the trade winds, causing mist and vapors to collect above the
landmasses. Magellan's lookout had seen this effect when he first
spied land and was unable to distinguish the island of Guam from the
surrounding clouds. Even the underside of clouds contained valuable
information because they reflected the color of the ocean directly
below. If the underside happened to be tinged with jade, it was likely
reflecting the greenish shallow water covering an atoll or reef.

They also sensed patterns in the placement of the islands, which
tended to be scattered in long archipelagos; if they found one island,
they would know approximately where to look for others.

For celestial navigation, island tribes employed a significantly different system than the Europeans used. Instead of relying on instruments, they developed a so-called star compass, a mental construct
in which points along the endless, undifferentiated horizon were
determined by places where stars and constellations rose and set.
With this construct, island navigators subdivided the horizon into
thirty-two segments, just as European compass bearings did. Rather
than rely on terms equivalent to north, south, east and west, the
island system named the points after the star or constellation.
Unlike the European system, the thirty-two segments migrated with

the stars, resulting in irregular bearings. In addition, the island navi-gator assumed that his *proa* was stationary, and the reference points on earth and in the sky were on the move. His reference point was the vessel, not landmarks, not even the stars. This custom may have derived from a common illusion experienced by sailors that their vessels seem to be motionless while landmarks slide by: hence the European sailor's tendency to say that an island falls astern of the boat, as if the island itself were on the move.

In the preliterate societies that Magellan encountered, the island system of navigation worked as well as, if not better than, the flawed European system, which still lacked the ability to determine longi-tude accurately.

*M*agellan set a westward course, journeying deeper into the unexplored reaches of the Pacific in his quest for the Moluccas. The revitalized fleet enjoyed another marvelous week of sailing down-wind, making seven or eight knots—top speed.

On March 16, a lookout spied the mountains of a large island rising from the sea in pale majesty. The fleet had reached the east-ern edge of the Philippine archipelago—over three thousand islands, most of which cover less than a square mile. Today, the two largest are known as Luzon and Mindanao. Magellan's lookout had glimpsed the third largest island in the archipelago, Samar. The Philippines are situated almost directly south of Japan, and north of Borneo. Magellan sensed he was getting close to the Spice Islands, but he did not realize how close.

*A*ccounts of Philippine history begin abruptly in 1521, with Magellan's arrival. But centuries before, these islands were well known to Chinese and Arab traders, who, with their superior sailing technology, profitably trafficked among them and developed sophis-ticated trading networks with the native societies. Archaeological

evidence suggests that trade between mainland Asia and the Philippines had become highly evolved as early as A.D. 1000. Chinese
junks, distinguished by their three tall, featherlike sails stiffened
with battens, became a familiar and welcome sight in the Philippines. The prevalence of commerce in the Philippine archipelago
brought islanders out of their isolation and spread Asian cultural
influences, especially writing, along with their goods. By the time
Magellan arrived, Filipinos who dwelled near oceans and inland
waterways had long been literate.

Chinese exploration of the Philippines reached its commercial
peak during the years 1405 to 1433, when the Treasure Fleet ruled
the South Pacific and the Indian Ocean. Its immense ships ranged as
far as the east coast of Africa to collect precious items and tributes
for the emperor. They were eight or nine times longer than Columbus's ships and five or six times longer than any in Magellan's
armada. For sheer size, the Treasure Fleet was unrivaled until the
zenith of the British navy in the nineteenth century. Despite its
importance and unique character, the Treasure Fleet is little known
in the West, even today. It was the creation, in many respects, of one
man whose accomplishments rivaled and in some ways surpassed the
more celebrated exploits of Columbus and Magellan: Cheng Ho.

*I*n 1381, the Chinese army seized control of the mountainous
province of Yunan, in southern China, and captured a young boy
named Ma Ho, the son of a devout Muslim. Along with other
young prisoners, he was castrated at the age of thirteen, a common
practice in China, where eunuchs engaged in servant occupations.
Ma Ho won an appointment as a servant to the fourth son of the
Chinese emperor, Prince Zhu Di. Tens of thousands of eunuchs
held such positions, which became so coveted that the Chinese eventually prohibited self-castration to discourage the overwhelming
number of office seekers. In this competitive environment, Ma Ho
rose to the rank of officer on the strength of his military and diplo-

matic skills. Later, the prince conferred the name Cheng Ho on his loyal and capable servant, and as Cheng Ho he played his pivotal role at the height of the Ming dynasty. (In the Pinyin transliteration of Chinese, he is now known as Zheng He, but he is still usually called Cheng Ho.) He was a giant of a man, seven feet tall, of considerable girth, endowed with a robust personality to match his stature and position. His complexion was said to be "rough like the surface of an orange," and "his eyebrows were like swords and his forehead wide, like a tiger's."

His star rose even higher when his patron, Zhu Di, became emperor in 1402. Zhu Di placed administrative authority in the hands of the eunuchs who had helped him come to power, among them Cheng Ho. Having rid his kingdom of enemies, the emperor decided to give himself a suitable name. He picked Yongle, which means "lasting joy." To pursue his goals of building an international commercial empire, Zhu Di named Cheng Ho as admiral, and placed him in charge of an ambitious and in some respects quite un-Chinese mission, that of building and leading a Treasure Fleet to explore the oceans.

Cheng Ho oversaw the operation of huge shipyards in Nanking, the planting of thousands of trees to provide wood for the ships, and the establishment of a school to train interpreters in foreign languages. Cheng Ho hastened to complete a fleet consisting of fifteen hundred wooden ships, including the largest sail-powered vessels ever built. They were extraordinarily luxurious, with staterooms, gold fittings, bronze cannon (for display rather than combat), and silk furnishings. Their seaworthiness was greatly enhanced by bulkheads, watertight compartments whose design was inspired by the chambers of the bamboo stalk. It would be several centuries before Western ships incorporated the same technology.

The Treasure Fleet assembled along the Yangtze River in Nanking in readiness for its first epic voyage, beginning in 1405. It consisted of 27,800 men, in contrast to the 260 men that made up the Armada de Molucca. Each of the Treasure Fleet's larger ships—

some as long as five hundred feet—carried nearly a thousand men. Others were devoted solely to transporting horses; still others carried water, or troops, or weapons, should they be needed to defend the fleet. Some ships carried only food, in case the crew found nothing to eat on distant shores; others carried large tubs of soil to cultivate fruits and vegetables. The luxury might have been responsible for preventing scurvy among the crew.

Unlike the Armada de Molucca, the Treasure Fleet did not conquer or claim distant lands. Although the Chinese considered themselves culturally superior to the outside world, they had no interest in establishing a colonial or military empire. Rather, the goal was to establish trade and diplomatic relations with the "barbarians" beyond their borders and to conduct scientific research. The unique Chinese philosophy of exploration received eloquent expression in a tablet said to have been written by the emperor himself at the height of the Treasure Fleet's activity.

> We rule all under heaven, pacifying and governing the Chinese and the barbarians with impartial kindness and without distinction between mine and thine. Extending the way of the ancient sage emperors and the enlightened kings so as to accord with the will of heaven and earth, we desire all distant countries and foreign domains each achieve its proper place under heaven.

At sea, the ships of the Treasure Fleet remained in touch with one another through a system of flags and lanterns, similar to the techniques employed by Magellan; they also used bells, gongs, and even carrier pigeons for communication. They measured time with burning graduated incense sticks. They navigated with compasses. Chinese pilots also employed a measuring instrument known as a *qianxingban* to fix their latitude, using the Southern Cross as their reference point. Cheng Ho often consulted a twenty-one-foot-long nautical chart that he unrolled section by section as his journey pro-

gressed. Like the portolan charts employed by Spanish and Portuguese navigators a century later, it contained landmarks, compass
bearings, and detailed directions for sailing from one point to
another. Chinese navigators also learned to steer by the stars, relying
on maps of the heavens to supplement their master charts. Chinese
constellations differed from those traditionally used in the West;
their great reference points were known as the Lantern and the
Weaving Girl.

In rough weather, Chinese sailors prayed fervently to be spared
from drowning, just as Magellan's crew did. In their case, the Chinese prayed to the Celestial Spouse, a Taoist goddess. Deliverance
from storms came in the form of the same Saint Elmo's fire that signaled salvation to Magellan's crew. Like Magellan's sailors, the Chinese mariners considered the appearance of the spectral light a sign
of safety conferred on their commander by divine forces.

The Treasure Fleet's first important destination was Calicut, on
India's southwestern coast. Chinese explorers had reached this city
overland eight centuries before, but the Treasure Fleet's arrival
prompted an outpouring of generosity from Calicut's ruler, who
conferred lavish gifts in the form of sashes made of finely spun gold,
pearls, and precious stones.

While in Calicut, the men of the Treasure Fleet became aware of
an unusual legend in which characters named Moses and Aaron fig-
ured prominently, along with a golden calf. The mysterious legend
was recorded by the Treasure Fleet's official chronicler, Ma Huan,
who performed approximately the same function for Cheng Ho as
Antonio Pigafetta did for Magellan. He wrote of a "holy man
named Mouxie [Moses], who established a religious cult; the people
knew that he was a true [man of] heaven, and all revered and followed him." As it happened, the holy man had a younger brother
with "depraved ideas." According to the story, he made a "gold calf
and said, 'This is the holy lord; everyone who worships it will have
his expectation fulfilled.' He taught the people to listen to his bid-
ding and to adore the gold ox, saying, 'It always excretes gold.' The

people got the gold, and their hearts rejoiced; and they forgot the way of Heaven." Later, when Mouxie returned, "He saw the multitude, misled by his younger brother . . . corrupting the holy way; thereupon he destroyed the ox and wished to punish his younger brother; [and] his younger brother mounted a large elephant and vanished." This was, of course, a modified version of the biblical account of Moses and Aaron, but the Chinese did not understand its true origins. They assumed it came from India because that was where they first heard it.

Cheng Ho returned from the Treasure Fleet's first voyage as a national hero, and he was soon making plans for future voyages. He stayed in China for the second voyage, and returned to sea for the third, commanding a fleet of forty-eight ships and thirty thousand men. With an eye to the future, they established trading posts and warehouses wherever they went. So it went for three more voyages, each lasting approximately two years as the Treasure Fleet established and maintained the first international maritime trading network. The Treasure Fleet explored the African coast all the way south to Mozambique, the Persian Gulf, and many other points throughout Southeast Asia and India. The lure and romance of ocean exploration spread throughout China. "We have beheld in the ocean huge waves like mountains rising sky-high," Cheng Ho wrote, "and we have set eyes on barbarian regions far away hidden in a blue transparency of light vapors, while our sails, loftily unfurled like clouds, day and night continued their course like that of a star, traversing the savage waves as if treading a public thoroughfare."

In 1424, the emperor, Zhu Di, died. His funeral was as excessive as his life, involving ten thousand mourners who watched as he was buried along with sixteen of his concubines. The unfortunate women had been hanged or ordered to take their own lives in preparation for the event. Their tomb was surrounded by a mile-long line

This anonymous sixteenth-century portrait of Magellan is one of the few accurate likenesses of the explorer.

The Guadalquivir River depicted in this sixteenth-century painting gave life to Seville's shipping industry but proved dangerous for ships to navigate.

Sixteenth-century Seville was a prosperous, tumultuous city and the starting point for Spanish expeditions to the Indies, including Magellan's.

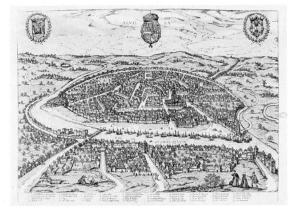

L EMPEREVR

A youthful Charles I by Orley van Barent. The king was just eighteen when he commissioned Magellan's expedition.

Idealized portrait of Manuel I of Portugal. This strange and withdrawn ruler, jealous of potential rivals, repeatedly refused Magellan's requests for backing a voyage to the Spice Islands. In frustration, the explorer turned to Portugal's main rival, Spain.

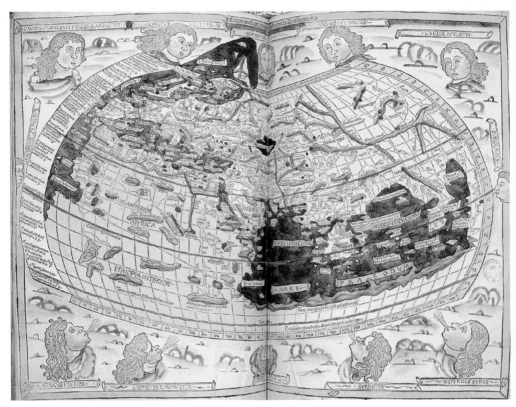

Ptolemaic map of the world, 1486. Maps based on the works of Claudius Ptolemy, the Greco-Egyptian astronomer and mathematician, distorted or omitted significant parts of the globe.

A T in O map, depicting a scripturally inspired view of the earth. In the T in O model, the ocean encircles three continents, Europe, Asia, and Africa, divided by waterways.

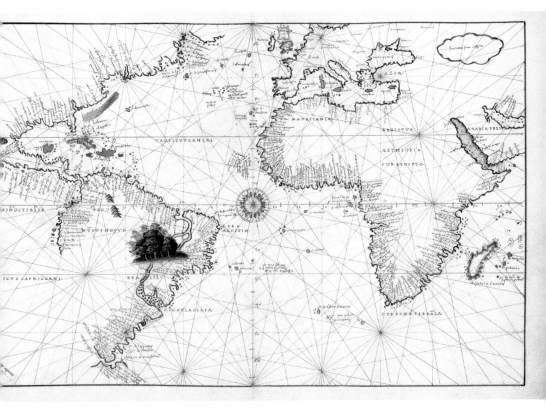

A portolan chart of the Atlantic Ocean by Battista Agnese, 1544. Agnese was a Genoese chartmaker known for his beautiful presentation copies of maps showing the newest discoveries.

Pope Leo X in a portrait by Raphael.

The *strappado* as depicted by Domenico Beccafumi in this sixteenth-century ink and charcoal rendition. Magellan subjected at least one of the participants in the Port Saint Julian mutiny to its agonies.

Cannibalism as practiced by the inhabitants of the New World, in a popular engraving by Theodore de Bry, late sixteenth century.

A romantic nineteenth-century view of Magellan's greatest accomplishment, the discovery of the strait that later bore his name.

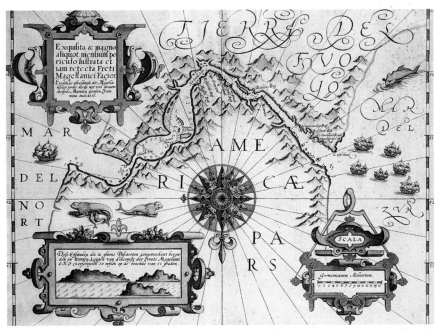

Map of the Strait of Magellan, 1606. Upon exiting, Magellan expected to reach the Spice Islands within a short span of time. Instead, he embarked on a grueling ninety-eight-day passage to his first Pacific landfall.

The Strait of Magellan in winter, as seen by NASA's *SeaStar* spacecraft. Magellan's fleet entered the eastern mouth of the strait, on the right, threaded its way through more than three hundred miles of frigid waters, and exited the western mouth, on the left, into the Pacific Ocean.

The Canary Islands off the coast of West Africa, as they appeared to NASA's *SeaStar* spacecraft. These sparkling volcanic islands served for centuries as stopovers for expeditions setting out and returning to the Iberian Peninsula.

The intricate estuaries and mysterious landscape of the strait look very much the same in this photograph from January 2002 as they appeared to Magellan and his armada nearly five hundred years ago.

The mesmerizing blue glaciers of the Strait of Magellan dwarfed the ships of the Armada de Molucca in search of the Western Sea—the Pacific Ocean.

The French illustrator Gustave Doré captured the psychological travail of a Pacific crossing during the Age of Discovery in this engraving published in 1878. It accompanied Samuel Taylor Coleridge's narrative poem "The Rime of the Ancient Mariner."

The baptism of Humabon, the king of Cebu, in April 1521, as it might have appeared.

Only weeks after the triumphant baptism of Humabon, the inhabitants of a neighboring island, Mactan, hacked Magellan to pieces in the harbor after he made a rash decision to burn their village.

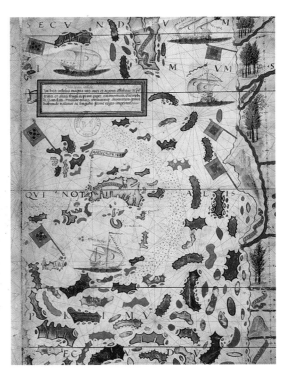

A pilot's chart of the Spice Islands created by the Portuguese in 1519. In the Age of Discovery such charts were considered top secret.

Antonio Pigafetta's chronicle of Magellan's voyage shows the Marianas, Magellan's first landfall in the Pacific, and the speedy, highly maneuverable outrigger canoe known as a *proa*, which occasionally bedeviled the fleet.

LEFT: Clove trees, the source of the rare and precious commodity for which Magellan and his fleet risked all.

The marvelous nutmeg, shown in an eighteenth-century French print.

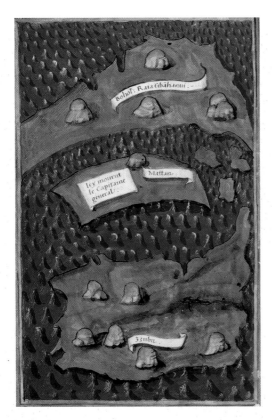

Pigafetta's illustration of
the island of Mactan, in the
Philippines. *"Icy mourut le
Capitaine general,"* reads the
inscription: This is where the
Captain General died. Unlike
many crew members, Pigafetta
revered Magellan as a
courageous explorer and
visionary.

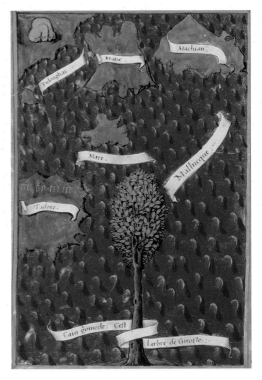

Antonio Pigafetta's depiction
of the Spice Islands, illustrated
with a clove tree. Believed to be
indigenous to the Spice Islands,
clove trees thrived in their
volcanic soil and drenching
rainfall.

Sailors endure the rigors of the icy Southern Passage as imagined by Gustave
Doré in this engraving published in 1878. Magellan's sailors protected
themselves against the cold with animal skins they obtained en route.

Oval World Map showing the track of Magellan's circumnavigation, by Battista Agnese, 1543–45.

Facsimile of the signature of the navigator Fernam de Magalhães at a Council of War at Cochim, 1510.

Magellan's signature to a letter to the Emperor Charles V, dated, Seville, October 24th 1518.

Magellan's signature. The example on top is from 1510, when he was in the service of Portugal, and the bottom from 1518, contained in a letter to King Charles.

of stone carvings representing soldiers, beasts, and officials. His son, Zhu Gaozhi, canceled all future voyages for the Treasure Fleet. Like other rulers during the Ming dynasty, Zhu Gaozhi was caught between followers of the Confucian traditions, who urged him to look inward and disdain traffic with foreigners, and the eunuchs, who encouraged international trade and grew rich off the proceeds. Zhu Gaozhi allied himself with the Confucians, and the admiral Cheng Ho, once the most powerful man in China next to the emperor, was reassigned to Nanking. The great shipyards, where thirty thousand men once toiled, fell silent as shipbuilding ceased.

That would have been the end of the Treasure Fleet, had Zhu Gaozhi lived. But he died a few years later, and his twenty-six-year-old son—Zhu Di's grandson—turned to the palace eunuchs, who quickly restored the Treasure Fleet to its former glory. In 1431, on its seventh voyage, the fleet consisted of 300 ships and 27,500 men. Cheng Ho was charged with restoring peaceful relations between China and the kingdoms of Malacca and Siam. After completing the mission, part of the fleet sailed on and probably reached northern Australia. This much has been strongly suggested by Chinese artifacts recovered in Australia and by the oral traditions of the Aborigines. The remarkable journey turned out to be the Treasure Fleet's last adventure; Cheng Ho, who inspired the enterprise, died on the voyage home.

The emperor mothballed the Treasure Fleet, shut down the Nanking shipyards, and destroyed records documenting its accomplishments. Chinese science and technology, especially regarding exploration, fell into decline. By 1500, an imperial edict made it a capital offense for a ship with more than two masts to put to sea; in 1525, officials set about destroying the larger ships of the Treasure Fleet. China abandoned the huge transoceanic trading empire created by the Treasure Fleet and, guided by Confucian precepts, turned inward, never to explore the ocean again.

Cheng Ho's voyages demonstrated that China was once the most powerful nation in the world, a seagoing empire that Spain or Portugal would have feared and envied, had they known of its reach. The reputation of the Treasure Fleet never made it to European shores. Portuguese and Spanish explorers sailed into the vacuum of power left by China. Like the Chinese, they came in search of wealth, but quite unlike them, they fiercely battled for territory, for commercial and political advantage over one another, and for religious conquest.

Driven by these imperatives, European progress into the regions formerly under the spell of Chinese commerce was swift. In 1498, Vasco da Gama and his men came across evidence of the vanished Chinese presence in East Africa: natives wearing green silk caps adorned with fringe. The inhabitants spoke of white ghosts wearing silk: a distant memory of the Treasure Fleet, which had visited these shores eighty years before. Now, in 1521, Magellan's Armada de Molucca arrived in the Philippines, claiming vast territories renounced by China. Magellan, like other Europeans, had no direct knowledge of the Treasure Fleet, but he and his men kept stumbling across artifacts of the vanished Chinese empire: silk, porcelain, writing, and sophisticated weights and measures were everywhere in evidence.

The Chinese experiment in maritime diplomacy and trade lasted for a single generation, but the rapacious and daring Europeans were here to stay. By the time Magellan arrived in the Philippines, Chinese influence was rapidly waning, and even a modest fleet such as the Armada de Molucca could have a major impact on the region. The era of Chinese colonization had ended; the era of Spanish colonization was just beginning.

*T*he sprawling Philippine archipelago did not exist on European maps, and neither Magellan nor his pilots knew what to make of their discovery. Magellan led his ships closer to the island of Samar, but within a mile or two of the shore, he found only unforgiving cliffs rising from the water, and nothing resembling a safe harbor.

He changed course once more, heading for diminutive Suluan, where the armada dropped anchor for a few hours' respite.

It was the fifth Sunday in Lent, with Easter fast approaching. Appropriately, Lent is dedicated to Lazarus, risen from the dead, and like him, the surviving crew members had overcome illness to regain their strength and persevere. Magellan decided to name the archipelago after Lazarus, but his inspiration did not last. Twenty-one years later, another European explorer, Villalobos, renamed it Las Islas Filipinas— the Philippines—after King Philip II of Spain.

Magellan's next landfall proved more satisfying than Samar. Homonhon Island did have a safe harbor, and Magellan, with tremendous relief, finally gave the order to drop anchor. He led his men ashore, to an oasis of dense rain forest, palm trees, and abundant water, where they erected two sheltering tents. At last they were free of the stench of the ships' holds. Instead, their nostrils twitched with the mingled fragrances of palm trees, wet sand, and decaying vegetation. They slaughtered a sow they had brought from Guam and prepared a great feast for themselves. For a time, their bellies were full, and the long-suffering sailors content.

On Monday, March 18, they saw a boat bearing nine men approach from the direction of Suluan. Calculating the risks and rewards inherent in their second encounter with the peoples of the Pacific, Magellan made certain that arms were at the ready; at the same time, he assembled a different sort of arsenal: shiny trinkets, in case the encounter turned out to be peaceful.

This time, Magellan handled the situation confidently. "The Captain General ordered that no one should move or say anything without his leave," Pigafetta wrote. "When those people had come to us in that island, forthwith the most ornately dressed of them went toward the Captain-General, showing that he was very happy at our coming. And five of the most ornately dressed remained with us, while the others who stayed at the boat went to fetch some who were fishing, and then they all went together. Then the captain, seeing that these people were reasonable, ordered that they be given food and drink, and he presented them with red caps, mirrors, combs, bells ... and other things. And when those people saw the captain's fair dealing, they gave him fish and a jar of palm wine, which they call in their language *vraca*, figs more than a foot long [bananas] and other smaller ones of better flavor, and two coconuts.... And they made signs with their hands that in four days they would bring us rice, coconuts, and sundry other food."

Perhaps they had found Paradise, after all, or at least a respite from an expedition well into its second year. Each day Magellan fed coconut milk supplied by the generous Filipinos to the sailors still suffering from scurvy. Pigafetta meanwhile became intrigued with the Filipinos' method for fermenting palm wine. "They make an aperture into the heart of the tree at its top ... from which is distilled along the tree a liquor ... which is sweet with a touch of greenness. Then they take canes as thick as a man's leg, by which they draw off this liquor, fastening them to a tree from the evening until next morning, and from the morning to the evening, so that the said liquor comes little by little."

Perhaps under the influence of too much Filipino palm wine, Pigafetta marveled at the coconut and all its uses. "This palm bears a fruit, named *cocho*, which is as large as the head or thereabouts, and its first husk is green and two fingers thick, in which are found certain fibers of which those people make the ropes by which they bind their boats. Under this husk is another, very hard and thicker than that of a nut.... And under the said husk there is a white marrow of a finger's thickness, which they eat with meat and fish, as we do bread, and it has the flavor of an almond.... From the center of this marrow there flows a water which is clear and sweet and very refreshing, like an apple." The Filipinos taught their visitors how to produce milk from the coconut, "as we proved by experience." They pried the meat of the coconut from the shell, combined it with the coconut's liquor, and filtered the mixture through cloth. The result, said the chronicler, "became like goat's milk." Pigafetta was so moved by the coconut's versatility that he declared, with some exaggeration, that two palm trees could sustain a family of ten for a hundred years.

Their idyll lasted a week, each day bringing with it new discoveries and a growing intimacy with their genial Filipino hosts. "These people entered into very great familiarity and friendship with us, and made us understand several things in their language, and the name of some islands which we saw before us," Pigafetta commented. "We took great pleasure with them, because they were merry and conversable."

*B*ut Magellan nearly destroyed the idyll when he invited the Filipinos aboard *Trinidad*. He incautiously showed his guests "all his merchandise, namely cloves, cinnamon, pepper, walnut, nutmeg, ginger, mace, gold, and all that was in the ship." Clearly he felt he was no longer among thieves. His trust was amply rewarded when the Filipinos appeared to recognize these exotic and precious spices and tried to explain where they grew locally, the first indication that the armada was approaching the Spice Islands. Magellan's reaction can be easily imagined. Perhaps he would reach the Moluccas after all.

He then did his guests a signal honor, or so he thought, by ordering his gunners to discharge their "artillery"—the awkward arquebuses. The roar shattered the silence and reverberated against the distant hills of Homonhom, terrifying the Filipinos who, afraid for their lives, "tried to leap from the ship into the sea." This might have been a gaffe, an excess of enthusiasm. Or was Magellan trying to impress these defenseless islanders, and himself, with the power of his weapons? At the very least, the display was a cruel practical joke on a tranquil tribe that had only helped and protected him and his men. Magellan quickly reassured the frightened Filipinos and coaxed them into remaining on board; at the same time, he could not fail to notice that his weapons conferred absolute power over the islanders, should he ever feel the need to exert it.

*A*fter a week in Homonhom, Magellan gave the order to weigh anchor on Monday, March 25, while light rain dappled the water's surface. As the three black ships were about to head out of the harbor on a west southwest course, deeper into the Philippine archipelago, toward the Moluccas, Pigafetta committed a rare lapse of judgment.

"I went to the side of the ship to fish, and putting my feet upon a yard leading down into the store room, they slipped, for it was rainy, and I fell into the sea, so that no one saw me. When I was all but under, my left hand happened to catch hold of the clew-garnet of the mainsail, which was dangling in the water. I held on tightly, and began to cry out so lustily that I was rescued by a small boat. I was aided, not, I believe, indeed through my merits, but through the mercy of that font of charity"—by which he meant the Virgin Mary. Had Pigafetta not been rescued, he would have drowned on the spot, or been rescued by the Filipinos, and would have spent the rest of his life with them, unable to tell his incredible tale.

The following night, the crew spied an island distinguished by a dull red glow, the unmistakable sign of campfires, and they knew they were not alone. In the morning, Magellan decided to risk approaching, and in a now familiar ritual, they were greeted by another small boat, this one bearing eight warriors with unknown intentions.

Magellan's slave, Enrique, addressed them in a Malay dialect, and to Magellan's astonishment, the men appeared to understand him and replied in the same tongue. No one, not even Magellan, knew how Enrique managed to converse with the islanders, but the slave's background provides some valuable clues. Magellan had acquired Enrique ten years earlier in Malacca, where he was baptized, and he had followed his master ever since across Africa and Europe. If Enrique had originally come from these islands, been captured as a boy by slave raiders from Sumatra, and sold to Magellan at a slave mart in Malacca, the chain of circumstances would account for his understanding the local language. But beyond that, it meant that Magellan's servant was, in fact, the first person to circle the world and return home.

As the islanders "came alongside the ship, unwilling to enter but taking a position at some little distance," the Captain General attempted to entice them with a "red cap and other things tied to a bit of wood." Still, they remained at a distance. Finally, Magellan's peace offerings were set out on a plank pushed in the canoe's direction. The men in the boat enthusiastically seized the gifts and paddled back to shore, where, Magellan presumed, they displayed their trophies to their ruler.

"About two hours later we saw two *balanghai* coming. They are large boats . . . full of men, and their king was in the larger of them, being seated under an awning of mats. When the king came near the flagship, the slave spoke to him. The king understood him, for in those districts, the kings know more languages than the other people. He ordered some of his men to enter the ships, but he always remained in his *balanghai,* at some little distance from the ship, until his own men returned; and as soon as they returned he departed." Magellan tried to conduct himself as a gracious visitor, but he was outdone by the generosity of the king, who proffered a "large bar of gold and a basketful of ginger." Magellan politely but firmly refused to accept this tribute, but he remained on such friendly terms with the natives that he moved his ships' anchorage closer to the king's hut for the night, as a symbol of their newfound allegiance.

This encounter with indigenous people was shaping up as the armada's most peaceful and successful since their delirious layover in Rio de Janeiro. A king willing to give gold and ginger might have other resources, and perhaps even women, but experience had shown Magellan that opening gestures could be deceptive, if not outright dangerous.

$\mathcal{T}$he next day, Good Friday of 1521, Magellan put his relationship with the islanders to the test. He sent Enrique ashore on the island of Limasawa. Even today, as part of southern Leyte in the Philippines, Limsawa is a remote, inaccessible island consisting of six square miles of sand. The island is remarkable for its broad, clean, inviting beaches, occasionally interrupted by unusual rock formations and caves. Although Magellan was the first European explorer to reach Limasawa, he was not the first outsider to find safe harbor here. Without realizing it, he had arrived at an important trading post. Chinese traders had been calling at the island for five centuries, their junks bearing sophisticated manufactured items such as porcelain, silk, and lead sinkers; the islanders traded for these items with products from their beaches and forests: cotton, wax, pearls, betel nuts, tortoiseshells, coconuts, sweet potatoes, and coconut leaf mats. The Limasawans enjoyed a reputation for hospitality and, more important, honesty. In 1225, Chau Ju Kuo, a Chinese merchant, described the orderly process of trading; the Limasawans, he said, efficiently carried away the Chinese goods they had been given and always returned with the arranged payment. So the appearance of the armada, while unusual, was not wholly unanticipated by the islanders, who were prepared to engage in trade with their guests.

Once he was ashore, Enrique asked the Limasawan ruler, Rajah Kolambu, to send more food to the fleet, for which payment would be rendered. As instructed, he added "that they would be well satisfied with us, for he [Magellan] had come to the island as friends and not as enemies." The king responded favorably to the request and came himself, along with "six or eight men," all of whom boarded the flagship. "He embraced the Captain General to whom he gave three porcelain jars covered with leaves and full of raw rice and two very large *orades*"—the dorado, a fish. In return, Magellan "gave the king a garment of red and yellow cloth made in the Turkish fashion, and a fine red cap.... Then the Captain General had a collation spread for them, and told the king through a slave that he desired to be *casicasi* with him. The king replied that he also wished to enter the same relations with the Captain General."

This was a strong statement. To be *casicasi* meant that Magellan wished to become blood brothers with the island king, a ceremony requiring the mingling of their blood. "Both cut their chests," said de Mafra, "and the blood was poured in a vessel and mixed together with wine, and each of them drank one half of it."

Magellan's attitude toward indigenous people had undergone a revolution. Where he had been content to convert, kidnap, and, when it suited his whim, even kill the giants of Patagonia, he felt a genuine kinship with this Filipino ruler. He took the king into his confidence and was soon trying to explain how the Armada de Molucca had navigated its way across the globe. "He led the king to the deck of the ship that is located above at the stern and had his sea-chart and compass brought. He told the king how he had found the strait in order to voyage thither, and how many moons he had been without seeing land, whereat the king was astonished."

The understanding nearly unraveled when Magellan decided to ask one of his gunners to demonstrate an arquebus, and the spectacle, all smoke and fire and noise, made the "the natives . . . greatly frightened." Recent experience should have warned Magellan that a show of force was courting disaster, but he could not resist the urge to impress the king with the power of European weapons.

Magellan gave an even more astonishing demonstration as he brought out one of his men, who was dressed in armor from his knees to his neck; then three other Europeans, "armed with swords and daggers . . . struck him on all parts of the body." As the blows fell and glanced off the armor, the clank of metal on metal echoing across the water, "the king was rendered speechless." The king seemed to think that these visitors possessed superhuman powers. No man could have withstood the shower of blows, yet the armored soldier had done just that.

Gratified by the king's reaction to the swordplay, Magellan instructed Enrique, his slave and translator, to tell the king that "one of those armed men was worth one hundred of his own men" and boasted that his armada brought with it two hundred warriors equipped with armor and weapons—swords, halberds, and daggers. The message was plain: A wise leader would do well to keep Magellan as an ally rather than antagonize him. Recovering from the shock of what he had seen, the king hastily agreed that a single warrior in armor was worth one hundred natives.

Magellan's Armada de Molucca carried enough weaponry to equip a small army. The sheer number of weapons reflected the growing reliance on arms in Spain and Portugal. Both nations depended on gunpowder, which had appeared in Europe at the beginning of the fourteenth century. Gunpowder was slow to reach the Iberian peninsula, but once it caught on, the Spanish and the Portuguese embarked on an arms race with a sense of deadly urgency. Local gunpowder works sprang up all over Spain; and eventually a government-sponsored gunpowder plant appeared in Burgos. The demand for gunpowder grew along with the demand for guns and cannon, and the number of foundries across Spain and Portugal increased as both countries armed themselves to compete for global dominance. It was only a matter of time before weapons found a place aboard the ships of both nations, at first to defend their harbors, and later to protect crews on voyages of exploration.

The most powerful weapons aboard Magellan's ships were the three *lombardas*. This was a cannon made of wrought iron. Designed for use at sea, it was equipped with rings to lift it on and off ships. Aboard the deck of a ship, the *lombarda* rested in a wooden cradle to which it was securely lashed. It could fire almost anything—stones, iron, and lead projectiles, but the most lethal shot consisted of an iron cube covered with a leaden sheath. To fire a *lombarda*, the gunner held a flaming taper to a touch hole leading to a small chamber holding priming powder; this in turn set off the main charge, expelling the shot with a great concussive roar as the *lombarda* shuddered in its massive cradle. The *lombarda* was not accurate, but its heavy projectile could inflict considerable damage on a hull. The fleet also carried seven breech-loading guns called *falcones*. They were smaller than the *lombardas*, and light enough for sailors to carry them into the longboats. The fleet also carried three *pasamuros*, another type of gun; nearly sixty *versos*, a crude rifle that could fire stone shots; fifty shotguns; three tons of gunpowder; and at least that weight in cannonballs.

Although these firearms could be exceedingly effective, they were also unreliable. Each time a gunner fired a weapon, he risked injury or death. Guns and cannon were liable to blow up or sputter harmlessly. The arquebus posed special dangers. It employed a matchlock, a small pan holding the gunpowder beside the gun barrel; its nine-foot-long match, or fuse, had to be lit at all times, which eliminated surprise in night combat. To maintain the match's length, the gunner pulled it by hand, risking injury. Even if a dexterous gunner managed to get off a shot, the bullets could not penetrate armor, and their effective range was less than a few hundred feet. At that moment, gun manufacturers were phasing out the awkward matchlock in favor of the wheel lock, which produced a spark, but the improvement came too late for Magellan's gunners to take advantage of it. If his expedition had left only a year later, he would have carried more advanced guns with him, and the outcome of his voyage might have been very different.

For real fighting, the weapons that mattered most were more traditional swords, knives, and pole weapons, which Spain had brought to a high level of refinement. The ships carried nearly one thousand spears (four for every member of the crew), several hundred steel-tipped javelins and pikes, and a dozen lances. They also carried halberds—an especially nasty weapon consisting of a blade mounted on a shaft with two handles. Properly used, a halberd could slice a man in two. There were at least sixty crossbows, and hundreds of arrows to supply them.

To complement the weapons, the fleet carried one hundred sets of armor (rather than the two hundred Magellan had claimed), consisting of corselets, cuirasses, helmets, breastplates, and visors. Magellan brought his own deluxe armor, which included a coat of mail, body armor, and six swords. His helmet was topped with bright plumage. With their firearms and armor, the men of Magellan's fleet believed they were the masters of all they surveyed. As far as Magellan was concerned, the combination of firepower and armor gave the armada unequaled power over the people of the islands, a belief that would cost him dearly.

*O*nce Magellan finished his military display, he formally requested that two emissaries inspect the island's huts and food stores. The king rapidly assented, and the Captain General chose Pigafetta and another crew member, whose name the attention-loving chronicler ignored. After his months at sea, keeping out of harm's way during the mutiny and narrowly avoiding disaster when he fell overboard, this was Pigafetta's great opportunity to distinguish himself as a diplomat.

The moment he stepped ashore, he encountered luxury the likes of which he had not seen since leaving Spain. "When I reached the shore, the king raised his hands toward the sky and then turned toward us two. We did the same toward him as did all the others." He made a regal spectacle, "very grandly decked out," and "the finest looking man that we saw among those people." His hair, "exceedingly black," hung to his shoulders, and he wore two large golden earrings. "He wore a cotton cloth all embroidered with silk, which covered him from the waist to the knees. At his side hung a dagger, the haft of which was somewhat long and all of gold, and its scabbard of carved wood. He had three spots of gold on every tooth, and his teeth appeared as if bound with gold." Tattoos covered every inch of his glistening, perfumed body. The women, Pigafetta noticed, "are clad in tree cloth from their waist down, and their hair is black and reaches to the ground. They have holes pierced in their ears which are filled with gold." Gold was everywhere, in jewelry, goblets, and dishes; it was evident throughout the king's dwelling. The precious metal, Pigafetta learned, was readily mined on the island in "pieces as large as walnuts and eggs."

Everyone, it seemed, chewed constantly on a fruit that resembled a pear. "They cut that into four parts, and then wrap it in the leaves of their tree, which they call betel. . . . They mix it with a little lime and when they have chewed it thoroughly, they spit it out. It makes the mouth exceedingly red. All the people in those parts of the world use it, for it is very cooling to the heart, and if they ceased to use it they would die."

Pigafetta had little time to gape. "The king took me by the hand; one of his chiefs took my companion, and they led us under a bamboo covering, where there was a *balanghai* as long as eighty of my palm lengths. . . . We sat down upon the stern of that *balanghai,* constantly conversing with signs. The king's men stood about us in a circle with swords, daggers, spears, and bucklers. The king had a plate of pork brought in and a large jar filled with wine. At every mouthful, we drank a cup of wine. . . . The king's cup was always kept covered, and no one else drank from it but him and me. Before the king took the cup to drink, he raised his clasped hand toward the sky, and then toward me; and when he was about to drink, he extended the fist of his left hand toward me. At first, I thought he was about to strike me. Then he drank. I did the same toward the king. They all make those signs toward one another when they drink."

Dinner was announced, and a royal feast it was. "Two large porcelain dishes were brought in, one full of rice and the other of pork with its gravy." Out of respect for the king, Pigafetta, an observant Catholic, forced himself to overlook one of his own religious customs. "I ate meat on Holy Friday," he confessed, "for I could not help myself." During the meal, Pigafetta gave the king a presentation that made almost as large an impression as Magellan's show of force: it was the power of the written word. Pigafetta coaxed the king to name various objects surrounding them, and recorded a phonetic transcription. "When the king and the others saw me writing, and when I told them their words, they were all astonished."

After the demonstration, "We went to the palace of the king"—in reality, a "hay loft thatched with banana and palm leaves. It was built up high from the ground on huge posts of wood and it was necessary to ascend it by means of ladders." Once everyone had clambered inside the flimsy structure, "The king made us sit down there on a bamboo mat with our feet drawn up like tailors. After a half-hour a platter of roast fish cut in pieces was brought in, and ginger freshly gathered, and wine. The king's eldest son, who was the prince, came over to us, whereupon the king told him to sit down near us, and he accordingly did so." More feasting ensued; Pigafetta claimed that he held his own, but "my companion became intoxicated as a consequence of so much eating and drinking." Eventually, the king, his appetite sated, retired for the night, leaving the prince behind. Pigafetta and the besotted prince slumbered in the rickety palace on bamboo mats "with pillows made of leaves."

In the morning, the king returned, took Pigafetta "by the hand" once more, and offered him another lavish meal, but before the feasting could resume, the longboat came to fetch the Europeans. Magellan had finally had enough; in addition, Easter was fast approaching. Pigafetta's reluctance to return to the fetid, barracks-like surroundings of *Trinidad* can be imagined. "Before we left, the king kissed our hands with great joy, and we his. One of his brothers, the king of another island, and three men came with us. The Captain General kept him to dine with us, and gave him many things."

*E*arly on the morning of Sunday, the last of March, and Easter Day, the Captain General sent the chaplain ashore to celebrate mass," wrote Pigafetta of that holiday. They explained the importance of the occasion to the king, so that he would not feel it necessary to feed everyone again, but neither he nor his royal brother could resist, and they sent two freshly slaughtered pigs to the Europeans. And then some of the islanders decided to worship alongside them.

Once the mass began, the islanders gradually fell under its incantatory spell, barely comprehending the rite's significance, but, to judge from Pigafetta's description, feeling its spiritual power nonetheless. "When the hour for mass arrived, we landed with about fifty men, without our body armor, but carrying our other arms, and dressed in our best clothes. Before we reached the shore with our boats, six weapons were discharged as a sign of peace. We landed; the two kings embraced the Captain General and placed him between them. We went in marching order to the place consecrated, which was not far from the shore. Before the commencement of mass, the Captain General sprinkled the entire bodies of the two kings with musk water. The mass was celebrated. The kings went forward to kiss the cross as we did, but they did not participate in the Eucharist. When the body of our Lord was elevated, they remained on their knees and worshiped Him with clasped hands. The ships fired all their artillery at once when the body of Christ was elevated, the signal having been given from the shore with muskets. After the conclusion of the mass, some of our men took communion."

After the solemn observance, it was time to celebrate. To amuse and impress his hosts, Magellan organized a fencing tournament, "at which the kings were greatly pleased." Next, Magellan ordered his men to display the cross, complete with "nails and the crown," and explained to the kings that his own sovereign, King Charles, had given these objects to him, "so that wherever he might go he might set up those tokens." Now he wished to set up the cross on their island, "for whenever any of our ships came from Spain, they would know we had been there by that cross, and would do nothing to displease them or harm their property." Magellan wanted to place the cross "on the summit of the highest mountain," and he explained the many benefits of displaying it as he proposed. For one thing,

"Neither thunder, nor lightning, nor storms would harm them in the least," and for another, "If any of their men were captured, they would be set free immediately on that sign being shown." The kings gratefully accepted the cross as a totem, without having any idea of what it actually meant.

Magellan inquired about the islanders' religious beliefs. "They replied that they worshiped nothing, but that they raised their clasped hands and their faces to the sky; and that they called their god 'Abba.'" Magellan indicated that their god sounded reassuringly familiar, "And, seeing that, the first king raised his hands to the sky, and said that he wished it were possible for him to make the Captain General see his love for him."

The discussion turned to politics. Magellan asked if the king had any enemies; if so, Magellan would "go with his ships to destroy them and render them obedient." By doing so, he hoped to strengthen their bond, and establish a permanent Spanish presence in the newly discovered archipelago. As it happened, the king said there were "two islands hostile to him, but . . . it was not the season to go there." Hearing this, Magellan turned warlike: "The Captain General told him that if God would again allow him to return to those districts, he would bring so many men that he would make the king's enemies subject to him by force." This was a curious offer because nothing in Magellan's charter from King Charles mentioned fighting tribal wars or mass conversions to Christianity; he was supposed to "go in search of the Strait," demonstrate that the Spice Islands belonged to Spain, and return in ships laden with spices. Now he put aside his commercial goals in favor of conversions and conquest. Magellan ordered his men back into formation; they fired their guns into the silent sky as a farewell gesture, "and the captain having embraced the two kings, we took our leave."

Magellan and the crew members returned briefly to their ships to retrieve the cross, and then made an exhausting ascent to the summit of the highest mountain in the area. "After the cross was erected in position, each of us repeated a Pater Noster and an Ave

Maria, and adored the cross; and the kings did the same. Then we descended through their cultivated fields, and went to the place where the *balanghai* was. The kings had some coconuts brought in so that we might refresh ourselves."

*C*onsidering his work done, Magellan announced his intention to depart in the morning. Despite all the pigs and rice and wine the kings had bestowed on them, the Captain General declared he needed even more food, and the kings recommended the island of Cebu as a convenient place to forage. Magellan's decision to sail on to Cebu troubled Pigafetta, who described it as "ill-fated." But Cebu itself did not pose any danger to Magellan; rather, it was his determination to form an alliance with friendly local rulers by making war on their enemies. Looking for trouble, he was sure to find it eventually.

Magellan asked the king for local pilots to escort the fleet to Cebu, and the king happily complied, but in the morning, he asked "for love of him to wait two days until he should have his rice harvested and other trifles attended to. He asked the Captain General to send him some men to help him, so that it might be done sooner; and said that he intended to act as our pilot himself." Magellan agreed, "But the kings ate and drank so much that they slept all day. Some said to excuse them that they were slightly sick."

Their departure delayed for forty-eight hours, Magellan fell to trading with the islanders, but he immediately ran into obstacles. "One of those people brought us ... rice and also eight or ten bananas fastened together to barter them for a knife which at the most was worth three *catrini*"—a Venetian coin of little value. "The Captain General, seeing that the native cared for nothing but a knife, called him to look at other things. He put his hand in his purse and wished to give him one real." The native refused the valuable coin. "The Captain General showed him a ducado, but he would not accept that, either." Magellan kept offering coins of increasing value, but met with the same reaction; the native "would take nothing but a knife." Finally, Magellan relented and gave it to him. Later, when a crew member went ashore to fetch water, he was offered a large crown made of gold in exchange for "six strings of glass beads," but Magellan blocked the trade, "so that the natives should learn that at the very beginning that we prized our merchandise more than their gold." The gold was far more valuable than the glass beads, but Magellan did not want the islanders to know how precious the Europeans considered gold. He instructed his men to treat it as just another metal. The ruse worked, and the armada, trading iron for gold, pound for pound, acquired vast riches. The gold they had acquired so easily would be worth a fortune in Spain, but the spices Magellan expected to find were even more valuable than the gold.

*T*he armada resumed its wanderings through the Philippine archipelago, dodging reefs so treacherous that even their native pilots hesitated. Along the way—it is impossible to know precisely where—they called at an island Pigafetta named Gatigan. Ashore, the crew members were fascinated by the profusion of bats; "flying foxes," they called the creatures as they swooped low over the ships and darted into the dense jungle in search of their main nourishment, fruit. The flying foxes reached astonishing proportions; Pigafetta claimed that they were as large as eagles. The fearless sailors even captured one of the creatures and ate it. The bat flesh, he claimed, tasted like that of a fowl.

Leaving Gatigan unscathed, the fleet continued on to Cebu. In his log, Francisco Albo traced their course as they threaded their way through the enchanted island realms: "We left Limasawa and went north toward the island of Seilani, after which we ran along the said island to the northwest as far as 10 degrees. There we saw three rocky islands, and turned our course west for about 10 leagues where we came upon two islets. We stayed there that night and in the morning went toward the south southwest for about 12 leagues, as far as 10 and one-third degrees. At that point we entered a channel between two islands, one of which is called Mactan and the other Cebu. Cebu, as well as the islands of Limasawa and Suluan, extend north by east and south by west. Between Cebu and Seilani we spied a very lofty land lying to the north, which is called Baibai. It is said to contain considerable gold and to be well stocked with food, and so great an extent of land that its limits are unknown."

Albo warned that the route, for all its lovely scenery, concealed hazards. "From Limasawa, Seilani, and Cebu, from the course followed to the south, look out for the many shoals, which are very bad. On that account a canoe which was guiding us along that course refused to go ahead. From the beginning of the channel of Cebu and Mactan, we turned west by a middle channel and reached the city of Cebu."

As a succession of warm, humid days and passionate nights in the Philippines passed, discontent among the crew subsided. For once, there was no talk of mutiny. All the crew members were aware of the armada's achievements. They had conquered an immense ocean and dispelled a thousand years of accumulated misconceptions about the world. They had sailed all the way from the West to the East, demonstrating that the earth was a globe. And they were beginning to savor the available women, exotic food, and tantalizing hints of the Spice Islands of which they had dreamed for so long. Yet a shadow hung over Magellan. Even if the rest of the expedition went flawlessly, and he did not lose another ship or sailor in his quest for spices, there would be hell to pay when he returned to Spain for marooning Cartagena and the priest. He could never return home with honor, and so he pressed on, a fugitive from society and a captive to the winds of fate.

CHAPTER X

The Final Battle

Are those *her* ribs through which the Sun
Did peer, as through a grate?
And is that Woman all her crew?
Is that a DEATH? and are there two?
Is DEATH that woman's mate?

As the Armada de Molucca approached the sandy, palm-shaded shores of Cebu, the crew members could see that the island was home to the most prosperous people they had encountered so far on the voyage. They watched village after village emerge as if by magic from the obscurity of the jungle; the inhabitants looked placid and well fed and not particularly startled by the strange ships. Their huts, rising on stilts in groups of five or six, resembled homesteads or even small estates. Overhead, tall palm trees blotted out the sky and cast wide swaths of shade. In front, extending from the water's edge, fishing lines crisscrossed the shallow water, and, a little farther from land, speedy *proas*, some powered by brightly colored sails, others by paddle, traveled out to greet the arriving fleet. No longer did the men of the armada have to contend with nomadic giants or wandering tribes living at the end of the world. Here was civilization, or at least a semblance thereof. "At noon on Sunday, April 7," Pigafetta recorded, "we entered the port of Cebu, passing by many villages, where we saw many houses built upon logs. On approaching the city, the Captain General ordered the ships to unfurl their banners. The sails were lowered and arranged as if for battle, and all the artillery was fired, an action which caused a great deal of fear to those people."

Once the ships dropped anchor, Magellan dispatched his illegitimate son, Cristóvão Rebêlo, "as ambassador to the King of Cebu," along with the slave Enrique to serve as an interpreter. Arriving on land, Rebêlo and Enrique "found a vast crowd of people together with the king, all of whom had been frightened by the mortars." To reassure the distraught inhabitants, Enrique explained that it was the fleet's custom to discharge their weapons "when entering such places, as a sign of peace and friendship." His words had their intended effect, and soon the local chieftain was asking what he could do for them.

Enrique stepped forward again and announced that his captain owed allegiance to the "greatest king and prince in the world, and that he was going to discover the Moluccas." His captain had decided to pass this way "because of the good report which he had of him from the king of Limasawa and to buy food." Impressed, the king welcomed the visitors, but he advised, "It was their custom for all ships that entered their ports to pay tribute." Only four days before, a junk from Siam "laden with gold and slaves" had called on the island and paid its tribute. To back up his story, the king produced an Arab merchant from Siam who had remained behind. The merchant explained that it was necessary to pay tribute to the local rulers in exchange for safe passage, and he urged Magellan to follow his example.

Magellan scorned the Arabs' live-and-let-live approach to the islanders and refused to pay anyone. He saw the local populace as prey, as helpers, and as heathen, not as equals, and he intended to claim their territory for Spain and their souls for the Church. Negotiations between Magellan and the king of Cebu broke down when Magellan—through Enrique—insisted that his king was the greatest

in all the world, and the Armada de Molucca would never pay tribute to a lesser ruler. He ended by declaring, "If the king wished peace
he would have peace, but if war instead, [then he would have] war."

At this point, the merchant from Siam uttered a few words that
Pigafetta took to mean, "Have good care, O king, what you do, for
these men are of those who have conquered Calicut, Malacca, and
all India the Greater. If you give them good reception and treat
them well, it will be well for you, but if you treat them ill, so much
the worse it will be for you." Enrique seconded the merchant's
advice; if the king refused to yield, the Captain General "would send
so many men that they would destroy him."

The king shrewdly replied that he would confer with his chieftains and return the next day. As a sign of his peaceful intentions, he
offered the landing party "refreshments of many dishes, all made
from meat and contained in porcelain platters, besides many jars of
wine" and sent them happily stumbling back to their waiting ships,
where they told Magellan (and the ever-present Pigafetta) the
details of the exchange. Despite his belligerent words, Magellan possessed one diplomatic asset, the king of Limasawa, who had come
along on this leg of the journey, and was pleased to "speak to the
king of the great courtesy of our Captain General."

The local king's soothing words had the desired effect, and on
Monday morning, the armada's notary, accompanied by Enrique,
held a formal meeting with the king of Cebu—"Rajah," or King
Humabon, in Pigafetta's transcription. This time, Humabon offered
to pay tribute to the most powerful king in the world, rather than
demanding it for himself. The impasse was broken. Magellan
acknowledged Humabon's generous offer and announced he would
"trade with him and no others." Prompted by the king of Limasawa,
the Cebuan ruler offered to become blood brothers with Magellan;
the Captain General had only to send "a drop of his blood from his
right arm, and he would do the same as a sign of the most sincere
friendship." Almost despite himself, Magellan had found a home in
Cebu.

The next day, Tuesday, Magellan had more good news: The Limasawan king announced that Humabon was preparing a great feast to send to the ships "and that after dinner he would send two of his nephews with other notable men to make peace with him." After gratefully receiving the food, Magellan decided to make another show of force and trotted out an armor-clad seaman, whose demonstration of European-style combat predictably alarmed the Cebuan emissary, "who seemed more intelligent than the others." Once again, Magellan turned the situation to his advantage: "The Captain General told him not to be frightened, for our arms were soft toward our friends and harsh toward our enemies; and as hand-kerchiefs wipe off sweat so did our arms overthrow and destroy all our adversaries and the enemies of our faith." The lesson had its intended effect.

*O*nce Magellan had alternately impressed and intimidated the Cebuans, relations between the two proceeded like a tale in a story-book. The king's nephew came aboard *Trinidad*, accompanied by a retinue of eight chieftains, to swear loyalty. Holding court, Magellan played the part of a magnanimous potentate with gusto: "The Cap-tain General was seated in a red velvet chair, the principal men of the ships on leather chairs, and the others on mats upon the floor. The Captain General asked them through an interpreter . . . whether that prince . . . had the authority to make peace. . . . The Captain General said many things concerning peace, and that he prayed to God to confirm it in heaven. They said they had never heard such words, but that they took great pleasure in hearing them. The Cap-tain General, seeing that they listened and answered willingly, began to advance arguments to induce them to accept the faith."

Rising from his special chair, Magellan abruptly changed the subject, wanting to know who would succeed the king after his death. "They replied that the king had no son, but many daughters, and that this prince who was his nephew had as wife the king's eldest daughter, and for love of her he was called prince. And they said moreover that, when the father and the mother were old, no more account was taken of them, but the children commanded them." This state of affairs struck Magellan as contradictory to the Commandments, and he proceeded to explain some basic tenets of the Bible. "God made the sky, the earth, the sea, and everything else and that He had commanded us to honor our fathers and mothers, and that whoever did otherwise was condemned to eternal fire; that we are all descended from Adam and Eve, our first parents; that we have an immortal spirit." His oratory must have been highly persuasive because "all joyfully entreated the Captain General to leave them two men, or least one, to instruct them in the faith." Magellan explained that he could not leave anyone behind with them, but if they wished, the armada's priest, Father Valderrama, would gladly baptize the Cebuans, and when they returned, they would bring priests and friars to instruct them. Pigafetta reported that the chieftains, Magellan, and the onlookers all became so excited by the prospect that everyone "wept with great joy." What significance the highly emotional baptismal rite held for the Filipinos can only be guessed, but it meant something very specific to Magellan. Baptism, a word derived from the Greek *baptismos,* meant immersion and carried with it the idea of cleansing the soul of sin and rebirth into the Christian faith.

Before he began in earnest, Magellan cautioned the Cebuans not to convert to Christianity simply to win his favor, and promised not to "cause any displeasure to those who wished to live according to their own law." But, he said, the Christians would get preferential treatment. "All cried out with one voice that they were not becoming Christians through fear or to please us," Pigafetta recorded, "but of their own free will." Magellan was so encouraged by this response that he promised to leave behind a suit of armor—just one—in gratitude.

He also raised the highly sensitive subject of sex between his men and the Cebuan women. "We could not have intercourse with their women without committing a very great sin, since they were pagans; and he assured them that if they became Christians, the devil would no longer appear to them, except in the last moment at their death." Magellan implied that it was a lesser sin to become intimate with Cebuan women who had been baptized, and the crew, as ravenous for sex as they were for food, immediately took advantage of the loophole. But there is no suggestion that he became intimate with any of the Cebuan women; he found fulfillment of a more spiritual nature. "The Captain embraced them weeping, and clasping one of the prince's hands and one of the king's between his own, said to them that he would give them perpetual peace with the king of Spain."

*A*fter mutual assurances and reassurances had been exchanged, it was time for another feast. Once again, Magellan was the fortunate recipient of island hospitality in the form of "rice, swine, goats, and fowls," all given with profuse apologies for their inadequacy.

The Cebuan women performed an elaborate consecration before slaughtering the hogs. The ceremony began with the sound of gongs, after which the celebrants appeared with three serving dishes, two holding rice cakes and roast fish wrapped in leaves, the other a coarse fabric made from palm trees. The women then spread the cloth on the ground, whereupon two elderly women, each holding a bamboo trumpet, wrapped themselves in it. "One of them puts a kerchief with two horns on her forehead, and takes another kerchief in her hands, and dancing and blowing on her trumpet, she thereby calls out to the sun. She with the kerchief takes the other standard, and lets the kerchief drop, and both blowing on their trumpets for a long time, dance about the bound hog." The dancing and music continued for quite some time, until one of the old women, after taking ritual sips of wine with her artificial horn, sprinkled the residue on the hog. "She is given a lance, and while dancing and clasping a lighted torch in her mouth, thrusts the instrument four or five times through the heart of the hog, with sudden and quick strokes." After the slaughter, the women unwrapped themselves, and, with other women they selected—no men allowed—devoured the contents of the three dishes. "No one but old women consecrate the flesh of the hog, and they do not eat it unless it is killed this way."

In return for this elaborately consecrated food, Magellan conferred a bolt of white linen, a red cap, strings of glass beads, and a gilded glass cup on the prince. ("Those glasses are greatly appreciated in those districts," Pigafetta commented.) There was more; Magellan asked Pigafetta to give Humabon a "yellow and violet silk robe, made in Turkish style, a fine red cap, some strings of glass beads, all in a silver dish, and two gilt drinking cups." By the time the feast ended, the Cebuans regarded Magellan as something more than a man; he was a powerful and beneficent god. The adulation rubbed off on the volatile Captain General, who increasingly believed himself to be divinely inspired, his expedition a manifestation of God's will. It was a dangerous delusion.

When Magellan finally left *Trinidad* to make his triumphal entry into Cebu, the occasion proved every bit as majestic as he could have wished. A delegation from the ships, including an excited Pigafetta, landed on Cebu to greet Humabon, dressed in regal splendor to greet his guests: "When we reached the city we found the king in his palace surrounded by many people. He was seated on a palm mat on the ground, with only a cotton cloth before his private parts, and scarf embroidered with the needle about his head, and necklace of great value hanging from his neck, and two large gold earrings fastened in his ears set round with precious gems. He was fat and short, and tattooed with fire in various designs. From another mat on the ground he was eating turtle eggs, which were in two porcelain dishes, and he had four jars full of palm wine in front of him covered with sweet-smelling herbs and arranged with four small reeds in each jar by means of which he drank. Having duly made reverence to him, the interpreter [Enrique] told the king that his master thanked him very warmly for his present, and that he sent this present not in return for his present but for the intrinsic love which he bore him. We dressed him in the robe, placed the cap on his head, and gave him the other things; then kissing the beads and putting them on his head, I presented them to him. He, doing the same, accepted them. Then the king had us eat some of those eggs and drink through those slender reeds.... The king wished to have us stay to supper with him, but we told him that we could not stay."

Impressive as they were, these exchanges were a mere prelude. The excitement began when the prince escorted Pigafetta and several others to his raised hut. They climbed ladders and within found "four young girls were playing—one, on a drum like ours, but resting on the ground; the second was striking two suspended gongs alternately with a stick wrapped somewhat thickly at the end with a palm cloth; the third, one large gong in the same manner; and the fourth two small gongs held in her hand, by striking one against the other, which gave forth a sweet sound. They played so harmoniously that one would believe that they possessed good musical sense."

The Europeans noticed more than musical ability; the girls were bare-breasted and extremely alluring. "Those girls were very beautiful and almost as white as our girls and as large. They were naked except for palm cloth hanging from the waist and reaching to the knees. Some were quite naked and had large holes in their ears with a small round piece of wood in the hole.... They have long black hair, and wear a short cloth about the head, and are always barefoot. The prince had three quite naked girls dance for us." Reluctant to contradict Magellan's prohibition against intercourse with native women until they converted to Christianity, Pigafetta refrains from describing the frolicking and lovemaking with the female musicians, but he leaves no doubt about the evening's outcome.

All around them similar celebrations of European-Cebuan amity involving ordinary villagers and sailors were taking place that night. The one question is whether Magellan participated as well, but, given his restraint and self-denial throughout the voyage, it is unlikely that he yielded to the temptations of the flesh, even on this occasion.

*W*hen they returned to the ships that night, the four emissaries were greeted with sobering news: Two shipmates lay near death. The next morning, April 10, Martín Barreta, a passenger, succumbed to the lingering effects of the scurvy he had suffered during the ninety-eight days of the Pacific crossing. Hours later, Juan de Areche, a sailor, breathed his last.

In the morning, Pigafetta and Enrique returned to the island to make arrangements for Christian burials for both men, which meant consecrating a cemetery on Cebu, complete with a cross. The king, as accommodating as ever, said he wished to worship the cross as soon as it was erected. Magellan turned the occasion into a religious lesson for the islanders' benefit. "The deceased was buried in the square with as much pomp as possible, in order to furnish a good example. Then we consecrated the place and in the evening buried another man."

Pigafetta spent enough time on Cebu to become familiar with local burial customs, and was impressed by their sophistication and parallels to European practices. He found that women took the leading role in the rites, which started simply and then grew in power. "The deceased is placed in the middle of the house in a box. Ropes are placed about the box in the manner of a palisade, to which many branches of trees are attached. In the middle of each branch hangs a cotton cloth like a curtained canopy. The most principal women sit under those hangings, and are all covered with white cotton cloth, each one by a girl who fans her with a palm-leaf fan. The other women sit about the room sadly. Then there is one woman who cuts off the hair of the deceased very slowly with a knife. Another who was the principal wife of the deceased lies down upon him and places her mouth, her hands, and her feet upon those of the deceased. When the former is cutting off the hair, the latter weeps; and when the former has finished the cutting, the latter sings." After five or six days of mourning, "They bury the body in the same box which is shut in a log by means of wooden nails and covered and enclosed by wooden logs."

A few days later, Pigafetta confided to his diary that he, along with other men of the fleet, had been intimate with the women of Cebu. That was not surprising in itself; far more extraordinary were the bizarre sexual customs practiced by both sexes, especially *palang,* or genital stretching.

"The males, large and small, have their penis pierced from one side to the other near the head, with a gold or tin bolt as large as a goose quill," Pigafetta observed, scarcely believing his eyes. "In both ends of the same bolt, some have what resembles a spur, with points upon the ends; others are like the head of a cart nail. I very often asked many, both old and young, to see their penis, because I could not credit it." Fascinated by the devices, Pigafetta studied them closely. "In the middle of the bolt is a hole, through which they urinate. The bolt and spurs always hold firm."

Pigafetta naturally wondered how the women of the island tolerated *palang* during sexual intercourse. Surely the bolts injured or hurt them. Not at all, the Cebuan men told him. "Their women wish it so, and said that if they did otherwise, they would not have communication with them." And they proceeded to explain precisely how *palang,* in their experience, actually enhanced sexual gratification for both men and women. In the process, Pigafetta received a graphic lesson in the art of love, Cebuan style. "When the men wish to have communication with their women, the latter themselves take the penis not in the regular way and commence very gently to introduce it [into the vagina], with the spur on top first, and then the other part. When it is inside, it takes its regular position; and thus the penis always stays inside until it gets soft, for otherwise they could not pull it out."

Palang was not confined to men. Women also used it, starting in infancy. "All of the women from six years and upward have their vaginas gradually opened because of the men's penises," he learned. Having sexual intercourse with *palang* prolonged the act; the bolts and spurs discouraged sudden movements; and it was believed to intensify the pleasurable sensations experienced by both sexes. One of the most difficult things for the Europeans to understand was that *palang* was intended to enhance female pleasure by stimulating a variety of sensations in the vagina. Intercourse using *palang* lasted as long as a day, or even more, as the two lovers remained locked in an embrace of passion.

Pigafetta's clinical description contained enough detail to suggest that he observed the islanders having intercourse, and he came away both excited and dismayed by what he saw. "Those people make use of that device because they are of a weak nature," he decided, equating weakness with pleasure-loving. He went on to explain that "they have as many wives as they wish, but one of them is the principal wife." Both the practice of *palang*, with its emphasis on increasing pleasure, and polygamy, which Pigafetta associated with it, ran counter to Catholic teachings. For all these reasons, Pigafetta found *palang* disconcerting and, to prove his point, he insisted, "All the women loved us very much more than their own men," presumably because the unadorned Europeans lacked the cumbersome accessories.

For the Armada de Molucca and the Spanish expeditions that followed, *palang* was just one of many unacceptable customs practiced by the islanders. It was said that the Filipino families of the ruling class resorted to infanticide by burying the victims or hurling them into the sea. Also, unmarried women regularly underwent abortions to make it easier to find a husband. Virginity was actually considered such a serious liability that "professional deflowerers" could be engaged to take care of the problem. The Filipinos emphasized female sexual pleasure, and women even had access to artificial

penises to assuage their lust. The Spanish, especially the clergy who
came after Magellan, were intent on eliminating the practice, which
they felt was nearly as repugnant as *palang* itself.

It can be said that Magellan's do-or-die emphasis on conversion
interfered with precious cultural traditions, but he saw matters quite
differently: He was engaged in a mission to rescue a benighted people from barbarism in this world and perdition in the next. In contrast to his pragmatic crew members, who considered themselves
travelers through an alien landscape, Magellan conducted himself as
if he were an instrument of the Lord. He believed that Providence
had sent him to the Philippines to bring Christianity to the heathen
and considered the local customs as grave social ills. In Magellan's
mind, Christianity offered the best, and the only cure.

$\mathscr{M}$agellan found that the Cebuans were organized and skillful
in their barter practices; they relied on a remarkably accurate system of weights, measures, balances, and scales. Accordingly, he
ordered his men to bring ashore their merchandise and open up for
business. The Europeans offered their usual assortment of metal
and glass objects, knives and beads and nails, and the islanders
rushed to offer gold in return. However, "The Captain General did
not wish to take too great a quantity of gold, so that the sailors
might not sell their share in the merchandise too cheaply, because of
their lust for gold, and he should therefore be constrained to do the
same with his merchandise, for he wished to sell it at as high a price
as possible."

Meanwhile, Pigafetta recorded the local language. The Cebuan
dictionary he compiled was even more detailed and thorough than
his primitive effort with the Patagonian giants. He took the trouble
to include Cebuan names for parts of the body, the sun and stars,
common plants and objects, and, for the first time, numerals. As
before, Pigafetta worked in a vacuum, guided by the dictates of his

intuition and common sense, since there was very little precedent and absolutely no professional standard for the ambitious task of writing down words and definitions for an entire oral tradition. Despite the obstacles, he managed to devise a phrase book that might be useful for subsequent expeditions that happened to pass through Cebu.

*T*he man... *lac.* The woman... *perampuan.* The youth... *benibeni.* The married woman... *babai.* The chin... *silan.* The spine... *lieud.* The navel... *pussud.* Gold... *boloan.* Silver... *pilla.* Pepper... *malissa.* Cloves... *chiande.* Cinnamon.... *manna.* A ship... *benaoa.* A king... *raia.*

*O*n Sunday morning, April 14, King Humabon's baptismal ceremony unfolded with all the pageantry that Magellan could muster. The day before, crew members had constructed a platform in the village square, festooned with palm branches and other decorative vegetation. A complement of forty seamen, including Pigafetta, clambered into the longboats. Two wore gleaming armor and stood just behind the king of Spain's banner as it waved benignly in the gentle ocean breeze. Once again, Magellan planned to fire his artillery, but this time he took the precaution to explain to the king that "it was our custom to discharge them at our greatest feasts." Having given fair warning, the crew fired their weapons at the moment they disembarked, marking the formal commencement of the event.

Magellan appeared, Humabon approached, and as the two embraced, the Captain General revealed that he had bent the rules of protocol in the king's favor. "The royal banner was not to be taken ashore except with fifty men armed as were those two, and with fifty musketeers; but so great was his love for him that he had thus brought the banner." Pigafetta wrote little about this banner, but it was probably the Royal Standard of the Catholic Kings, in use since 1492, during the reign of Ferdinand and Isabella. It featured the eagle of Saint John with inverted wings, and it might have included symbols of the kingdoms of Spain—León, Aragón, Castile, Sicily—as well as arrows and possibly a scroll. The reverence that Magellan accorded the flag, along with his familiarity with exactly how it was to be displayed—accompanied by fifty soldiers in armor and fifty musketeers—showed his devotion to King Charles, even here, on these distant shores, and how misplaced were the long-standing suspicions that he remained secretly attached to Portugal.

Once the priest baptized Humabon, the king took the name of Charles, after Magellan's sovereign. Next, the king of Limasawa took the name of John. Even the Siamese merchant was swept up in the religious fervor and decided that he would convert, too, taking the name of Christopher. The baptism was more successful than Magellan had dared to hope.

"Then all approached the platform joyfully," Pigafetta wrote. "The Captain General and the king sat down on chairs of red and violet velvet, the chiefs on cushions, and the others on mats. The Captain General told the king through the interpreter that he thanked God for inspiring him to become a Christian; and that he would more easily conquer his enemies than before." The king declared that although he wished to become a Christian, his chieftains still resisted the idea.

The Captain General instantly summoned the recalcitrant chieftains and, as Pigafetta tells it, warned that "unless they obeyed the king as their king, he would have them killed, and would give their possessions to the king." This was a nearly complete reversal of Magellan's earlier declaration, when he insisted that no one would be forced to become a Christian, though he might give preferential treatment to the converts. This declaration ran contrary to the Church's doctrines concerning the baptism of adults; they were supposed to be voluntary, for one thing, and, more important, based on faith, not fear. Magellan might have been bluffing so that the conversion could proceed rapidly; it is difficult to imagine him, or his men, staging a massacre of the generous and good-natured islanders they had just befriended. In any event, the chieftains swiftly agreed to obey Magellan and converted.

Gratified, Magellan announced that when he returned from Spain, he would bring so many soldiers with him that the king would be recognized "as the greatest king of those regions, as he had been the first to express a determination to become a Christian." Swept along by Magellan's fervor, the king lifted his hands to the sky, profusely thanked the Captain General, and even asked that some of his sailors stay behind to instruct the others in Christianity. This time Magellan relented and said he would appoint two men to stay here with the king, but in return, he wished to take "two children of the chiefs with him" to visit Spain, learn Spanish, and describe the wonders of that country on their return to Cebu.

At last the general baptism was ready, and Magellan, dressed in splendid white apparel, presided over the throng. "A large cross was set up in the middle of the square. The Captain General told them that if they wished to become Christians as they had declared on the previous days, they must burn all their idols and set up a cross in their place. They were to adore that cross daily with clasped hands, and every morning after their custom, they were to make the sign of the cross (which the Captain General showed them how to make); and they ought to come hourly, at least in the morning, to that cross, and adore it kneeling." Magellan also explained that he was dressed in white "to demonstrate his sincere love toward them"—his recent threat to kill them notwithstanding. He continued to bestow Christian names on the converts. "Five hundred men were baptized before mass," Pigafetta reports.

The ceremony ended on a solemn note, with the king and the other chieftains, now Christians, declining Magellan's offer of dinner aboard *Trinidad*, but embracing as brothers in the same faith, while the ships discharged their artillery and the jarring blasts reverberated throughout the island kingdom.

*A*fter dinner, the women took their turn at conversion, and their ceremony proved to be even more emotional. Father Valderrama, along with Pigafetta and several crew members, returned to the island to baptize the queen, who brought a retinue of forty women. She made a regal impression on the Europeans. "She was young and beautiful," Pigafetta noted, "and was entirely covered with a white and black cloth. Her mouth and nails were very red, while on her head she wore a large hat of palm leaves in the manner of a parasol, with a crown about it of the same leaves, like the tiara of the pope."

The women now participated in a very different sort of ceremony. "We conducted her to the platform, and she was made to sit down upon a cushion, and the other women near her, until the priest should be ready. She was shown an image of our Lady, a very beautiful wooden child Jesus, and a cross. Thereupon, she was overcome with contrition, and asked for baptism amid her tears. We named her Johanna, after the Emperor's mother [Juana the Mad]; her daughter, the wife of the prince, Catherina; the queen of Limasawa, Lisabeta; and the others each [received] a distinctive name. Counting men, women, and children, we baptized eight hundred souls."

As more conversions occurred spontaneously in the following days, the entire population of Cebu embraced Christianity, and soon the inhabitants of other islands were making their way to Father Valderrama for the same reason. In all, 2,200 souls converted, without a shot being fired in anger.

The scenes of conversion seemed touching and inspiring at first glance, but on closer inspection, they were incongruous and improbable. Theater had won the day. The rapidity with which the Cebuans accepted Christianity was suspect, but neither Magellan nor Pigafetta saw beyond the outward signs of faith to the lack of sincerity, conviction, and understanding that lay beneath. Thousands of islanders had converted to Christianity, but for how long? A tribe that converted so easily could readily accept another religion, or none at all.

$\mathcal{B}$y mid-April 1521, Magellan's trajectory as an explorer reached its zenith. He had quelled vicious mutinies, made good on his promise to discover the strait, navigated uncharted reaches of the Pacific Ocean, and claimed the Philippines, among other lands, for Spain, converting thousands of islanders in the process. But his erratic behavior—sometimes beneficent, sometimes menacing, occasionally both—suggests that his accomplishments had gone to his head and caused him to take an increasingly zealous approach to religious matters. Throughout the voyage, he had displayed a penchant for piety, but he now went further, threatening to kill those who defied his crusade. This time, Magellan intended to carry out his threat.

"Before that week had gone," Pigafetta wrote, "all the persons of that island, and some of the other islands, were baptized." But there were holdouts. Magellan sent word to the recalcitrant chieftains that if they did not convert immediately and swear allegiance to King Charles, he would confiscate their property, a European concept that was nearly meaningless to the islanders, and he vowed to punish them with death, a threat they understood but chose to ignore. To demonstrate his seriousness, Magellan sent a band of his men to wreak havoc. "We burned one hamlet which was located on a neighboring island, because it refused to obey the king or us. We set up a cross there, for those people were heathen," Pigafetta said, without a trace of remorse as the smoldering ashes sent a sickening plume into the sky.

The neighboring island was called Mactan.

$\mathcal{A}$s the Mactan hamlet burned, and all its inhabitants fled, Magellan forced the potentates of Cebu to adopt more authoritarian and hierarchical methods of exercising power, in the Spanish mode. First, he gathered various chieftains and coaxed them into swearing obedience to Humabon, who in turn had to swear loyalty to the king of Spain. "Thereupon, the Captain General drew his sword before the image of our Lady, and told that king that if anyone so swore, he should prefer to die rather than break such an oath." Next, Magellan endowed Humabon with a red velvet chair, "telling him that wherever he went he should always have it carried before him by one of his nearest relatives; and he showed him how it ought to be carried." Humabon, in return, presented Magellan with a special gift: two large earrings made of gold, two gold armlets, and two gold bands to be worn above the ankles. But the king was mistaken if he thought Magellan regarded those precious tokens as equal in importance to the power symbolized by the velvet chair.

For all his apparent success in bringing the islanders to Christianity, Magellan was troubled by signs that the conversions were incomplete, and might be undone. Despite his orders, for example, they had failed to burn their idols; in fact, they continued to make sacrifices to them, and he demanded to know why. Everywhere Magellan looked, there seemed to be an idol mocking him; they were even arrayed along the shore, and their appearance was disturbing to European sensibilities. "Their arms are open and their feet turned up under them with the legs open," wrote Pigafetta. "They have a large face with four huge tusks like those of a wild boar, and are painted all over."

In their defense, the islanders explained that they were propitiating the gods to aid a sick man; he was so sick that he had been unable to speak for four days. He was not just any man, he was the prince's brother, considered the "bravest and wisest" on the entire island. But Christianity could not help him, for he had not been baptized.

Magellan seized on the illness to demonstrate the healing power of Christian faith. Burn your idols, he commanded, believe in Christ, and only Christ, and, if the sick man is baptized, "he would quickly recover." Magellan was so adamant that if the sick man failed to recover, he would allow Humabon to "behead him, then and there." In fact, he would insist. Humabon, compliant as always, "replied that he would do it, for he truly believed in Christ." Magellan was convinced that his life depended on the outcome of the baptism, and it did. If the sick man failed to recover, the cause of Christianity would lose all credibility, and Magellan, undone by his fanaticism, would very likely lose his head.

He prepared carefully for the ordeal, relying on a show of power and a display of ritual to preserve the sick man's life. Once again, Pigafetta was in the thick of things: "We made a procession from the square to the house of the sick man with as much pomp as possible. There we found him in such condition that he could neither speak nor move. We baptized him and his two wives, and ten girls. Then the Captain General asked him how he felt. He spoke immediately and said that by the grace of our Lord he felt very well. That was a most manifest miracle. When the Captain General heard him speak, he thanked God fervently. Then he made the sick man drink some almond milk, which he had already prepared for him." The miraculous healing made a tremendous impression on the trusting islanders, who now revered Magellan as they would a god. He was more powerful than their idols, yet he walked among them.

Magellan made the most of his victory, revealing a tenderness and compassion he had previously held in abeyance, and thus won even more glory and adulation from the Cebuans. "Afterward, he sent him a mattress, a pair of sheets, a coverlet of yellow cloth, and a pillow. Until he recovered his health, the Captain General sent him milk, rosewater, oil of roses, and some sweet preserves. Before five days the sick man began to walk. He had an idol that certain old women had concealed in his house burned in the presence of the king and all the people." In the following days, Magellan, inflamed with biblical fervor, destroyed other idols arrayed along the shore, and incited the agitated islanders to follow his example. "The people themselves cried out, 'Castile! Castile!' and destroyed those shrines." The campaign to rid the island of idols consumed Magellan and the Cebuans, who vowed to burn all they could find, even the idols concealed in Humabon's dwelling.

*F*or a brief time, Magellan made his peace. All the hamlets on Cebu and the neighboring islands paid him tribute, presumably gold, and gave his men food in exchange for Christianity and faith healing. Life seemed tranquil, for a change, and the men, enjoying their nights with their island lovers, reveling in the exotic sexual practices of Cebu, were reluctant to leave.

In the midst of this serenity, one ominous sign—a "jet black bird as large as a crow"—appeared over the island huts around midnight each night, and "began to screech, so that all the dogs began to howl; and that screeching and howling would last four or five hours." The Europeans found it impossible to get a decent night's sleep ashore while the racket lasted, and they earnestly inquired about the disturbance, "But those people would never tell us the reason for it." To superstitious sailors, the relentless screeching might have been a portent of impending disaster.

*O*n April 26, the island of Mactan beckoned the armada. Its chief, Sula, sent one of his sons to Cebu, where he presented Magellan with an offering of two goats. He would have brought more, he explained, but the king with whom he shared the island, Lapu Lapu, had thwarted him. Lapu Lapu was the chieftain who had stubbornly resisted converting to Christianity, and whose village Magellan had burned to the ground.

Caught between two intransigent warriors, Lapu Lapu and Magellan, Sula tried his hand at diplomacy. He told Magellan that Lapu Lapu was married to his sister and would cooperate in the end, but Lapu Lapu remained adamantly opposed to the European invader. Sula abruptly reversed himself and offered to place his soldiers at

Magellan's disposal to fight Lapu Lapu. The combined forces might
be able to get rid of Lapu Lapu altogether. Magellan refused the
offer and said he wanted to "see how the Spanish lions fought" without any help.

Turning the situation to his advantage, Sula asked Magellan for a
boatload of armored warriors to fight against Lapu Lapu's men.
Magellan, never one to back down, declared that he would send not
one but three longboats filled with warriors. Thanks to Magellan's
belligerence, Sula came out the clear winner; rather than placing his
soldiers at Magellan's disposal, Magellan now placed his men at
Sula's service.

And so the battle lines were drawn.

*T*he decision to fight threw the armada into a state of alarm. Magellan's inner circle immediately recognized that they had reached
another turning point in the expedition. For the first time since
their arrival in these lush islands, they seriously questioned Magellan's judgment, if not his sanity. "A man who carried on his shoulders so momentous a business had no need to test his strength,"
Ginés de Mafra observed. "From victory . . . he would benefit little;
and from the opposite, the Armada, which was more important,
would be set at risk." Juan Serrano, the captain of unlucky *Santiago*,
passionately argued against entering into a needless battle. Converting natives was all to the good, but their primary mission was to
reach the Spice Islands; that was what their orders from King
Charles commanded them to do. He reminded Magellan that they
had already suffered many casualties and could not afford more loss
of life. Assembling a force large enough to face the islanders meant
the ships would stand nearly empty and thus become vulnerable to
attack; in the worst-case scenario, they might lose the battle and
their ships. Even Pigafetta, among the most fervent believers in
Magellan, cautioned the Captain General against taking drastic and
unnecessary measures against Lapu Lapu. But no matter how many

times they all implored him to follow a peaceful and practical strategy, Magellan refused to back down. "We begged him repeatedly not to go, but he, like a good shepherd, refused to abandon his flock."

In the face of criticism, Magellan did make two minor concessions. He reduced the number of men to a bare minimum, and he ordered his ships to keep far from shore. These crucial strategic decisions would place the entire enterprise at a tremendous disadvantage as the battle unfolded.

Without realizing it, Lapu Lapu had done just the right thing to incite Magellan to battle. He could not resist a challenge and thrived on confrontation. Throughout the voyage, he had put down mutinies, braved storms, navigated the strait, and crossed the Pacific, all with single-minded determination. He had even offered his head if a converted islander failed to recover. Each time, Magellan had succeeded, and he was convinced that the battle of Mactan would fit the same pattern. He would emerge victorious, not because of superior manpower or strategy, but because it was God's will. His officers, however, did not share his faith in divine intervention. They had no choice but to go along with the Captain General, and so they did, for the sake of form. At the same time, they planned to remain at a safe distance. If Magellan wanted to try to take Mactan virtually single-handed, with only a few amateur warriors at his side, so be it. His officers would leave him to his fate.

The Captain General gave the order to prepare for attack, and his men donned armor, this time for actual combat, not for show. Their ranks included Pigafetta; Magellan's slave, Enrique; and his illegitimate son, Cristóvão Rebêlo; along with a cadre of Cebuans in their own small vessels. The Cebuans were under orders not to fight, but merely to observe the "Spanish lions" hunt their prey.

"At midnight, sixty of us set out armed with corselets and helmets, together with the Christian king, the prince, some of the chief men, and twenty or thirty *balanghai*. We reached Mactan three hours before dawn." Magellan declared that he did not wish to fight, which must have come as a relief to his apprehensive men; instead, he sent a message to Lapu Lapu that if he would simply "obey the king of Spain, recognize the Christian king as their sovereign, and pay us our tribute, he [Magellan] would be their friend; but that if they wished otherwise, they should wait to see how our lances wounded." This was the same arrangement Magellan had offered the other islanders, who had readily accepted either of their own free will or after a brief show of force. Based on his recent experiences, Magellan anticipated a ragged band of nearly naked warriors who would flee the moment he fired his artillery, and whose flimsy bamboo spears would be useless against impenetrable Spanish armor.

Lapu Lapu refused to yield and sent back a message boasting of his weapons' strength; his lances, he said, were made from stout bamboo, and his stakes "hardened with fire." At the same time, Lapu Lapu asked Magellan to postpone his attack "until morning, so that they might have more men." At first, Lapu Lapu's absurd request baffled Magellan's men, but later it was revealed as a delaying tactic. "They said that in order to induce us to go in search of them; for they had dug certain pits between the houses in order that we might fall into them."

As he considered the request and the possible motives behind it, Magellan lost precious time, along with the advantages of darkness and a favorable tide. The shallow water meant that the longboats had to keep away from the beach, which was bad enough; worse, the big ships had to stay even farther back, in deep water. The increased distance from the longboats to the shore meant that Magellan's men would be completely exposed to Lapu Lapu's spears for a much longer period of time as they waded to land, and it meant that the ships would be so far from the scene of battle that their crossbows and artillery would be useless.

*B*y the time Magellan ordered his men to charge, the sky had already begun to glow with the approaching dawn. "Forty-nine of us leaped into the water up to our thighs, and walked through the water for more than two crossbow flights before we could reach the shore," Pigafetta wrote. By this reckoning, the distance was about two thousand feet, nearly half a mile, and a very dangerous half mile that was, because the men were completely unprotected. "The other eleven men remained behind to guard the boats." Meanwhile, the Cebuan king, prince, and soldiers, confined to their light, maneuverable *balanghai*, looked on, powerless to affect events. They acted under orders from Magellan himself, who had repeatedly warned them to stay clear of the fighting.

As Magellan's men awkwardly waded through the water to the beach, they were confronted by armed warriors prepared for a battle to the death. The Mactanese emerging from the jungle numbered not in the dozens, as expected, but, according to Pigafetta's reckoning, fifteen hundred—all from the village Magellan had just destroyed. The ratio of Mactanese fighters to Europeans was thirty to one. Magellan had boasted that just one of his armored men was worth a hundred island warriors; his estimate was about to be put to the test.

"When they saw us, they charged down upon us with exceeding loud cries, two divisions on our flanks and the other on our front. When the Captain General saw that, he formed us in two divisions, and thus did we begin to fight. The musketeers and crossbowmen shot from a distance for about a half hour, but uselessly; for the shots only passed through the shields, which were made of thin wood and the [bearers'] arms." The artillery failed to have any effect on the enemy; the Europeans' predicament grew worse, and the battle intensified. "The Captain General cried to them, 'Cease firing! Cease firing!' but his order was not heeded. When the natives saw that we were shooting our muskets to no purpose...they redoubled their shouts. When our muskets were discharged, the natives would never stand still, but leaped hither and thither, covering themselves with their shields. They shot so many arrows at us and hurled so many bamboo spears (some of them tipped with iron) at the Captain General, besides pointed stakes hardened with fire, stones, and mud, that we could scarcely defend ourselves."

The besieged Europeans, protected by their armor, awkwardly made their way through this deadly gauntlet to the shore. "The beach where they landed is very low," de Mafra recalled, "so they left the skiffs very far from the shore. Reaching it, they saw a big village in a palm grove, but there was nobody to be seen." Magellan, instead of rethinking the situation, ordered the men to do the one thing that was most likely to incite the Mactanese: "Burn their houses in order to terrify them," in Pigafetta's words. Predictably, "When they saw their houses burning, they were roused to greater fury."

The Europeans set fire to one house, driving fifty warriors armed with swords and shields from their hiding place into the open. "They charged down upon our men," said de Mafra, "striking them with their swords. In the midst of this skirmish, one of those heathens slashed a Galician [crew member] with his sword, cutting his thigh, and he later died as a result. Our men, wanting to avenge this, charged against the heathens, who beat a retreat, and as our men were chasing them, they came out of a path at the backs of our men, as if it had all been planned as an ambush, and, with earsplitting shouts, pounced on our men and began to kill them."

As the mayhem grew, the Europeans suffered more casualties. "Two of our men were killed near the houses, while we burned twenty or thirty houses," Pigafetta reported. Even their armor failed to protect the men against all the arrows flying in their direction. "So many of them charged down upon us that they shot the Captain General through the right leg with a poisoned arrow." It was only now, too late, that Magellan realized the gravity of his situation. He finally gave the order to retreat, even though his men were stranded

far from their longboats. More than forty of the Europeans scattered as best they could, while six or seven diehards, Pigafetta
included, stuck by the wounded Captain General as the Mactanese
pressed the attack: "The natives shot only at our legs, for the latter
were bare; and so many were the spears and stones that they hurled
at us, that we could offer no resistance. The mortars in the boats
could not aid us as they were too far away. So we continued to retire
for more than a good crossbow flight from the shore, always fighting
up to our knees in water. The natives continued to pursue us, and
picking up the same spear four or six times, hurled it at us again and
again. Recognizing the Captain General, so many turned on him
that they knocked his helmet off his head twice, but he always stood
firmly like a good knight, together with some others. Thus did we
fight for one hour, refusing to retire farther."

All this time, no one came to the aid of Magellan and his small
band fighting for their lives—no Cebuans in their *balanghai,* and no
reinforcements from the ships. Pigafetta explains that the "Christian
king"—faithful Humabon—"would have aided us, but the Captain
General charged him before we landed not to leave his *balanghai* but
to stay to see how we fought," an order that Humabon was only too
glad to obey. At last, some Cebuan converts showed up in their
balanghai, but by then it was too late. Friendly fire from the ships
felled many of them before they came to Magellan's aid; perhaps
the seamen mistook them for adversaries rather than allies. Meanwhile, Magellan was rapidly weakening from the effects of the poisoned arrow in his leg, as the implacable Mactanese closed in and
the two sides fought hand to hand.

"An Indian hurled a bamboo spear into the Captain General's
face, but the latter immediately killed him with his lance, which he
left in the Indian's body. Then, trying to lay his hand on his sword,
he could draw it out but halfway, because he had been wounded in
the arm with a bamboo spear. When the natives saw that, they all
hurled themselves upon him. One of them wounded him on the left
leg with a large cutlass, which resembles a scimitar, only larger."

The wounded leader "turned back many times to see whether we were all in the boats," Pigafetta took care to note, and without that concern, "Not a single one of us would have been saved in the boats, for while he was fighting, the others retired to the boats." Meanwhile, the scimitars' repeated blows took their mortal toll. "That caused the Captain General to fall face downward, when immediately they rushed upon him with iron and bamboo spears and with their cutlasses, until they killed our mirror, our light, our comfort, and our true guide. Thereupon, beholding him dead, we, wounded, retreated as best we could to the boats, which were already pulling off."

At that moment, the Cebuan warriors finally came to the Europeans' aid. They charged into the water, brandishing their swords, and drove off the Mactanese, who displayed little desire to make war on their neighbors. When the water had cleared, the Cebuans dragged the exhausted survivors into their *balanghai* and delivered them to the armada's longboats, which remained curiously distant from the scene of battle.

This was not the dignified, pious ending that Magellan had envisioned for himself during those pressured months of preparation in Seville. No paupers would say prayers in his memory, no alms would be distributed in his name, no masses would be said for him in the churches of Seville. Not one *maravedí* from his contested estate would go to his wife or young son, or to his illegitimate older son, who had been killed in battle at his side in Mactan harbor. He would not be buried in the tranquil Seville cemetery he had picked out for himself; none of the plans he had carefully set out in his will would come to pass. Instead, pieces of his body, driven by the winds and tide, washed up on the sands of Mactan.

*I*n Magellan's death, Pigafetta, who had fought at his side, saw a shining example of nobility, heroism, and glorious acceptance of fate. In the most emotional, eloquent entry of his entire diary, he

memorialized his slain leader, whom he had revered. "I hope that...
the fame of so noble a captain will not become effaced in our times.
Among the other virtues that he possessed, he was more constant
than anyone else in the greatest adversity. He endured hunger better than all the others, and more accurately than any man in the
world did he understand sea charts and navigation. And that his was
the truth was seen openly, for no other had had so much natural tal-
ent nor the boldness to learn how to circumnavigate the world, as he
had almost done...." *Almost*... perhaps the saddest and most telling
word in Pigafetta's eulogy.

"That battle was fought on Saturday, April 27, 1521," he concluded. "The Captain General died on a Saturday because it was the
day most holy to him. Eight of our men were killed with him, and
four Indians who had become Christians and who had come afterward to aid us were killed by the mortars of the boats. Of the
enemy, only fifteen were killed, while many of us were wounded."
The dead included Cristóvão Rebêlo, Magellan's illegitimate son
and constant companion on the voyage; Francisco Gómez, a seaman;
Antonio Gallego, a cabin boy; Juan de Torres, a man-at-arms;
Rodrigo Nieto, who had been Cartagena's servant but had switched
his loyalty to Magellan; and Anton de Escovar, who lingered for two
days after the battle.

Pigafetta's eulogy makes clear that he was genuinely devastated.
He had left Europe as a young man of literary inclination, eager to
explore the world as Magellan's guest, and now his Captain General
was dead, and the identity of his successor uncertain. What Pigafetta
had experienced of the world beyond Europe could only alarm him.
Instead of monsters, magnetic islands, boiling seas, and mermaids,
he had encountered fierce storms, cruelty and suffering, and widely
scattered humans living in conditions unimaginable to him, people
who were as likely to kill him as assist him. Most frustrating of all,
the armada had come all this way, halfway around the world, sacrificed dozens of men, including Magellan, and had yet to reach the
Spice Islands.

*I*n death, Magellan was not a hero to everyone, not even to those who had admired his daring and skill. His loyalists believed he had courted death by picking an unnecessary quarrel with the Mactanese, who held all the military advantages. In his misguided quest for glory, Magellan had squandered lives and the resources of the armada; his reckless conduct grieved other crew members, but more than that, it angered them. In de Mafra's judgment, Magellan's final campaign amounted to "madcap foolhardiness which the unfortunate Magellan attempted . . . when he could have done some much better things instead." In the name of King Charles, Magellan had pillaged and betrayed his hosts, and paid the ultimate price.

The circumstances leading to Magellan's spectacular, gory death were not, as has often been suggested, an aberration, the result of an unusual tactical error or inexplicable lapse of judgment. Rather, it was the direct outcome of his increasingly belligerent conduct in the Philippines, where he burned the dwellings of people who could easily have been converted to Christianity by diplomacy rather than force. Through frequent displays of his military might, Magellan convinced the islanders—and himself—that he was omnipotent. It was only a matter of time until he provoked a confrontation with enemies who held a decisive advantage from which faith alone could not protect him. His thirst for glory, under cover of religious zeal, led him fatally astray. In the course of the voyage, Magellan had managed to outwit death many times. He overcame natural hazards ranging from storms to scurvy, and human hazards in the form of mutinies. In the end, the only peril he could not survive was the greatest of all: himself.

Magellan's death may also have been the result of one final mutiny by his own disenchanted sailors. Although Pigafetta and other eyewitnesses provide a detailed account of the Captain General's actions during the fight in Mactan harbor, the whereabouts and actions of his backup is open to question—and to suspicion. During his amphibious landing, Magellan and his coterie expected

the gunners aboard his ships to cover them with fire that would dis-
perse the island warriors. Pigafetta, a gentleman, not a soldier or a
seaman, believed the tide made it impossible for their ships to
anchor close enough to the raging battle to be effective, but even
after several hours of fighting, they failed to dispatch reinforcements
in their longboats; indeed, the most striking element of Pigafetta's
account of the battle of Mactan concerns the inexplicable isolation
of Magellan and his small band. The Cebuans eventually intervened, but not Magellan's own men, a circumstance that makes no
sense, unless the crew members refused to come to the Captain
General's aid or their officers ordered them to stay put. From the
standpoint of the men in the ships, this mutiny had the advantage of
being easy to disguise; the revolt consisted of what they failed to do
rather than what they did. In effect, they allowed the Mactanese to
do the dirty work for them; they left Magellan to die the death of a
thousand cuts in Mactan harbor.

Antonio Pigafetta was among the few men of the armada who
saw the Captain General's death in a different, more glorious light.
He was no tyrant and engendered no anger or disloyalty; his end
embodied the Portuguese ideal of submission to fate, no matter how
tragic, in the service of a noble principle. Magellan seemed even
greater because he was doomed; he had become a martyr to a cause
greater than himself. Even so, the chronicler's own record tells a
more ambiguous story, one in which light and shadow are virtually
inseparable, and in which Magellan is both heroic and foolish, per-
spicacious yet blind, a man of his time who was trying to escape his
time, a visionary whose instincts outran his ideals.

Magellan was generally at his best, and a far more sympathetic
character, when he was the underdog. At such moments, his best
qualities came to the fore: tenacity, cunning, and courage. When his
plan to reach the Spice Islands was turned down by the king of Portugal—not once, but many times—Magellan successfully assembled and promoted a mission to the king of Spain. When mutineers
seized three of his ships in Port Saint Julian (and nearly captured a

fourth), Magellan immediately, and with little assistance from others, managed to reclaim the vessels, one by one, to end the mutiny. When his officers doubted the existence of the strait, Magellan found it; when they quailed at the prospect of entering the Pacific Ocean, he proceeded to sail into its roiling waters. And it took the massed forces of fifteen hundred men to kill him.

*A*fter the furious battle ended, the hacked pieces of the explorer's corpse drifted aimlessly in the water near the beach at Mactan until the victorious warriors claimed them. That afternoon, Magellan's distraught loyalists urged Humabon to send a message to Lapu Lapu, requesting the remains of Magellan and the other victims of the battle of Mactan; they even offered to pay as much as the victors wanted in exchange for the nine fallen soldiers.

Lapu Lapu's reply was shockingly arrogant: "They would not give him for all the riches in the world. . . . They intended to keep him as a memorial." That might have been the case, but nothing of Magellan was ever recovered, not even his armor.

*T*oday, in the Philippines, the tragic encounter between Magellan and Lapu Lapu is seen from a radically different perspective. Magellan is not regarded as a courageous explorer; instead, he is portrayed as an invader and a murderer. And Lapu Lapu has been romanticized beyond recognition. By far the most impressive sight in Mactan harbor today is a giant statue of Lapu Lapu, his bamboo spear at the ready, as he gazes protectively over the Pacific. There is no other record of Lapu Lapu or his reign; were it not for his battle with Magellan, his name would be lost to history.

Within the harbor, a white obelisk commemorates the ferocious battle between the Europeans and the Filipinos, and it offers two sharply varying accounts of the events. One face presents the European point of view: "Here on 27th April 1521 the great Portuguese navigator Hernando de Magallanes, in the service of the King of Spain, was slain by native Filipinos." The other portrays the conflict from the Filipino perspective: "Here on this spot the great chieftain Lapu Lapu repelled an attack by Ferdinand Magellan, killing him and sending his forces away." This version is naturally more popular in the Philippines, where the name Magellan is often regarded with loathing and even gloating at the circumstances of his death. Every April, Filipinos restage the battle of Mactan on the beach where it occurred, with the part of Lapu Lapu played by a film star, and Magellan by a professional soldier. Thousands turn out to witness the reenactment between the nearly naked Filipino warrior and the armor-clad invader who eventually falls face down into the surf.

Book Three

Back from the Dead

CHAPTER XI

Ship of Mutineers

The naked hulk alongside came,
And the twain were casting dice;
"The game is done! I've won! I've won!"
Quoth she, and whistles thrice.

*T*he officers and seamen of the armada had long anticipated Magellan's death. "As soon as the Captain General died," wrote Pigafetta, "the four men of our company, who had remained in the city to trade, had our goods brought to the ships." With the precious trading cargo—the bells, beads, and fabrics designed to entice islanders—safely stowed away, the survivors held an election to select the next admiral of the Armada de Molucca. They sought a man who would, above all, avoid high-risk endeavors similar to those that had endangered and claimed the lives of so many, and who would rededicate the fleet to its primary commercial goal: spices.

There was no discussion of disbanding the fleet or turning back. They had come too far and suffered too much for that. Nor was there any shortage of candidates to succeed Magellan; the ranks swelled with rivals and would-be admirals who had long been waiting for this moment. Although the loss of the Captain General was tragic—no one, not even his detractors, begrudged Magellan his courage—his death brought a palpable sense of relief that the ordeal of sailing under him had at last ended. When completed, the voting produced an unusual result, electing not one but two men: Duarte Barbosa, Magellan's brother-in-law, and Juan Serrano, the Castilian captain. Even now, the sailors maintained a balance of power between the Spanish and Portuguese presences in the fleet.

Even so, this careful outcome did not satisfy everyone. Sebastián Elcano, the Basque mariner who had played a leading role in the mutiny against Magellan at Port Saint Julian, believed that Serrano was miscast as co-commander. Elcano thought Serrano a competent pilot but nothing more. Implicit in Elcano's judgment was the conviction that he was better equipped to lead the expedition.

Magellan's loyal servant, Enrique, was even more bitterly opposed to the new leadership of the fleet. Enrique had rendered valuable service with his ability to interpret the Malay tongue, a skill now more necessary than ever, but he refused to leave *Trinidad* as ordered, claiming that he was suffering from battle wounds. He remained in his bunk, wrapped in a blanket, loudly proclaiming that he was free now that his master was dead. He was correct on this point; in the event of Magellan's death, his will provided for Enrique's liberty along with a bequest of 10,000 *maravedís*, but the new leaders of the expedition, accustomed to the slave's unquestioning subservience and still in need of his linguistic and diplomatic skills, insisted that he continue to obey orders. Enrique, coming into his own after years of servitude, stubbornly refused to yield to anyone's authority.

A loud argument between Enrique and Barbosa ensued, which Pigafetta recorded. "Duarte Barbosa, commander of the Captain General's flagship, told him in a loud voice that, although his master was dead, he would not be set free or released, but that, when we reached Spain, he would still be the slave of Madame Beatriz, the wife of the deceased Captain General. And he threatened that if he did not go ashore he would be driven away." Pigafetta's rendition of Barbosa's threats likely disguised a considerable amount of verbal abuse. But Sebastián Elcano left a more complete account of the confrontation. According to him, Serrano, and not Barbosa, abused Enrique. "Serrano, being unable to do anything without this inter-mediary, reprimanded him with bitter words, telling him that in spite of his Master, Magellan, being dead, he was still a slave and that he would be whipped if he did not obey everything that he [Serrano] commanded. The slave became enraged by Serrano's threats. Ire overtook his heart."

The harsh words succeeded in rousing Enrique from his stupor, and he furiously stalked off the ship.

Pigafetta believed that Enrique, on leaving the ship, sought out Humabon, "the Christian King," as he was called, to scheme against the armada, even though the Cebuan leader seemed to be the Euro-peans' staunch ally. On hearing of Magellan's death, he had wept copious tears, obviously undone by the tragedy he had tried so hard to avoid. Despite these strong emotional ties, Enrique "told the Christian King that we were about to depart immediately"—this much is true—"and that, if he would follow his advice, he would gain all our ships and merchandise. And so they plotted a conspir-acy. Then the slave returned to the ships, and he appeared to behave better than before."

Elcano told much the same story: Enrique "secretly spoke with the master of Cebu"—Humabon—"telling him that the Castilians were endlessly greedy and that . . . they would come back and arrest him." To Elcano's way of thinking, "The slave convinced the king that because the Castilians had been plotting against them, there was no other solution for the Cebuans than to plot back against the Castilians." With these arguments, Enrique launched his betrayal of Magellan's memory. Enrique's motives were powerful and probably quite complex. Perhaps he resented being a slave for his entire adult life, perhaps the rediscovery of his Filipino origins awakened long-suppressed feelings of loyalty and kinship, or perhaps he failed to realize the drastic effect his words would have on Humabon, who found himself in a desperate situation. Magellan, to whom he had been loyal, was dead, and the crew was about to depart, terminating the protection Humabon had enjoyed. In the absence of the armada, Humabon would have to contend with Lapu Lapu, whose victory over Magellan had emboldened the local chieftain. Because Humabon had sided with Magellan, it was only a matter of time before Lapu Lapu, seeking revenge, came after him. Pressure on Humabon to retaliate against the Europeans came from yet another direction. Many of the island men resented the way their women had been treated by the Europeans. For all these reasons, plotting against Magellan's men was the most effective way for Humabon to demonstrate his loyalty to his own people and save his own neck.

*O*n Wednesday, May 1, Humabon requested that the armada's leaders attend a feast. The invitation, presumably delivered orally by Enrique, promised a lavish meal accompanied by gifts of jewels and other presents, which Humabon wished the fleet to carry across the waters as tribute to the king of Spain. The Christian king hoped that as many people as possible would partake of his hospitality and generosity. In all, around thirty men, most of them officers, decided to accept.

This was a large contingent, approximately a quarter of the entire crew; their number included Barbosa and Serrano, the new co-commanders, as well as their astrologer and astronomer, Andrés de San Martín. Antonio Pigafetta was also invited to the feast, but, as he later explained, "I could not go, because I was all swollen from the wound of a poison arrow that I had received in the forehead." He had sustained his injury at Magellan's side, during the battle of Mactan.

This banquet promised to be another occasion for Humabon's guests to fill their bellies and get drunk on the island's palm wine. But shortly after the officers went ashore, Pigafetta, recovering on board *Trinidad,* was startled to hear João Lopes de Carvalho, the Portuguese pilot, and the master-at-arms, Gonzalo Gómez de Espinosa, returning unexpectedly. Pigafetta listened apprehensively as they "told us that they had seen the man cured by a miracle"—

the prince's brother healed by Magellan—"leading the priest into his house, and for this reason they had departed, fearing some evil chance." The sight of Father Valderrama entering a Cebuan hut hardly seemed sinister, but in this charged atmosphere it was enough to send the two Europeans scurrying back to the ships for safety.

*N*o sooner had those two spoken their words than we heard great cries and groans," said Pigafetta. "Then we quickly raised the anchors, and, firing several pieces of artillery at their houses, we approached nearer to shore."

What they saw exceeded their worst imaginings; it was worse, even, than the massacre of Juan de Solis. Ginés de Mafra, among those who had remained behind, described the murderous chaos engulfing the sailors on shore:

> As the banquet was about to end, some armed people emerged from the palm grove and attacked the invitees, killing twenty-seven of them, and captured the priest who had remained there and Juan Serrano, the pilot, who was an old man; others, although there were few of them, swam to the ships and, helped by those aboard, cut the cables and set sail; the barbarians, gorging on the killing and anxious to steal whatever was in the ships, brought their armada to the sea and, in order to stop our men while they were preparing to leave, also brought Juan Serrano to the shore and said that they wanted to exchange him for ransom. The old man implored our men with words and tears to feel sympathy for his old age and not to become accomplices, lest his last days end in the hands of such cruel barbarians, but to strive so that at least he could spend what little life he had left amidst his kin.
>
> Our men told him that they would do as they could. The ransom was discussed and they asked for an iron gun,

which is what they fear the most; this was sent to them on a skiff, and upon seeing it, the Indians asked for more, and no sooner would our men grant their request than the Indians would reply asking for more, and this continued until, realizing their intention, those aboard the ships did not want to remain there any longer and said to Juan Serrano that he himself could very well see what was going on, and how the Indians' words were all but a pretence.

Serrano pleaded for his crew members to come to his rescue, but they refused to leave the safety of their ships for fear that they would be massacred, as well. "Then Juan Serrano, weeping, said that as soon as we sailed he would be killed," Pigafetta recorded. "And he said that he prayed God that at the day of judgment he would demand the soul of his friend João Carvalho." Serrano's desperate words fell on deaf ears, and his friend Carvalho refused to intervene. Pigafetta was appalled by this cowardice, but there was nothing that he, as a supernumerary, could do.

Hoarse cries from the ships floated to land. Had the worst happened? Were the men ashore all dead? Could it be possible? Summoning his last reserves of strength, Serrano, stranded on the shore, confirmed that the other men, including Barbosa and San Martín, were dead, slaughtered during Humabon's banquet. He then watched the ships weigh anchor, preparing to abandon him to bloodthirsty island warriors seeking to reclaim their lost honor and dignity. "I do not know whether he is dead or alive," Pigafetta wrote in anguish as the ships sailed. Left behind by his own men, Serrano eventually met the same fate as his crew members. Enrique's revenge on the Europeans had been bloodier than anyone could have foreseen.

The three black ships of the Armada de Molucca raised anchor, set sail, and headed out of Cebu harbor with all the speed they could muster. No thought was given to sending a rescue party to stop the massacre, to recover bodies or search for survivors, or even to punish Enrique for his betrayal. Only 115 men remained of the 260 who had left Spain, and as they fled to safety, their last sight of Cebu was of enraged islanders tearing down the cross on the mountaintop and smashing it to bits.

*T*he May 1 massacre claimed many of the ablest and most prominent crew members. The victims included Duarte Barbosa, who had served as co-commander for just three days; Serrano; Andrés de San Martín, the fleet's cautious astrologer; Father Valderrama; Luis Alfonso de Gois, who had succeeded Barbosa as the captain of *Victoria;* two clerks named Sancho de Heredia and León Expeleta; a barrelmaker by the name of Francisco Martín; Simón de la Rochela, a provisioner; Francisco de Madrid, a man-at-arms; Hernando de Aguilar, who had been the servant of the mutineer Luis de Mendoza, whom Espinosa had executed; Guillermo Feneso, who operated the *lombardas;* four sailors; two cabin boys; three ordinary seamen; a servant attached to Serrano; and four men described in the roster as "servants of Magellan."

According to some accounts, eight of these crew members survived, but were imprisoned and sold off as slaves to the Chinese merchants who regularly visited Cebu, but the rumors were impossible to confirm. Enrique, whose treachery had set the stage for the ambush, disappears from history at this point, as does the wily Humabon. Such was the tragic conclusion of what had begun as a highly promising experiment in the Philippines.

*F*ive days later, and half a world away, a weatherbeaten vessel tied up at the harbor in Seville.

The arrival of a ship from distant lands was hardly an unusual event in Seville, but she was not just any vessel, this was *San Antonio,* part of the Armada de Molucca. It was May 6, 1521, and the event marked the first news of the fleet since it left Sanlúcar de Barrameda on September 20, 1519.

No one ashore knew what to make of her arrival, for the fleet had not been expected to return for months. They would soon learn that Magellan had found the fabled strait, after all, but before he could traverse it, *San Antonio* had been commandeered by mutineers fleeing Magellan's cruelty and excessive daring. She carried her captain, Estêvão Gomes; his chief co-conspirator, Gerónimo Guerra; and fifty-five men, including Magellan's cousin, Álvaro de Mesquita, whom the mutineers had stabbed and kept in irons throughout the return journey.

Gomes had skillfully piloted the ship across the Atlantic to Spain. There had been talk of returning to Port Saint Julian to rescue Cartagena and the priest whom Magellan had marooned, but in all likelihood the ship never attempted to rescue them. Instead, *San Antonio* made for the coast of Guinea to find water.

Despite having braved the Atlantic Ocean alone, *San Antonio's* captain and crew felt no joy on seeing Seville's familiar cathedral because they were returning in disgrace, mutineers who would face the prospect of an official inquiry, incarceration, and even punishment by death. They could console themselves in the knowledge that Guerra was related to Cristóbal de Haro, who had financed the expedition. They could also draw strength from Magellan's lack of popularity in Spain, and planned to destroy the Portuguese Captain General's reputation with tales of his poor judgment and brutal mistreatment of Spanish officers. But these stories had to be compelling because their lives depended on convincing the authorities that the mutiny had been necessary and justified. Of course, Magellan would not be present to plead his case or contradict their assertions. The only one likely to speak up on his behalf was Álvaro de Mesquita, whose wounds offered eloquent evidence of the mutineers' tactics. And Mesquita had used the long sea journey home to prepare for an inquiry because his life also depended on how persuasively he argued his case.

So began a fierce, tangled battle between the two competing versions of the mutiny.

The moment King Charles heard that *San Antonio* had returned, he ordered the Casa de Contratación to restore all the merchandise and equipment aboard the ship to Cristóbal de Haro, to whom Charles, perpetually strapped for cash to fund his empire, was deeply in debt. The Casa was to sell off anything of value "and after the sale," the king instructed, "send me an account . . . of what you have sold so that the said Cristóbal de Haro can make an accounting of it so that we can know what our share will be." Anything over ten thousand ducats would be remitted to the crown. The order crackled with the young king's eagerness to get his hands on the money, if there was any.

As it happened, there was none. The Casa's detailed inventory of the ship's contents listed tarnished combs, crumbling paper, rusty knives, scissors, bent sewing needles, beads, crystals, pearls, a velvet-covered chair, a bolt of decaying altar cloth for celebrating mass, iron, mercury, copper, an oven, a scale, pots, a green moth-eaten cloth, decaying barrels, two compasses, and a small bag of fishhooks, but no spices—nothing, in fact, of any great value. Furthermore, the ship was much the worse for wear after eighteen months at sea. The heat and humidity had taken their toll, to say nothing of the termites boring into the hull. Eventually, the authorities in Seville realized that *San Antonio* had not made it to the Spice Islands, after all. The king's dreams of claiming the Moluccas for the glory of Spain would have to wait.

No one aboard *San Antonio* knew what had happened to Magellan. They assumed—or perhaps they hoped—that his recklessness and secret loyalty to Portugal had caused him to perish at sea, somewhere over the edge of the world, and the Casa was inclined to believe them. "They believe he must have been double-dealing," a representative of the Casa reported to the king, "so they had no hope at all of his returning." The weather-beaten *San Antonio* and her mutinous, ragtag crew of fifty-five were presumed to be the only survivors of what had once been the glorious Armada de Molucca.

*W*ithin days of their return, the mutineers delivered their finely honed accounts to the Casa de Contratación. Fifty-three out of the fifty-five members of the crew gave depositions, and the sudden activity threw the Casa's clerks into a frenzy. "Since the morning of the Feast of the Ascension, we have been asking questions and tak-ing the declarations, without forcing them, in the presence of two clerks," wrote Juan López de Recalde, an accountant at the Casa, to Archbishop Fonseca on May 12, just six days after the ship's return. The task of gathering and squaring fifty-three separate accounts was exhausting and daunting. "We had with us the lawyer Castroverde, legal counsel of this House, and until last night, Saturday, for three days now, we have not been able to take the declarations of more than twenty-one of them. Half a day is needed to take down their account from the day they left this place till their return."

Mesquita, meanwhile, went directly from confinement aboard ship to jail on land. He was now "in the custody of the Admiral, where he is well-guarded." The Casa's representatives insisted they were only protecting Mesquita from the others, but the deposed captain believed he had been singled out for unfair treatment.

The Casa did a remarkably thorough job in uncovering the details of the mutiny at the strait in the allotted time. Their report included an elaborate description of the initial confrontations between Magellan and Cartagena shortly after the armada's depar-ture from the Canary Islands. There was even an account, inaccurate and inflammatory, of the homosexual behavior that had sent Magel-lan into a rage and sparked so much resentment among the crew: "It seems that on *Victoria* captained by Luis de Mendoza, a sailor attacked a cabin boy in an act contrary to nature and they told Ma-gellan about it. One calm day, he had the boy thrown into the sea."

As the report unfolded, a strong anti-Magellan bias became increasingly apparent. "It seems that the captains and officers, seeing that they were moving along the coast instead of going ahead to search for Cape Horn"—the southernmost promontory of South America—"decided to require Magellan to follow His Majesty's instructions, which were for them to continue their voyage with the agreement, counsel and opinions of the captains, officers and pilots of the armada." In fact, Magellan's orders were to "go in search of the strait," not Cape Horn, and despite what the mutineers later claimed, he had made a point of calling a formal conference and soliciting in writing the opinions of his captains and pilots, just as his orders required. Although he had not accepted their recommendation to turn back, he was not obliged to heed their advice. This was not a democracy, it was an armada, and he was the admiral.

Not surprisingly, the mutineers rearranged events at Port Saint Julian to suit their cause. To hear them tell it, they aroused Magellan to fury simply by asking him to obey the king's orders, or at least their interpretation of them. "One night Gaspar de Quesada with certain companions went from his ship, *Concepción*, to *San Antonio*, which was commanded by Álvaro de la Mesquita. He asked for the said Álvaro de la Mesquita and took him prisoner and told the men of the ship, in the presence of Juan Cartagena... that they already knew how Magellan had treated him [Cartagena]; and that Magellan would have him killed because he had asked Magellan to comply with the orders of His Majesty.... They demanded that Magellan comply with the orders of His Majesty; and for their having done this, not to be maltreated by him.... If he did this, they were and would be at his command." They would even, they claimed, "call him Your Lordship and kiss his feet and hands."

The mutineers delivered a wildly distorted rendition of the meeting to which they had tried to lure Magellan. In reality, Magellan had spurned their invitation to attend a conference aboard the rebel ship for fear of his life. But to hear the mutineers tell it, "Magellan sent word for them to come to his ship and that he would hear them out and do whatever was right. They replied that they dare not go to his ship for fear that he would punish them, that instead, he should come to *San Antonio,* where all of them could meet and they would do what he ordered them."

The mutineers remained oblivious to Magellan's successful effort to sabotage the revolt. In their account, Cartagena and Quesada ordered the rebel ships to sail out of Port Saint Julian, an act that meant confronting Magellan, whose flagship, *Trinidad,* blocked their path to freedom. "*San Antonio* raised two anchors and started to steady itself with one. Quesada agreed to set his prisoner Álvaro de Mesquita free and sent him to Magellan so there would be peace between them." This was fiction, as Mesquita knew, but the muti- neers invented still more incidents with Mesquita playing a critical part. For example, as the rebel ships sailed past the flagship, Mesquita supposedly asked Magellan not to fire on them so they could "iron out their differences, but before they could move from where they were, in the middle of the night while the men slept, the flagship fired heavy and light volleys at their ship." This was a good story, but the truth was that *San Antonio,* carried along by a power- ful current and dragging her anchor, had approached *Trinidad* quite unintentionally in the middle of the night because her cable had parted, not because Quesada had given the order to sail. The befud- dled mutineers were left telling tales of *San Antonio* somehow slip- ping past the flagship in the middle of the night... of Magellan's loyal cousin temporarily siding with the mutineers... of the leaders of the mutiny offering to kiss the hands and feet of the Captain General they obviously despised. None of this made sense unless their stories were seen for what they were: rather obvious attempts to exculpate themselves.

Inevitably, the mutineers recast the climactic struggle at the strait in their favor. In their version, Mesquita provoked the rebel- lion by stabbing Gomes in the leg, and Gomes retaliated by stabbing Mesquita's left hand. (In reality, of course, Gomes had stabbed

Mesquita first.) They also insisted that the trip home had been unspeakably difficult because each man was limited to a ration of three ounces of bread a day. This was another doubtful assertion because *San Antonio* carried provisions for the entire fleet, more than enough food to fill the mutineers' bellies.

While the mutineers spun their tales for the Casa's representatives, Gomes and Guerra were held in custody, as was Mesquita, despite his claims that he was the mutiny's principal victim, not its perpetrator. "We receive a thousand complaints every hour from them, insisting that they should not be imprisoned," Recalde complained, "that they be given a chance to see Your Majesty to tell Your Majesty what had happened in the said voyage." But they never got the chance. From his jail cell, Mesquita insisted, truthfully, that he had been tortured into signing a confession that he had tortured Spanish officers, that it was spurious, and that he had acted loyally to Magellan and the king of Spain. Nevertheless, suspicion fell more heavily on Mesquita than on anyone else.

Mesquita's account, so different from the mutineers' exculpatory version, received little attention and even less credence at the Casa de Contratación. In his defense, Mesquita presented the Casa with the documents he had kept when he presided over the prolonged mutiny trial in Port Saint Julian. The dossier recorded the rebellious actions of every accused crew member, the sentences they received, and Magellan's clemency, all to no avail. Mesquita was ordered to remain in prison, while the mutineers went free. The ringleaders, Gomes and Guerra, even had their travel expenses to and from court reimbursed, while Mesquita, considered guilty until proven otherwise, was ordered to pay travel costs out of his own, threadbare pocket.

*I*n their depositions, the crew members skillfully played on Spanish fears that Magellan was a Portuguese tyrant after all, a cunning agent of his native land who skillfully assembled the Armada de

Molucca at Spain's expense merely to destroy it and to dupe King Charles. They embellished this stereotype with fresh horrors: Magellan was a murderer who tortured honorable Spanish officers with connections in the highest possible place, the Church. They told the tragic tale of Cartagena—a Castilian officer!—who, through no fault of his own, was left to rot on a remote island by Magellan. As if that were not wicked enough, the Captain General left a priest to the same miserable fate.

This was an accomplished argument, but it was not flawless. For one thing, the mutineers had difficulty explaining why they had not rescued Cartagena as they retraced their route home. Fortunately, they generated enough shock and anti-Magellan hysteria in Seville that their inconsistent behavior was overlooked, for the present. Instead, the authorities focused on the accusation that Magellan had tortured loyal Spanish officers at Port Saint Julian, and not only abused them, but dismembered and disemboweled them, and placed his victims' heads on stakes.

*O*n May 26, Archbishop Fonseca—Cartagena's father—delivered his response to the depositions, and it became apparent that the mutineers' conspiracy to distort the truth had worked as planned. The bishop expressed shock and dismay at Magellan's treatment of Cartagena and Quesada. It seemed incredible that Spanish officers would be capable of mutiny, and there was no excuse for drawing and quartering one man and marooning the other. So the mutineers went free, for now, though a taint of suspicion clung to them, and they did not receive the back pay they claimed was due them. "We told the officers and seamen . . . to look for a means to earn a living without wasting more time," Recalde noted. "They have begun to look for work. We request Your Majesty to let us know what to do regarding said salary."

In Magellan's absence, his wife, Beatriz, became an object of suspicion, as if she were somehow involved with events at the other end of the world. The Casa de Contratación cut off her financial resources, and in a memorandum to the king suggested a convenient excuse for not paying her. "The wife of Ferdinand Magellan, as authorized by Your Majesty, has 50,000 *maravedís* in this House due the said Magellan as captain.... We doubt whether we should pay these claims considering the outcome of the voyage.... Inasmuch as we do not have the funds right now to pay them the first trimester of this year, we shall not pay them until Your Majesty advises us what to do."

The vindictive Archbishop Fonseca had even more punitive measures in store for Magellan's family. He ordered Beatriz and their young son to be placed under house arrest; they were forbidden to return to Portugal while the inquiry continued. Of course, she had no way of knowing that her husband had died only weeks before, on April 27, in the battle of Mactan, followed by her brother, Duarte Barbosa, who died on May 1 in the massacre at Cebu. And so throughout her captivity she waited, Penelope-like, for them to return home from their wanderings.

But Fonseca was almost as suspicious of the mutineers as he was of Magellan loyalists. He ordered Gomez, Guerra, and several other ringleaders to be brought to him in custody, insisting that they travel separately because they might continue to conspire. He told them he was making plans to send a caravel to Port Saint Julian to retrieve Cartagena and the priest. How the mutineers must have regretted their hasty decision to leave those two in the wilderness. Had Cartagena, who had always despised Magellan, returned to Spain, he would have done more than anyone else to blacken Magellan's reputation and to win vindication and even honor for the mutineers.

No one besides Mesquita spoke up on Magellan's behalf. The Spanish officials clearly planned to prevent him from returning in triumph to claim the lands, titles, and riches promised him by King Charles. But they had no way of knowing that their precautions were unnecessary, that Magellan was already dead. Mesquita, whose chief crime was being Magellan's cousin, remained confined in jail for another year, during which he frequently proclaimed his innocence, to no avail.

*T*he inquiry into the mutiny of *San Antonio* consumed six months, and in the end, Guerra and Gomes were set free in addition to all the sailors; Gomes even received a royal appointment to another expedition, a sure sign of rehabilitation.

Those who sided with Magellan fared much worse. His wife and son remained under house arrest, and now his father-in-law, the well-connected and prominent Diogo Barbosa, was ordered to give up property that Magellan had given to him before the fleet left Seville. This shabby treatment roused him to fury, and he spoke to the king in defense of Magellan's conduct during the mutiny: "He had to take great care so that it would be to your advantage and not against your honor," Barbosa said, pointing out that "when the men he brought with him mutinied with three of the principal ships, he did not punish them severely when he could have, and he pardoned many who later proved to be ungrateful." In addition, "The captain [Mesquita] was taken to Seville as a prisoner and later to Burgos until the time Your Majesty arrived in Spain. Prior to this, he was never given a chance to air his side, nor was he shown any justice." Barbosa recklessly lectured King Charles about the principles at stake. "These [events] serve as bad examples which discourage those who wish to do what they should and give greater encouragement to those who do otherwise." Barbosa was not fighting to clear Magellan's name alone; the official disgrace extended to Barbosa's daughter, his grandson, and to himself. For all their sakes, he offered a rousing if solitary defense of the Captain General, yet Barbosa's impassioned arguments in Magellan's defense came off as special pleading, and because they challenged Fonseca, worked against his own interests. As a Portuguese, Barbosa was seen as treasonous rather than honorable, and his star fell along with Magellan's.

One other Magellan loyalist, the brilliant but unstable cosmologist Ruy Faleiro, remained at large. After the Armada de Molucca had left Spain, he returned to Portugal, only to be imprisoned. He suffered a breakdown in jail, but he eventually regained his strength and was released. He then returned in secret to Seville, where he gained some sympathy at the Casa by displaying the marks made by the shackles he had worn during his time in prison. Out of pity, and to keep him away from Portugal, where he might have some value, the Casa gave him (and his brother Francisco) separation pay "because they had arrived worn out and penniless from Portugal; besides, they are here by order of Your Majesty." Faleiro, the prime mover behind the Armada de Molucca, lived out his days in obscurity.

During the clamor arising from the unexpected return of the mutineers, not a word was heard from the young king who had authorized the expedition two years earlier, despite the petitions and correspondence pleading for his attention. Charles had not lost interest in the enterprise, but ever since the ships had sailed from Seville, he had been engulfed in political turmoil. His mother, Juana the Mad, lived on, hopelessly insane. She was said to have kept the body of her late husband, Philip I, the Handsome, who died suddenly at the age of twenty-eight, next to her bed for years in the belief that he would eventually return to life on the anniversary of his death. After his death, she always wore black and refused to clean herself. Meanwhile, the young monarch, encouraged by his backers, was still making every effort to become the next emperor of the Holy Roman Empire, the most powerful political entity in Europe.

The Holy Roman Empire was founded on Christmas Day, A.D. 800, when Charles the Great (Charlemagne), the ruler of the Franks, an affiliation of Germanic kingdoms, was crowned emperor. His coronation unified France, much of Germany, the Netherlands, Belgium and Luxembourg, and northern Italy. Although Charle-

magne's line of male descendants died out within a century, he was
an ancestor of many European ruling dynasties. Over time, the
Holy Roman Empire became so fragmented that by the eighteenth
century Voltaire remarked that it was "neither holy, nor Roman, nor
an empire." Nevertheless, it survived.

The Holy Roman Empire was an elective monarchy; its German
electors had the authority both to appoint the emperor and to control his actions thereafter. Charles's grandfather, Maximilian, while
emperor had elicited promises from the seven German electors to
appoint the boy as emperor, but promises alone were not enough to
assure Charles's succession. He faced competition from the king of
France, Francis I, who was eager to make a reputation, especially at
Spain's expense. It was true that Charles belonged to the House of
Hapsburg, which traditionally ruled the Holy Roman Empire, but
he needed money, lots of it, to clinch the deal. Charles had to pay
bribes, thinly disguised as tributes, to the electors and to representatives of the papacy if he wished to secure the title. Lacking
resources of his own, he borrowed heavily from various banking
houses, permanently placing himself in their debt. He eventually
paid the electors an astounding figure, 850,000 ducats, of which
540,000 ducats came from loans arranged with the Fugger banking
dynasty. Thus Charles was borrowing from German bankers to pay
German electors to win his largely German title, "Emperor of the
Holy Roman Empire." The Germans made a fortune from Charles's
imperial ambitions, and he expected Spain to pay the bill, incurring
wrath from one end of the Iberian peninsula to the other.

To complete his quest to become emperor of the Holy Roman
Empire, Charles required the blessing of Leo X, the Medici family
pope whose excesses helped to inspire the Reformation. According
to popular mythology, he was a free-spending libertine, but
Raphael's renowned portrait of Leo X, painted in 1518, depicts a
very different image, that of a pudgy, thoughtful scholar and aes-
thete averting his saturnine expression from the viewer. With his
puffy face and massive, fleshy nose, he presents an altogether

homely and unprepossessing figure, flanked by two young cardinals, who stand directly behind him, uncomfortably crowding him. Although all three are swathed in luxury, in damasks and velvets and silks, they look at odds with each other, as if their robes concealed sharp weapons. Raphael's portrait reflected a difficult and divisive time in Rome. The year before, Leo X had uncovered a plot among the younger cardinals to poison him. Cardinal Petrucci, who admitted to knowledge of the plot, was strangled in prison, and the other conspirators were exiled or executed. No wonder Raphael's portrait showed a careworn, abstracted Leo X surrounded by menacing cardinals.

There was another side to Leo X. When not presiding at Church functions, he displayed a sense of bonhomie, rich laughter, and an addiction to theater and music and art, and other secular pleasures such as banquets and hunting. "Let us enjoy the papacy, since God has given it to us," he declared. He dispensed papal largesse to his entourage without regard to the dwindling papal treasury. Leo X then tried to raise money with the same lack of discipline, indiscriminately selling titles, favors, and indulgences, the latter popularly understood as the promise of avoiding Hell in the afterlife, given in exchange for donations.

To discontented outsiders, the Church deteriorated into a spectacle of corruption, selfishness, and arrogance. In 1520, Martin Luther, in Wittenberg, Germany, wrote a furious and menacing letter to Pope Leo X. "Among those monstrous evils of this age," he wrote, "I am sometimes compelled to look to you and call you to mind, most blessed father Leo." Under the influence of Pope Leo X, the Church of Rome, "formerly the most holy of all Churches, has become the most lawless den of thieves, the most shameless of all brothels," Martin Luther wrote, and many agreed. "Not even the antichrist, if he were to come, could devise any addition to its wickedness." He ranted in this vein for many pages, inciting others to follow his example. The Reformation was in full cry.

In theory, the beleaguered Leo X could draw additional support—and funds—from the Holy Roman Empire, but that con-glomeration was in disarray. After the death of Maximilian, Leo X nominally supported Francis I, the king of France, over King Charles, but in reality the pope skillfully played one candidate against the other. More zealous and better funded, King Charles ultimately prevailed, and the pope reluctantly threw his weight behind the young man who had suddenly arrived at the summit of power in Europe. Or was it a precipice?

On July 28, 1519, less than a month before Magellan's fleet sailed down the Guadalquivir River to the Atlantic, King Charles, then in Barcelona, learned that he had been elected emperor of the Holy Roman Empire, but the title would not belong to him until he paid for it. He had counted on Spanish nobles for financial support, but they turned their backs on him. He remained in Europe, raising funds, and finally, on October 23, 1520, in the ancient city of Aachen, Germany, from which Charlemagne had once ruled the empire, Charles, now twenty-one years old, was crowned emperor. The occasion marked the formal alliance between a hesitant, cash-starved pope under siege from the forces of the Reformation and an untested, cash-starved monarch.

In Spain the nobility resented Charles even more fiercely now that he was emperor of the Holy Roman Empire. Despite his prom-ise not to appoint foreigners to government posts in Spain, Charles selected his former tutor, Adrian of Utrecht, as regent, and the choice confirmed the nobles' fears that Charles was essentially a German interloper plunked down in their midst. The city of Toledo responded by expelling its *corregidor,* as the royal administrative executive was known, and the Revolt of the Castilian Comuneros was under way. Cities and towns across Spain, Madrid and Sala-manca among them, joined in the Junta Santa de las Communidades to return political power to Spain. They showed their determination by raising militias that marched on Tordesillas, where they placed their trust in King Charles's mother, the mad Queen Juana, but she refused to come out of seclusion to offer support or even to sign a document expressing their grievances.

The insurrection spawned a counterrevolution in rural areas among those who despised the nobles; they now turned to King Charles for protection. He eagerly sought their support, promised to indemnify them against losses they incurred while fighting the rebellious nobles, and consented to appoint two Castilian noblemen to serve alongside Adrian of Utrecht as co-regents. He also showered titles and dukedoms on those who rallied to his side, and managed to bring the recalcitrant nobles around. Despite these victories, King Charles's position in Spain remained hotly contested as alliances between the *comuneros* and the royalists shifted constantly. Desperate to shore up his empire, King Charles paid scant attention to the controversy surrounding a rogue ship tied up in Seville. He remained abroad until July 1522, and in his absence Spain struggled to redefine itself as a nation and as part of the Holy Roman Empire.

Seville, the center of Spanish commerce, reflected the tensions afflicting the rest of the country and developed a reputation as a city in crisis. Criminal behavior flourished in the streets and alleys of its shabbier neighborhoods. Triana, the suburb across the Guadalquivir River, served as home to many underworld types, as well as to the sailors who manned Spanish ships. Gypsies, slaves, palm readers, beggars, itinerant thespians, and minstrels populated a rapidly expanding underworld. In time, its ranks came to include defrocked clergy, destitute nobles, and unemployed soldiers, as well as an assortment of con artists and dealers in questionable merchandise. With goods flowing into Seville from Africa and across Europe, smuggling became a major enterprise; the value of smuggled goods far outstripped that of legitimate merchandise. Chronically unemployed people masqueraded as disabled beggars; it was often difficult for their victims to distinguish them from mendicant orders of monks. Knife fights were common throughout Seville, as were bribery and prostitution. Each year, eighteen thousand prisoners entered the gates of the Royal Prison, further stressing the city's already burdened economy.

Meanwhile, Seville's titled oligarchy fattened itself with income derived from leasing lands to farmers or cashed in on their titles and prestige to engage in commercial pursuits, importing wine, oil, and soap. With the profits, they constructed impressive castles, gardens, and ravishing courtyards. Throughout Spain, Seville's wealthy nobility was renowned and envied even as the city's criminals were feared.

These two disparate sides of Seville met at the docks, where wealthy merchants jostled with sailors and dishonest middlemen seeking merchandise to peddle. Amid the chaos on the banks of the Guadalquivir, *San Antonio,* now stripped of her rigging and fittings, rode at anchor, a mute but eloquent witness to an expedition gone awry. In Seville no one knew that the Armada de Molucca had successfully navigated the strait, or crossed the immense Pacific Ocean. No one realized how close the survivors were to their ultimate goal, the Spice Islands. Everyone—from King Charles to the bureaucrats in the Casa de Contratación to the recently freed sailors looking for their next ship—assumed that the fleet was lost and the expedition a complete failure.

Everyone was wrong.

CHAPTER XII

Survivors

The helmsman steered, the ship moved on;
Yet never a breeze up-blew;
The mariners all 'gan work the ropes,
Where they were wont to do;
They raised their limbs like lifeless tools—
We were a ghastly crew.

*T*en thousand miles from Spain, in a remote corner of the Philippine archipelago, a ship was burning. The blaze turned night into day, and its reflection formed hypnotic patterns on the inky, swelling sea. As it hissed and sent a pungent vapor skyward, the blaze consumed the ship's timbers down to the water's edge. The dull red glow from the waterborne bonfire was visible for miles around. The next morning, thick smoke from the dying embers of the charred hull turned day into night.

The ship was *Concepción*, one of the three vessels that had escaped the massacre at Cebu the previous day. Since then, the survivors had tried to navigate the three large vessels around the uncharted shoals and islands of the Philippines, but they soon discovered that they were hopelessly shorthanded. To add to their problems, *Concepción's* master, Juan Sebastián Elcano, complained that shipworms infested the hull. Magellan, had he been alive, would have ordered the men to undertake arduous repairs, but the survivors adopted a more pragmatic approach and decided to burn the ship to prevent it from falling into the hands of an enemy who might use it against them. The crew transferred the contents of *Concepción*—her provisions, rigging, sails, fittings, weapons, and navigational devices—to the two other ships, *Trinidad*, still the flagship of the fleet, and *Victoria*. And then, on the night of May 2, 1521, the empty ship was set ablaze in symbolic, and wholly unconscious, expiation of the fleet's sins.

A hasty vote among the sailors placed Espinosa in command of *Victoria*, while João Lopes Carvalho, the Portuguese pilot, won election as the new Captain General. Elcano, the master of *Victoria*, silently cursed the new Captain General, who might be a talented pilot but was incapable of imposing discipline on the unruly fleet. In Brazil, Carvalho had attempted to bring his mistress on board; although he did not succeed, their child had been traveling with the fleet ever since. Elcano had no respect for a leader who set such a poor example for the others.

The new command placed Pigafetta in a vulnerable position. He had always identified himself as a Magellan loyalist, but the Captain General's inner circle—his slave, Enrique; his illegitimate son, Cristóvão Rebêlo; his cousin, Álvaro de Mesquita; and his brother-in-law, Duarte Barbosa—had all perished or disappeared. Only Pigafetta survived. He believed he would continue to serve as the expedition's chief chronicler, as well as its chief interpreter, because he alone had troubled to make a methodical study of the Malay tongue. He lacked Enrique's facility with it, but he knew how to make himself understood and obtain information. Equally important, he was familiar with Filipino customs, ranging from *casicasi* to *palang*, and could make himself useful as the expedition's emissary to the strange and changeable islanders all around them. Carvalho and the newly elected leaders of the expedition agreed, and Pigafetta's role in the post-Magellan era was, if anything, enhanced.

As for his diary, he continued to maintain it, and to keep its contents to himself.

After the multiple tragedies the armada had suffered in the Philippines, commercial considerations ruled their actions. Never again would they erect crosses or insist on mass conversions. Everything was different now. Knowing they were lucky to be alive, the men turned their attention to reaching the Spice Islands, where they hoped to find safety, supplies, and the precious commodity they had sailed halfway around the world to find.

Carvalho faced the task of leading the fleet's two remaining ships southward through the archipelago to the Moluccas, but the arrival of the rainy season in the Philippines and its storms often made navigation next to impossible. They had adapted to sailing over vast stretches of open water, but now they had to thread their way through a labyrinth of islands. For the short distances and intricate maneuvering involved, they needed a reliable map or, failing that, a guide familiar with these waters, but after their horrific experiences on Cebu and Mactan, the sailors were reluctant to call at strange islands and ask for help. Who could guess the real intentions of the islanders lurking in the shadows of the palm trees?

Occasionally, the fleet was approached by *balanghai* powered by rowers chanting in unison. Whenever possible, Pigafetta asked the rowers for directions to the Moluccas, but the others kept their relations with the islanders to a bare minimum.

*C*arvalho, aided by Albo, the pilot, veered from one island to another, following a meandering but generally southerly course through the labyrinth of the Philippine archipelago to the Moluccas. Albo's methodical record, barely mentioning the ambush at Cebu, tracked the fleet's wanderings, as if the ships were wounded beasts in search of a healing sanctuary.

They soon encountered an island populated by Negritos, aboriginal pygmies with dark skin, as their name indicates. After an unsuccess hunt for food, the fleet approached a towering island clad in dense foliage cut by steep channels and waterfalls flowing from hidden springs. Here and there the shore suddenly cleared to offer an inviting, if narrow, stretch of beach. This was Mindanao. The idyllic setting soothed the chastened yet hard-bitten crew, who dropped their guard long enough to establish friendly relations with a local ruler named Calanoa, who appeared eager to make peace. Calanoa, Pigafetta wrote, "drew blood from his left hand marking his body face, and the tip of his tongue with it as a token of closest friendship, and we did the same." Despite his offer of friendship, he was unable, or unwilling, to feed the crew.

After the ceremony, Calanoa invited Pigafetta ashore as a sign of respect, but Pigafetta does not explain why he alone received this honor. Perhaps his facility with the Malay language had impressed the chieftain, or perhaps the invitation gave him an opportunity to prove his usefulness to Carvalho and the other leaders of the expedition. Pigafetta boldly accepted the invitation, even after witnessing the recent massacre. One explanation for Pigafetta's sudden courage might be that Calanoa had put him at ease; another might be that he had no intention of returning to the fleet, that he had seen enough of death and disaster at sea and preferred to live out his days as an honored guest among the islanders and, especially, their beautiful women.

"We had no sooner entered a river than many fishermen offered fish to the king"—so food was available after all. "Then the king removed the clothes which covered his privies, as did some of his chiefs; and began to row while singing past many dwellings which were upon the river. Two hours after nightfall we reached the king's house. The distance from the beginning of the river where our ships were to the king's house was two leagues." Isolated from his crew mates, Pigafetta was now at the mercy of his hosts, but if he felt fear, he left no trace in his diary.

"When we entered the house, we came upon many torches of cane and palm leaves," he continued. "The king with two of his

chiefs and two of his beautiful women drank the contents of a large
jar of palm wine without eating anything. I, excusing myself as I
supped, would only drink but once." It was a scene familiar to
Pigafetta, the drinking, and feasting, and women; he might have
been back on Limasawa, in the days before the massacre. At his
ease, and inquisitive as ever, he observed food preparations: "They
first put in an earthen jar . . . a large leaf lining the entire jar. Then
they add the water and the rice, and after covering it allow it to boil
until the rice becomes as hard as bread, when it is taken out in
pieces." (In recording this recipe, Pigafetta became the West's first
guide to Oceanic cuisine.) After the meal, the chieftain offered
Pigafetta two mats for sleeping, one fashioned of reeds, the other of
palm leaves. "The king and his two women went to sleep in a separate place, while I slept with one of the chiefs."

In the morning, Pigafetta explored the island, devoting special
attention to huts, whose fittings gleamed with gold. Gold seemed to
be on display everywhere; there was, he said, "an abundance of gold.
They showed us certain small valleys, making signs to us that there
was as much gold there as they had hairs, but that they had no iron
or tools to mine it, and moreover that they would not take the trouble to do so."

Over a midday meal of rice and fish, Pigafetta courteously asked
Calanoa for an audience with the queen. The chieftain agreed, and
the two of them trudged up a steep hill to pay their respects to her.
"When I entered the house, I made a bow to the queen, and she did
the same to me, whereupon I sat down beside her. She was making a
sleeping mat of palm leaves. In the house there were a number of
porcelain jars and four bells . . . for ringing. Many male and female
slaves who served her were there."

If Pigafetta had ever considered seeking refuge on this island
with its abundant gold, the temptation waned. After his audience
with the queen, he clambered aboard a waiting *balanghai,* along with
the chieftain and his retinue, and they glided along the serene river
toward the ocean. When he least expected it, the tranquil surround-

ings were disturbed by an appalling spectacle: "I perceived to the right, on a small hill, three men hanging from a tree which had its branches cut off." Once again, he was struck by the stark contrast between the splendor of the setting, the peaceful, generous, and open nature of the inhabitants, and the macabre reminders of brutality that lurked just out of sight. Who were these people, Pigafetta asked, and why did they meet such a gruesome ending?

"Malefactors and thieves," Calanoa grimly explained.

The *balanghai* approached *Trinidad,* and Pigafetta bid farewell to his hosts and rejoined the fleet. It had been, over all, a pleasant interlude, with the exception of the nightmarish vision of the men hanging from bare trees.

*S*till unable to pinpoint the Spice Islands, the fleet weighed anchor "and laying our course west southwest, we cast anchor at an island not very large and almost uninhabited." They were veering seriously off course, heading west into the Sulu Sea, toward China, rather than south to the Spice Islands. In its wanderings, the fleet called on the island of Caghaian, as Pigafetta designated it. Once again, he enthusiastically went ashore to establish relations with the islanders, but this time other crew members accompanied him. Their mission: to find enough food to restock their rapidly dwindling stores before they starved.

Only a short distance from their previous anchorage, the fleet encountered a far more predatory culture. "The people of that island are Moros"—Moors—"and were banished from an island called Burne"—Borneo. "They go naked as do the others. They have blowpipes and small quivers at their side, full of arrows and a poisonous herb. They have gold daggers whose hafts are adorned with gold and precious gems, spears, bucklers, and small cuirasses of buffalo horn." Fortunately, these menacing-looking warriors believed that the European intruders were "holy beings" and spared them from harm. But the ravenous sailors found no food to speak of and, growing desperate, the armada embarked on a twenty-five-league detour to the northwest, almost directly away from the Spice Islands.

The search for food grew more frantic. "We were often on the point of abandoning the ships in order that we might not die of hunger," Pigafetta wrote. At last they arrived at "the land of prom-ise, because we suffered great hunger before we found it." The island was called Palawan, and it divides the Sulu Sea from the South China Sea. Although the fleet was getting even farther from its goal, Palawan offered a tropical paradise to men who had endured so much for so long. "The winds are mild, the sun warm, the sea teem-ing with fish," wrote Samuel Eliot Morison of the island. "The land is so fertile that for more than half a year, after the main crops are gathered, people have nothing to do but enjoy themselves."

Their stomachs growling and their heads spinning from fatigue and hunger, the sailors rushed through another *casicasi* ceremony with the local chieftain and then gorged themselves with "rice, gin-ger, swine, goats, fowls," and "figs…as thick as the arm." Pigafetta declared these "figs," actually bananas, to be "excellent" fare. That was not all; the grateful crew members also sated themselves with coconuts, sugarcane, and "roots resembling turnips in taste." Pigafetta pronounced their wine, distilled from rice, to be exceed-ingly light and refreshing, far superior to the rough palm brew they had been drinking for weeks. Hours before, they had been so des-perate that they contemplated the prospect of relinquishing the safety of their ships to forage for food. Now they offered thanks to God for saving them from starvation.

When he had filled his belly, Pigafetta once again became an amateur anthropologist. He charmed his island hosts into displaying their exotic weapons for him: "They have blowpipes with thick wooden arrows more than one palmo long, with harpoon points, and others tipped with fishbones, and poisoned with an herb; while oth-ers are tipped with points of bamboo like harpoons and are poi-soned. At the end of the arrow they attach a little piece of soft wood, instead of feathers. At the end of their blowpipes they fasten a bit of iron like a spear head; and when they have shot all their arrows they fight with that." In this culture, Pigafetta found, the fascination with combat included their animals. "They have large and very tame cocks, which they do not eat because of a certain veneration they have for them. Sometimes they make them fight with one another, and each one puts up a certain amount on his cock, and the prize goes to him whose cock is the victor." The more closely he looked at cultures like these, the more he began to see disturbingly familiar suggestions of his own.

When the crew had rested and loaded provisions onto the ships—provisions for which their weeks in the Pacific had taught them to barter skillfully—they weighed anchor, and on June 21, 1521, prepared to leave Palawan. This time, they had on board a local pilot, a Negrito who gave his name as Bastião, and said he was a Christian, but he vanished just before the fleet left the harbor. In search of a replacement, Carvalho ordered the fleet to encircle a large *balanghai*. Feigning peaceful intentions, the armada captured all three of the *balanghai*'s pilots, believing that they would lead the way to the Spice Islands at last, but these pilots—all Arabs—complicated matters by directing the armada southwest, toward Brunei, an Arab stronghold, rather than southeast, toward the Moluccas.

This was a hazardous crossing, replete with shoals and sandbanks, and the fleet needed the pilots' assistance to reach Brunei. Even Albo, the resolute pilot, became agitated on this leg of the journey. "You must know that it is necessary to go close to land, because outside there are many shoals," he complained in a rare outburst, "and it is necessary to go with the sounding lead in your hand, because it is a very vile coast, and Brunei is a large city, and has a very large bay, and inside it and without it there are many shoals; it is necessary to have a pilot of the country." Reaching the mouth of the harbor, the fleet followed junks whose pilots were familiar with the route to safety. At last, they dropped anchor in the harbor of Brunei, in the midst of a realm of enchantment and luxury that would surpass anything they had previously experienced on the voyage.

*T*he next day, July 9, what appeared to be a *proa* appeared on the horizon, but as it approached, the crew realized it was a much larger vessel "whose bow and stern were worked in gold. At the bow flew a white and blue banner surmounted with peacock feathers." Trailing the ornamental *proa* were two smaller vessels. To add to the theatri-cal nature of the scene, musicians on board serenaded the shocked Europeans. "Some of the men were playing on musical instruments and drums," Pigafetta noted in disbelief.

The *proa*'s crew signaled with elaborate gestures that they wished to board, and "eight old men, who were chiefs, entered the ships and took seats in the stern upon a carpet. They presented us with a painted wooden jar full of betel and areca (the fruit which they chew continually), and jasmine"—a shrub whose white and yellow flowers released a soft, almost cloying scent into the sea-permeated air—as well as orange blossoms, whose sweet intoxicating perfume the crew members had not sampled since Seville. The old chiefs brought much more: bolts of yellow silk cloth, two cages filled with flapping fowl, jars filled with sublime rice wine, and bundles of sugarcane. After depositing their offerings aboard *Trinidad,* the chiefs did the same with *Victoria.*

Their generosity toward the armada likely stemmed from a case of mistaken identity. Most of these regions had been visited by the Portuguese, who, traveling a different route, had pioneered trading relationships with the local Arab rulers. Ginés de Mafra described the rajah of Brunei as a "friend of the Portuguese and an enemy to the Castilian, whom he hates." That made the Armada de Molucca an interloper, but many of the crew were Portuguese and appeared to be the latest emissaries of the Portuguese crown.

That night, the men, craving distraction from their trials, tasted the local rice wine, found it to their liking, and drank themselves into oblivion.

*T*he fleet remained anchored off Brunei for six peaceful days, allowing the men to recover, at least partially, from the violence that had marked the last few weeks. From the decks of their ships, the men could see an assortment of elevated houses constructed over a complicated series of waterways, piers, and boardwalks. Behind the city, tall palms stood as sentries. At night, dim fires flickered in the distance and sent slender plumes skyward. If the sailors listened carefully, they could hear faint voices from the shore echoing across the surface of the water, or even a kind of primitive music consisting of gongs and bells and chanting. It was a scene of domestic tranquility transplanted to an exotic setting, but the men were afraid to leave their ships and explore the unknown.

The fleet's isolation ended when their benefactor dispatched a convoy of *proas* to beguile and seduce them. Arriving "with great pomp," Pigafetta wrote, they "encircled the ships with musical instruments playing and drums and brass gongs beating. They saluted us with their peculiar cloth caps which cover only the top of their heads. We saluted them by firing our mortars without stones [bullets]. Then they gave us a present of various kinds of food, made only of rice. Some were wrapped in leaves and were made in somewhat longish pieces, some resembled sugarloaves, while others were made in the manner of tarts with eggs and honey. They told us that their king was willing to let us get water and wood, and to trade at our pleasure."

The king's messenger promised to help them with all their needs. "The messenger was an old man," de Mafra recalled, "handsome and well-dressed. He wore gold jewelry on his fingers, neck, and ears." He wanted to know where they were going, and when they spoke of the Moluccas, he scoffed; there was nothing there but cloves, he advised, but if they were determined to go, he would supply a pilot for each ship. "For this our men thanked him, and then asked whether there was in that land any pitch with which to caulk the ships." After months in tropical water, the hulls badly needed reconditioning. The messenger explained that "they caulked their own boats with a pitch they made with coconut oil and wax, for which they could send out some people to town, where they could find many things to buy." And again he invited the men to stay awhile and sample the pleasures of Brunei.

The repeated entreaties from the mysterious island ruler eventually had their intended effect, and the crew members reciprocated by sending a delegation consisting of Gonzalo Gómez de Espinosa, the master-at-arms, still on the job; Elcano, the would-be Captain General; two Greek sailors; Carvalho's illegitimate Brazilian son; Pigafetta; and one other sailor. The delegation transferred from *Trinidad* to the *proa*, bearing gifts salvaged from the wreckage of the fleet: "a green velvet robe made in the Turkish manner, a violet velvet chair, five *brazas* of cloth, a cap, a gilded drinking glass, a covered glass vase, three writing books of paper, and gilded writing case." The crew thoughtfully brought along separate tributes for the queen, should there be one: "Three *brazas* of yellow cloth, a pair of silvered shoes, and a silvered needle case full of needles."

After a short trip over water, the delegation reached an elaborate city "entirely built in salt water," said Pigafetta, "except the houses of the king and certain chiefs. It contains twenty-five thousand fires"—that is to say, hearths indicating family units. "The houses are all constructed of wood and built up from the ground on tall pillars. When the tide is high, the women go in boats through the settlement, selling the articles necessary to maintain life. There is a large brick wall in front of the king's house with towers like a fort, in which were mounted fifty-six bronze pieces, and six of iron." The gunpowder for these weapons was likely imported from China,

where it was invented. After months of drifting among more primitive (though none the less dangerous) tribes, the armada had finally
reached a civilization at least as advanced as their own.

After waiting in the *proa* for two hours, Pigafetta, Elcano, and the
others were rewarded with the spectacle of "two elephants with silk
trappings, and twelve men, each of whom carried a porcelain jar
covered with silk in which to carry our presents." The members of
the delegation were invited to mount the elephants, and from their
swaying eyries, they surveyed the landscape. Their grins can be readily imagined. The elephants lurched forward, carrying the members
of the armada toward the dwelling of the "governor," while "twelve
men preceded us afoot with the presents in the jars."

Reaching their destination, the elephants knelt, discharging their
astonished passengers, who immediately sat down to a great feast.
After they ate and drank their way into a state of pleasant stupefac-
tion, they were treated to "cotton mattresses, whose lining was of
taffeta and the sheets of Cambaia." It was the first night the men had
slept on mattresses and linen since they had left Seville, but few
remained awake long enough to savor the sublime comfort, because
they fell into a deep sleep. As they slept, servants constantly tended
large candles fashioned from white wax and oil lamps, adjusting the
wicks and finally snuffing them when the sun rose.

*A*t noon the next day, the men awoke and remounted the elephants and proceeded to the king's palace, while onlookers treated
them with a respect reserved for great dignitaries. "All the streets
from the governor's to the king's house were full of men with swords,
spears, and shields, for such were the king's orders." Dismounting,
they passed through a courtyard to a "large hall full of many nobles,"
perhaps as many as three hundred, and came upon an extraordinary
scene: "We sat down upon a carpet with the presents in the jars near
us. At the end of that hall there is another hall higher but somewhat
smaller. It was all adorned with silk hangings, and two windows,

through which light entered the hall . . . opened from it. There were three hundred foot soldiers with naked rapiers at their thighs to guard the king. At the end of the small hall was a large window from which a brocade curtain was drawn aside so that we could see within it the king seated at a table with one of his young sons, chewing betel. No one but women were behind him." They were cautioned not to speak directly to the king. Should they wish to say anything, they were to inform a servant, who would pass it on to a functionary of slightly higher rank, who would then tell the governor's brother, who would in turn whisper the message through a "speaking-tube" passing through the wall, where another servant would intercept it and relay it to the king. As if that were not sufficiently off-putting, they were instructed to kowtow. "The chief taught us the manner of making three obeisances to the king with our hands clasped above the head, raising first one foot and then the other and then kissing the hands toward him, and we did so, that being the method of the royal obeisance."

Once they had completed these formalities, Pigafetta explained that they wished only to make peace and to trade. The king, through his intermediaries, happily cooperated. Take water and wood, he offered, trade as you wish, and he ordered his minions to place a cloth made of gold and silk brocade on his visitors' shoulders. For a moment, they resembled their hosts, "all attired in cloth of gold and silk which covered their privies," and carrying "daggers with gold hafts adorned with pearls and precious gems," but then the ornamental cloth was quickly and mysteriously removed. Of greater importance, the king conferred samples of cinnamon and cloves, the spices his guests had been seeking for nearly two years. It appeared they were now on the Spice Islands' doorstep.

"That king is a Moro," or Muslim, Pigafetta observed, "and his name is Rajah Siripada. He was forty years old and corpulent. No one serves him except women who are daughters of the chiefs. He never goes outside of his palace, unless he goes hunting." No less than ten scribes wrote down his every action "on very thin tree bark." These people also had a written language, another indication of how advanced they were.

Ceremony pervaded every aspect of life in Brunei, and after the audience with Rajah Siripada, the Europeans were ceremoniously returned atop elephants to the "governor's house" accompanied by seven bearers carrying the presents bestowed on them by the ruler. When they dismounted, each man received his present, which the bearers carefully placed on the left shoulder, and in return, "We gave each of those men a couple of knives for his trouble."

That evening, nine servants came to the house, each man carrying a large tray, and "each tray contained ten or twelve porcelain dishes full of veal, capons, chickens, peacocks, and other animals, and fish." Pigafetta claims they dined on thirty-two different kinds of meat, in addition to the fish. "At each mouthful of food we drank a small cupful of their distilled wine from a porcelain cup the size of an egg. We ate rice and other sweet food with gold spoons like ours."

Even now, nearly a century after the era of the Treasure Fleet, Chinese wares were everywhere. Pigafetta mentions porcelain ("a kind of very white earthenware"); silk; and, amazingly enough, "iron spectacles." Eyeglasses are thought to have been invented in Venice, but it appears likely that the Chinese also developed techniques for grinding glass, and this technology had found its way to Brunei. Even the kingdom's currency revealed a pronounced Chinese influence. "The money coined by the Moors in those parts is of metal, pierced at the center for stringing. And it bears only, on one side, four marks, which are letters of the great king of China." All the men were curious to inspect two giant pearls "as large as eggs" owned by the king. "They are so round they cannot lie still on a table," Pigafetta marveled. After considerable negotiation, and even more tributes, the officers of the armada made their wishes known, and the king reluctantly displayed the two giant pearls.

After their second night ashore, the delegation rode by elephant back to the ocean, and boarded their crude and confined ships. The familiar creaks filled their ears, and the familiar reek of stagnant

water filled their nostrils. Not everyone returned, however. According to Ginés de Mafra, only four men made it back to the fleet while
three—the two Greek sailors and Carvalho's son—remained ashore.
(De Mafra forgot to mention that Elcano and Espinosa were also
among the missing.) The Europeans suspected that they were all
being held against their will and anxiously awaited their safe return.

$\mathcal{S}$hortly after dawn on July 29, more than one hundred *proas*,
organized into three groups, appeared out of nowhere, bearing down
on the armada.

For the first time since the massacre three months earlier, the
crew feared for their lives. They broke out their halberds, crossbows,
and arquebuses, knowing that they were badly outnumbered,
because each *proa* carried a full complement of warriors. To complicate matters, two great junks—de Mafra claims three—had
anchored just behind the armada during the night. No one aboard
Trinidad or *Victoria* noticed the junks at the time, but it now
appeared that the *proas* intended to drive the armada toward the
junks, whose crew would overwhelm the Europeans and take them
as prisoners, or worse.

"Upon catching sight of them, imagining that there was some
trickery afoot, we hoisted our sails as quickly as possible, abandoning an anchor in our haste," Pigafetta wrote. As the armada began to
gain speed in the water, some crew members jumped aboard the
junks and captured four warriors. The men-at-arms fired their
weapons at their adversaries, "killing many persons," according to
Pigafetta. Several of the menacing *proas*, frightened by the armada's
vehement response, veered away. De Mafra, a more cynical com-
mentator than Pigafetta, was bewildered by the battle. How would
such behavior lead to the recovery of the three lost crew members?
Nevertheless, the battle raged on, as the armada turned its guns on
one of the huge junks. They ordered the junk to drop sail, and when
her captain refused, the Europeans opened fire at the rudder; still

her crew refused to comply. The Europeans swarmed aboard the junk, where they discovered that her captain was not the murderous pirate they had imagined. "Their captain said that he served the king of Luzon and that while with a fleet to an island he had been cut off from the rest of the ships by a storm, and that being near the island he had resolved to call on it to repair his vessel, since the local king was a relative of Luzon's king." After that, Carvalho and the captain fell into secret conversation, to the dismay of the armada's officers, who had risked their lives to disable and board the junk. In hushed tones, the wily captain offered Carvalho jewels, two cutlasses, and a dagger "with golden hilts and guards inlaid with many diamonds," all for his personal use. The gifts had their intended effect: "Having received these presents," according to de Mafra, "our captain released the junk and its people, something which everyone later regretted because they saw that under their poor-looking cotton garments, most of those men were wearing silk clothes with gold embroidery."

Pigafetta recognized the transaction as a simple case of bribery, and his opinion of Carvalho, never high to begin with, fell several notches. Had they held the captain hostage, Pigafetta believed, Rajah Siripada would have paid a tremendous ransom for him, far more than the bribe that Carvalho had accepted. As Pigafetta interpreted local politics, the captain was needed to battle the heathens who threatened the rajah's Muslim empire.

The matter did not end there. The extent of the Europeans' confusion became apparent when Rajah Siripada revealed that the *proas* had no intention of attacking the armada. They were actually on their way to attack the Arabs' enemies when the armada got in the way and thwarted their battle plan. "As a proof of that statement, the Moros showed some heads of men who had been killed, which they declared to be the heads of heathens." Once they realized their mistake, the armada's officers awkwardly struggled to make amends with the rajah. At the same time, they requested that the detained men, including Carvalho's illegitimate son, be returned. But Rajah

Siripada refused. He had lately pampered the Europeans, treating them to elephant rides and mattresses, to feasts and gifts of precious jewels; he had even granted them a personal audience, and they had repaid his generosity by meddling in his internal affairs and letting the troublesome captain go. As a result, the rajah insisted on holding his hostages, at least for the present.

Carvalho responded with an insult of his own. He decided to keep sixteen of the prisoners they had captured at sea, as well as another prize, three extraordinarily beautiful women. He declared that he would present them to King Charles, a plan that the other officers enthusiastically seconded. Magellan had always forbidden the presence of women and slaves (his own slave excepted) aboard the ships because he believed that their presence would become divisive, and Carvalho's captives proved Magellan's belief correct. Soon everyone on board *Trinidad* was aware that Carvalho had turned the women prisoners into his personal harem, and he was busy taking liberties with all three. This behavior so incensed the other officers that they muttered threats to kill Carvalho, who bartered for his life, and his harem, with liberal gifts of gold and jewels from the loot he had received from the captain of the captured junk. In the end, Carvalho was spared, and he even kept his harem, but he lost all authority in the eyes of his men. As the officers realized, if they took bribes and maintained harems, they would become pirates themselves.

Carvalho's unscrupulous behavior made Pigafetta long for Magellan's icy sense of duty and discipline; without those driving forces, the expedition's sense of moral imperative melted away amid the luxuriant Indonesian heat.

At length, the rajah released two hostages, Elcano and Espinosa, whom the messengers promptly returned to the waiting fleet. They said they had been detained separately, "treated well," and knew nothing about the mysterious flotilla of *proas* bearing down on the armada. But where were the others? Elcano and Espinosa explained to Carvalho that the two Greek sailors had decided to desert. The story seemed unlikely, but there was no way to confirm it. Magellan, had he been alive, would have immediately launched a search for the deserters, but Carvalho did not lift a finger. He was naturally more interested in the fate of his young son; with long faces, Elcano and Espinosa said they had heard the boy had died ashore, but they did not know for certain.

That was only the beginning of Carvalho's misfortunes. On September 21, 1521, the other officers decided to replace him. The change of command did not amount to a mutiny, and Carvalho was neither attacked nor restrained; he was simply told to step down, and he did, returning to his former post as pilot.

The officers settled on an awkward triumvirate to command the fleet. The purser, Martín Méndez, became the fifth Captain General, and Gonzalo Gómez de Espinosa took over the captaincy of *Trinidad*, still the flagship. Elcano gnashed his teeth in frustration, having been bypassed yet again in favor of men with lesser skills but greater rank. No one could forget that he had participated in the mutiny against Magellan and served his time in chains. Since then, he had rehabilitated himself, but some stain of dishonor clung to him. Still, he could console himself with becoming the captain of *Victoria*. Because neither Espinosa nor Méndez had firsthand navigation experience, Juan Sebastián Elcano, the veteran Basque mariner, became the unofficial head of the expedition.

*T*o be a Basque meant, and still means, to be a historical anomaly. The Basques are the oldest ethnic group of Europe, a breed apart ever since Paleolithic times. In their province in northern Spain, next to the French border, the Basques speak a distinct language, actually, eight dialects of a distinct language. No direct link between the Basque tongue and another language has been identified. Over the centuries, various monarchs had attempted to annex the Basques, and although King Ferdinand finally conquered them in 1512, and Basques became fervent Catholics, the fiercely independent Basque culture persisted.

The sea loomed large in the lives of Basque men; they were born facing the sea, they lived by the sea, they died at sea. It was into this highly idiosyncratic and tenacious culture that Juan Sebastián Elcano was born in 1487 in the Basque province of Guipúzcoa. His name, usually given as Elcano or Del Cano, is said to have derived from *Elk-ano,* a Basque word for a district of fields. From his youth in Guipúzcoa, the center of the Basque fishing industry, Elcano was destined for the sea. Of his eight siblings, at least two brothers became mariners, and one sister married a pilot. At twenty, Elcano found work ferrying Spanish soldiers in ships, although he had undoubtedly gone to sea much earlier. Two years later, he found work aboard an expeditionary ship taking Spanish forces and matériel to Africa, where the king's soldiers engaged Arabs in battle; his duties included overseeing the ship's cargo—gold to pay the soldiers—and weapons. By the time he was twenty-three, Elcano became the owner and captain of his own ship, a large vessel weighing two hundred tons. He offered his services to Spain, which refused to pay him; the situation forced him to borrow to pay his crew members, and ultimately he had to sell the ship to pay his debts, which involved him in more trouble, for it was illegal to sell an armed Spanish ship.

Elcano took refuge in Seville, where he attended the Casa de Contratación's School of Navigation, receiving formal training as a pilot, probably from the boastful and controversial Amerigo Vespucci, who served as head of its board of examiners. Students received credit in the form of beans won from their instructor; if they successfully completed a course, they were awarded a dry bean; if unsuccessful, they received a shriveled pea. Under Vespucci's supervison, Elcano learned his navigation, was awarded his bean, and became a master pilot.

With his new credentials, he applied for a position as a pilot for the Armada de Molucca, but even here Elcano's business troubles continued to haunt him, for many of the officials of the Casa de Contratación were Basques, including the chief accountant, who came from the same little province as Elcano and might detect Elcano's old financial transgressions. As luck would have it, a relative who worked at the Casa and was willing to overlook Elcano's problems recommended him to Magellan, who in turn appointed Elcano master of *Concepción* at a salary of 3,000 *maravedís* per month. Even better, he received six months' pay in advance—18,000 *maravedís*, a small fortune for a young man from a modest Basque family. Although he would have to pay for his furnishings out of the advance, he would still have a considerable amount left over. By combining his salary with his share of the expedition's profits, he would become wealthy. Once Elcano accepted his position, he recruited other seamen for the voyage, and in the end, ten Guipúzcoans wound up on the armada's rolls, largely through Elcano's efforts.

Just before the fleet departed from Seville, Elcano was called to testify before a formal board of inquiry in Seville, where he testified that Magellan was a "discreet and virtuous man and careful of his honor." After that brief moment of prominence, Elcano blended into the background, and even though he was among the mutineers at Port Saint Julian, he made little impression on his fellow crew members. In his entire chronicle of the voyage, Pigafetta did not mention even once the name of the Basque mariner who now led the armada.

*A*fter thirty-five days in Brunei, the fleet was ready to make the final assault on the Moluccas. They had reason to believe they were approaching the Spice Islands at last, because they were now following the path of an earlier European traveler, Ludovico di Varthema of Bologna, who had published a popular account of his travels, including his visit to the Spice Islands, in 1510. (He had reached the

Spice Islands by traveling east along the overland route rather than west over water.) Varthema was a pioneer many times over, the first European to become wealthy by trading in gems in India, and among the first to gain a sustained look behind the veil of Islam. He even claimed to be the first nonbeliever to visit Mecca, at the risk of his life. Soon after, he arrived in the Spice Islands, where he was transfixed by the sight of the fabled clove tree. "The tree of the cloves is exactly like the box tree," he wrote, "that is, thick, and the leaf is like that of the cinnamon, but it is a little more round.... When those cloves are ripe, the said men beat them down with canes, and place some mats under the said tree to catch them." He observed how the people of Molucca traded in their precious resource, and was not impressed: "We found that they were sold for twice as much as nutmegs, but by measure, because these people do not understand weights."

Distracted and overextended, the surviving men of the fleet lacked Varthema's cunning and ability to blend into the surroundings. From the moment the fleet weighed anchor, the ships ran into serious navigational trouble. Sailing downwind out of the harbor of Brunei, *Trinidad* ran aground as she attempted to round a point; the shoal could have sliced the hull open. The accident was caused solely by the pilot's negligence, according to Pigafetta, "but by the help of God we freed it." Actually, they had to wait for four hours, praying that the hull would remain intact until the tide rose and freed the ship.

Shortly afterward, a sailor "snuffed a candle into a barrel full of gunpowder, but he quickly snatched it out without any harm." An explosion could have destroyed the ship and claimed many lives. Mishaps like these would never have occurred on Magellan's watch, and in each case, the undisciplined fleet had been lucky to survive a mistake of its own making, but how long would their luck hold?

*D*amaged by running aground, *Trinidad*'s hull needed repairs; in fact, both ships leaked rapidly, and the constant seepage meant that the men had to take exhausting turns at the pumps just to keep them afloat. It became apparent to all that they would have to recondition the fleet for the first time since the painstaking overhaul conducted during the grim winter in Port Saint Julian.

Arriving on the island of Cimbonbon, the armada spent the next forty-two days on repairs. Pigafetta describes their refuge as a "perfect port for repairing ships," for it was remote from waterborne traffic, and tranquil, but the work itself was difficult to perform efficiently, "as we lacked many things for repairing the ships." The difficult and exhausting task, made even more taxing by the Indonesian heat, was absolutely necessary if the ships were to be seaworthy. "During that time, each one of us labored hard at one thing or another. Our greatest fatigue, however, was to go barefoot to the woods for wood." Wandering in the shade, they were attacked by wild boar. They managed to kill one of the beasts as it was swimming across the harbor, pursuing it in a longboat. They also found a wide variety of fish and amphibious life, including "large crocodiles," giant oysters five or six feet long and weighing hundreds of pounds, and a curious fish with a "head like a hog and two horns. Its body consisted entirely of one bone, and on its back it resembled a saddle. And they are small." To judge from this description, this might have been the squamipen, or angelfish, the brightly colored, highly compressed fish found in the region.

Another natural marvel to be found on Cimbonbon was worthy of Pliny the Elder's *Natural History.* "trees . . . which produce leaves which are alive when they fall and walk. . . . They have no blood, but if one touches them, they run away." With childlike enthusiasm, Pigafetta managed to capture a specimen. "I kept one of them for nine days in a box. When I opened the box, the leaf went round and round it." These walking leaves have been identified as phyllium, insects whose flat, broad back resembles a leaf, including scars and stems; it is a remarkable example of camouflage. In flight, or when moving, these insects reveal bright colors, but when they rest in a tree, they melt into the shadows, and avoid the sharp-eyed birds that prey on them.

*O*nce the arduous renovations were completed, the fleet resumed its search for the Spice Islands on September 27. Days later, the fleet sighted a large junk from the island of Pulaoan, bearing the local ruler. "We made them a signal to haul in their sails, and as they refused to haul them in, we captured the junk by force, and sacked it. [We told] the governor if [he] wished his freedom, he was to give us, inside of seven days, four hundred measures of rice, twenty swine, twenty goats, and one hundred and fifty fowls." The governor tried to mollify the marauders with a liberal tribute of coconuts, bananas, sugarcane, and especially palm wine, all of which had their intended effect. The contrite Europeans returned the firearms and daggers they had taken from the governor, along with tributes of their own, cloth, a flag, a "yellow damask robe," and other trinkets. "We parted from them as friends," Pigafetta noted with satisfaction, and the search for the Spice Islands resumed.

Traveling southeast, they came upon a weird outcropping in the ocean. It seemed to Pigafetta that the sea was "full of grass, although the depth was very great." Passing the outcropping, he thought they were "entering another sea." Actually, they were still in the vicinity of Mindanao, traveling along its western coast until they arrived at another island Pigafetta calls Monoripa. "The people of that island make their dwellings in boats and do not live otherwise," he observed of the Bajau, the sea gypsies who were widely scattered throughout the area, adjusting their moorings to avoid the monsoon. Of all the tribes the armada encountered, the Bajau were among the most enigmatic.

They are thought to have flourished well before the armada's arrival, when the Chinese were exploring the region. The Bajau developed a brisk trade in a Chinese delicacy, *trepang*, or sea cucumber. This leathery echinoderm, normally a few inches in length, grew to extraordinary dimensions in the area, occasionally as long as three feet. It was considered an aphrodisiac, the ginseng of the sea.

Long after the Chinese presence faded, the Bajau remained. Each anchorage usually served an extended family that spread across several boats, as little as two or as many as six. They fished together, shared food, and maintained relationships with other families through intermarriage. The boats were only thirty feet from stem to stern and six feet amidships, but far more spacious than *proas* or *balanghai*. Their living areas were sheltered by poles supporting mats made from palm fronds, and each boat had its clay hearth for cooking.

Bajau fishermen employed handheld lines and spears to catch hundreds of other edible species in addition to *trepang*. On moonless nights, they fished by lantern. They preserved their catch much as Europeans did, by salting and drying. Their activities were confined almost exclusively to the sea; they owned no land, but they held small islands in common devoted to burials, and when necessary they went ashore for fresh water. They were not at all predatory; when attacked, the Bajau usually fled across the water. More conventional tribes on shore considered the waterborne, nomadic Bajau unreliable and not subject to any one set of laws or beliefs. Over time, many of them became Muslims, but they retained some of their earlier customs. They practiced trance dancing and called on mediums to purge the community of evil spirits or illness. The evil forces were led to a particular boat, which was set adrift in the open sea to wander eternally. The custom might serve as a metaphor for the entire Bajau culture, always adrift.

The crew was tempted to remain among the Bajau because the men heard that on two nearby islands they could find the best cinnamon grown anywhere. Next to cloves, cinnamon was the most valuable spice; the temptation to fill their ships with the fragrant spice proved almost irresistible. "Had we stayed there two days, those people would have laden our ships for us, but as we had a wind favorable for passing points and certain islets which were near the island, we did not wish to delay."

Just before they left, they got their first, tantalizing look at the fabled cinnamon tree: "It has but three or four small branches and its leaves resemble those of the laurel. Its bark is the cinnamon, and it is gathered twice a year." In Malay, Pigafetta noted, the unprepossessing tree was called *caiu* (sweet) *mana* (wood). The men conducted a quick, probably illicit transaction, exchanging two large knives for about seventeen pounds of cinnamon, worth enough on the docks of Seville to buy an entire ship. They expected to obtain far more cinnamon, along with nutmeg, pepper, mace, and many other precious spices, once they reached their goal.

Just when it seemed that a measure of order had returned to the fleet, they attacked a large *proa* to obtain information about the whereabouts of the Moluccas. In a bitter struggle, they slaughtered seven of the eighteen men on board the little craft. Pigafetta mentioned the matter only in passing, without remorse. In the past, the needless deaths of the Chamorros and the Patagonian giants had caused sorrow and guilt, but by now he had become desensitized to the business of killing, which he reported with less emotion than he would a passing storm. Pigafetta's lack of fellow-feeling reflected the entire crew's frame of mind. It is one of the outstanding ironies of the voyage that the closer they came to fulfilling their mission, the more they lost their sense of mission, which Magellan, for all his faults, had done so much to impart.

Before leaving the unlucky *proa* in their wake, the armada spared the life of one of its occupants, the brother of Mindanao's ruler, who insisted that he knew the way to the Moluccas. Making good on his promise, he guided the armada on a different course; they had been traveling northeast, but he took them to the southeast, toward the Moluccas. Along the way, they passed a cape inhabited by cannibals, and the crew studied these fabled creatures with rapt attention. The cannibals were every bit as frightening as their reputation: "shaggy men who are exceedingly great fighters and archers. They use swords one palmo in length and eat only raw human hearts with the juice of oranges and lemons." The crew members naturally kept their distance, and listened closely to their captured guide's account of the tribe as if they were tourists on safari. In all likelihood, they had encountered members of the Manobos tribe, who did on occa-sion practice a ritual cannibalism in which they devoured the heart or the liver of their enemies. But no European hearts were con-sumed that day.

The armada had just reached the southernmost part of Min-danao when the ships were swept by the strongest storm they had encountered since the life-threatening gales off the eastern coast of South America, but once again, they received brilliant supernatural reassurance that they would safely reach their goal. "On Saturday night, October 26, while coasting by Birahan Batolach, we encoun-tered a most furious storm. Thereupon, praying to God, we lowered all the sails. Immediately our three saints appeared to us and dissi-pated all the darkness. St. Elmo remained for more than two hours on the maintop, like a torch; St. Nicholas on the mizzentop; and St. Clara on the foretop. We promised a slave to St. Elmo, St. Nicholas, and St. Clara, and gave alms to each one." The storm passed, and the shaken crew members once again gave thanks for their lives, raised the sails, and the fleet recommenced its southeasterly voyage. They were only two hundred miles from the Spice Islands, yet they spent weeks zigzagging blindly throughout the Sulawesi and Maluku seas without knowing how to reach their destination.

At the island Pigafetta called Cavit, the crew members struck again, capturing two more pilots and ordering them to take the fleet to the Moluccas on pain of death. "Laying our course south south-west," Pigafetta tells us, "we passed among eight inhabited and unin-habited islands, which were situated in the manner of a street. Their names are Cheaua, Cauiao, Cabaio, Camanuca, Cabalizao, Cheai,

Lipan, and Nuza"—all members of the Karkaralong group, located at the southern tip of Mindanao.

Even now, as they approached their goals, they were bedeviled by misfortune. On November 2, Pedro Sánchez, a gunner aboard *Trinidad*, attempted to fire an arquebus; the weapon exploded, killing him, and two days after that, another *Trinidad* gunner, Juan Bautista, died in a gunpowder explosion.

Unable to sail close enough to the wind to pass a cape, the fleet had to double back and forth past the point until the wind changed. As they did, three of their captives, two men and a boy, jumped ship and swam for their lives toward a nearby island. "But the boy drowned," Pigafetta relates, "for he was unable to hold tightly to his father's shoulder."

*O*n the ships sailed, gliding past the islands of Sanguir, Kima, Karakitang, Para, Sarangalong, Siao, Tagulanda, Zoar, Meau, Paginsara, Suar, Atean: a string of emeralds set in gleaming sapphire. And then, on November 6, 1521, they saw four more islands shimmering on the horizon. "The pilot who still remained with us told us that those four islands were the Moluccas," Pigafetta recorded. After losing three ships and more than a hundred men—half the crew— they were finally on the doorstep of the Spice Islands . . .

     . . . Ternate . . .

        . . . Tidore . . .

           . . . Motir . . .

              . . . Makian . . .

They stretched from north to south, four small islands, each no more than six miles across. To the south lay a fifth Spice Island, Bacan, which was considerably larger.

The Moluccas actually comprise about one thousand islands of varying sizes, but for Europeans of the sixteenth century, the

Moluccas referred to just those five islands. The best-known among them were Ternate and Tidore, volcanic islands whose steep cones towered about a mile above the sea, imparting an impressive solidity to the tiny landmasses. Bartolomé Leonardo de Argensola, writing in 1609, described Ternate's volcano as a "dreadful burning of mountain flames." He guessed that winds "kindle that natural fire, or the matter that has fed it for so many ages. The top of the mountain, which exhales it, is cold, and not covered with ashes, but with a sort of light cloddy earth, little different from the pumice stone burnt in our fiery mountains." Volcanic ash enriched the soil on islands where the spices grew, and the moist climate also promoted lush growth; this combination made them unique sources for spices. The occasional volcanic eruptions terrified those who beheld them, and gave Ternate and the other islands a magical reputation. It would not have been more marvelous to see a dragon or the lost city of Atlantis rising from the depths of the sea than to witness an eruption in the Moluccas.

"Look there, how the seas of the Orient are scattered with islands beyond number," wrote the Portuguese poet Luís de Camões in *The Lusíads* about the spell cast by the Spice Islands:

> *See Tidore, then Ternate with its burning*
> *Summit, leaping with volcanic flames.*
> *Observe the orchards of hot cloves*
> *Portuguese will buy with their blood . . .*

All these exotic sights and more were now within the grasp of the Armada de Molucca. "So we thanked God, and for joy we discharged all our artillery," Pigafetta wrote. "And no wonder we were so joyful, for we had spent twenty-seven months less two days in our search for the Moluccas."

CHAPTER XIII

Et in Arcadia Ego

The harbour-bay was clear as glass,
So smoothly it was strewn!
And on the bay the moonlight lay,
And the shadow of the Moon.

On November 8, 1521, the Armada de Molucca entered the harbor of Tidore, firing a joyful salute. They dropped anchor in twenty fathoms and fired another round of artillery, the report of the guns echoing off the island's tranquil hills. In the humid climate, the strong scents of clove and cinnamon wafted across the water, reviving the weary crew members with the promise of riches.

The following day, an emissary from Tidore floated out to the ships in a luxurious *proa*, his head protected from the sun by a silk awning; his son, bearing a ceremonial scepter, was at his side. They were accompanied by a pair of ritual hand washers bearing sweet water in jars made of gold, and two other bearers carrying a gold casket filled with an offering of betel nuts. The emissary introduced himself as al-Mansur, a Muslim name, but the officers came to know him by the Spanish version, Almanzor. He appeared to be in his forties and rather rotund.

Almanzor's theatrical arrival was calculated to announce that he was an important personage: the king of Tidore and an enthusiastic astrologer. As intended, the officers recognized that gaining Almanzor's goodwill would be vital because he was the gatekeeper to the cloves, which they had come so far to find. But Almanzor's little kingdom was in constant peril, and he needed these visitors from afar as much as they needed him, or his spices.

From his resplendent *proa*, Almanzor enthusiastically welcomed the fleet. "After such long tossing upon the seas, and so many dangers, come and enjoy the pleasures of the land, and refresh your bodies, and do not think but that you have arrived at the kingdom of your own sovereign," he declared, according to Pigafetta. And then Almanzor startled them all by announcing that he had dreamed of their arrival, and they had fulfilled his prophecy.

Almanzor boarded *Trinidad* under the watchful eyes of the officers, who offered him the velvet-covered chair of honor. Almanzor lowered himself into it, but conveyed the impression that he was accommodating them by consenting to sit, after which he "received us as children" in Pigafetta's astonished words. For all his graciousness, Almanzor had a stubborn streak; he refused to bow or even to tilt his head even when it was necessary. When he was invited to enter *Trinidad*'s cabin, he refused to stoop, as her crew members routinely did. Instead, he mounted the upper deck and descended from above, his head rigidly erect.

In conversation, Almanzor revealed that he was familiar with Spain, and even with its great and powerful ruler, King Charles. He insisted that he and the people of Tidore fervently desired to serve the king and his kingdom, an assertion that immediately made the officers suspect that Almanzor had another agenda that involved switching his allegiance from the Portuguese to the Spanish. The officers were correct. A decade earlier, the father of the island's current ruler had encouraged the Portuguese to set up a trading station, in part because he wished to loosen the Arab stranglehold on the islands' crops.

The experience left a bitter legacy on both sides. The Portuguese came to detest the Moluccans with the passion of a jilted lover. At the outset, the Portuguese had hoped to break the Chinese and Arab monopoly on spices and grow fat on the proceeds, fatter even than their neighbor and rival, Spain. They would then assert control over the global economy. But the islanders turned out to be devious partners, murderous and slippery; most infuriating of all, they continued to sell spices to anyone with a ship capable of carrying them away. Portugal never got its monopoly and blamed the rulers and inhabitants of the islands.

João de Barros, a Portuguese court historian, expressed the official attitude toward the inhabitants of the Spice Islands: "In everything but war they are slothful; and if there be any industry among them in agriculture or trade, it is confined to the women," he declared, enumerating their failings. "Altogether, they are a lascivious people, false and ungrateful, but expert in learning anything. Although poor in wealth, such is their pride and presumption that they will abate nothing from necessity; nor will they submit, except to the sword that cuts them. . . . Finally, these islands, according to the account given by our people, are a warren of every evil, and contain nothing good but their clove tree." Barros came to consider the clove itself as the ultimate source of evil in this region. "Though a creation of God," he wrote, the spice was "actually an apple of discord and responsible for more afflictions than gold."

No wonder Almanzor had grown tired of the Portuguese; and no wonder he preferred Spaniards (although he did not realize that many of the crew were Portuguese). But there was more. Local politics also influenced Almanzor's thinking. At the time, Tidore was embroiled in a conflict with its island neighbor, Ternate, still in the Portuguese grip, and Almanzor thought these representatives of the Spanish crown could make powerful allies in the struggle.

The triumvirate of officers—Elcano, Espinosa, and Méndez— quickly made trading pacts with Almanzor and bestowed so many gifts that he asked them to restrain their overwhelming generosity because "he had nothing worthy to send to our king as a present, unless, now that he recognized him as a sovereign, he should send himself."

$\mathcal{O}$n November 10, Carvalho and a small detachment went ashore, and for the first time the men of the Armada de Molucca set foot on the Spice Islands.

Antonio Galvão, the Portuguese administrator who arrived at the Spice Islands a few years later, evoked the ethereal landscape that greeted the armada's crew as they looked at their surroundings: "The shape of most of these islands is that of a sugarloaf, with the base going downward into the water, surrounded by reefs at little more than a stone's throw; at ebb tide one can go there on foot. One can put into the islands through some channels in the reef which outside is very high; and there is no place to anchor except in certain small sandy bays: a dangerous thing! They look gloomy, somber, and depressing. That is always the way they strike the onlooker at first sight; for always, or nearly always, there is a large blanket of fog on their summits. And for the greatest part of the year the sky is cloudy, which makes it rain very often; and if it does not, everything withers but the clove tree, which prospers. And at certain intervals there falls a dismal, misty rain."

What made the islands seem alive to the first European explorers were the active and highly unpredictable volcanoes rising to the sky. "Some of these islands spit fire and have warm waters like hot springs. And they are so thickly crowded with groves as to look like one big mass of them, and they are therefore hiding places for evil doers," Galvão warned. As a result of the volcano's ejecta raining down on the islands, the soil "is black and loose; and in places there is clay and gravel, which is unstable because it lies on the rock where it does not take hold. And however much it may rain, the water stands only a while before it is absorbed."

Of supreme importance were the spices themselves, especially the cloves. The armada's men had seen cloves, smelled cloves, and tasted cloves, but only now did they find cloves growing in the wild—not just a few trees scattered here and there, but a dense, impenetrable forest of cloves. "The hills in these five islands are all of cloves," wrote Magellan's brother-in-law, Duarte Barbosa, after his visit to the Spice Islands in 1512. "[They] grow on trees like laurel, which has its leaf like that of the arbutus, and it grows like the orange flower, which in the beginning is green and then turns white, and when it is ripe it turns coloured, and then they gather it by hand, the people going amongst the trees."

On their first visit to Tidore, the armada's leaders reached an agreement with Almanzor recognizing Spain's sovereignty over the island, even though it violated the Treaty of Tordesillas. Once these formalities were over, the leaders wanted to obtain the spices as quickly as they could, before local strife drove them away. The men had seen too many warm receptions turn violent for them to believe that Almanzor would keep his word for very long.

For the Europeans of the armada, a treaty was, above all, a written document, but for the Tidoreans, only the oral expression carried the force of law. To record commercial transactions, the inhabitants of the Spice Islands occasionally wrote on palm leaves or paper imported from India, using a system borrowed from the Chinese, but when they made treaties, they relied on oral rather than written communication. Both sides managed to overcome their differences to seal the bargain, and with the treaty in force, the king of Tidore advised the armada's officers that he did not have enough cloves on hand to satisfy their needs, but he offered to accompany them to Bacan, where he assured them that they would find as much as they wanted. But before the officers began filling the ships with spices, they inquired after one of their own: Francisco Serrão, the author of the letters that had inspired Magellan's voyage to the Spice Islands.

None of the Europeans knew what had become of this legendary figure. The most recent information—and it was only gossip—was

that he and a small band of Portuguese adventurers arrived at Ternate, where they allied themselves with the island's ruler, Rajah
Abuleis. In the eyes of the authorities, Serrão and his band of Portuguese adventurers had become little more than mercenaries; like
Magellan, they were willing to switch loyalties to Spain in exchange
for a better deal. Now, Serrão's fate assumed great importance to the
armada, which was starved for leadership. It was possible that he was
still in the Spice Islands, and, if so, the armada's officers hoped to reunite with him. He might even take command of the fleet in Magellan's stead, if he were still alive.

The reunion was not to be. Almanzor revealed that Serrão had
died eight months before, about the time of Magellan's death, but
the king concealed the whole story behind Serrão's end. The facts
were these: After his arrival in the Spice Islands in 1512, Serrão had
chosen sides in a power struggle between the rulers of Tidore and
Ternate, and he served as admiral of the Ternate navy, such as it was.
The two island kingdoms battled for years, with Ternate, under Serrão's leadership, winning every time. To make peace, Serrão forced
Tidore to give up the sons of its rulers as hostages and forced
Almanzor to marry off his daughter to his enemy, the king of Ternate, whose child she bore.

Almanzor neither forgot nor forgave the terrible humiliations
Serrão had inflicted on him. "Peace having been made between the
two kings," Pigafetta relates, "when Francisco Serrão came one day
to Tidore to trade cloves, the king of Tidore had him poisoned
with...betel leaves. He lived only four days. His king wished to
have him buried according to his law"—meaning Muslim rites—
"but three Christians who were his servants would not consent to it.
He left a son and a daughter, both young, born by a woman whom
he had taken to wife in Java the Great, and two hundred barrels of
cloves. He was a great friend and a relative of our good and loyal
dead Captain General." The vendetta did not end there. Ten days
later, the king of Ternate, "having driven out his son-in-law, the king
of Bacan, was poisoned by his daughter, the wife of the said king,

under pretext of wishing to conclude peace between them." He lingered two days before he died.

The fleet's officers realized that Serrão's death contained disturbing echoes of Magellan's. Each had taken sides in a protracted struggle between two island kingdoms, and each had acted harshly in his
dealings with the enemy. Eventually, the warring tribes formed common cause, and the formerly heroic outsider paid for his bold deeds
with his life. These cautionary tales reminded the officers to resist
the temptation to fight anyone else's battles. Despite their sorry history, the unhappy inhabitants of these two islands hoped that the
distant but powerful king of Spain, about whom they had heard,
could bring lasting peace where their own efforts had failed.

On Monday, November 11, the rulers of Ternate began their
diplomatic offensive.

One of the king's many sons came out to the fleet in a *proa*,
accompanied by Serrão's widow, a Javanese, and their two children.
The sight of the approaching craft caused Espinosa to panic, for he
had cast his lot with Ternate's enemy, Tidore. What was he to do?
Almanzor, who remained near at hand, calmly advised Espinosa to
act as he saw fit.

Espinosa and the other officers aboard *Trinidad* stiffly welcomed
the visitors, bestowed gifts on them, and watched closely for signs of
trouble. Meanwhile, Pigafetta, drawing on his linguistic skills, fell
into conversation with a servant named Manuel, who said he served
a Portuguese governor named Pedro Alfonso de Lorosa, who had
come to the Spice Islands with Serrão and lived there still. Manuel
claimed that while considerable enmity still existed between the
kings of Tidor and Ternate, the rulers of Ternate were also in favor
of Spain, and he assured the officers that they were as welcome on
Ternate as they were on Tidore.

Taking the servant at his word, Pigafetta went ashore to see the
Spice Islands for himself. Always intrigued by the local sexual customs

and the women, he felt greatly disappointed by the females of Tidore, calling them "ugly," a word he rarely uses elsewhere in his chronicle. Both men and women went about naked, or wore only a scanty loincloth "made from the bark of trees," he noted. Tidore was not to be the scene of Filipino-style orgies, because the men "are so jealous of their wives that they do not wish us to go ashore with our drawers exposed for they assert that their women imagine that we are always in readiness." Pigafetta meant that the European-style breeches made the sailors appear to be erect.

Despite the apparent sexual exclusivity of the inhabitants, Pigafetta heard that the local rulers had fathered dozens of children. He wondered if there was any truth to the story, and found that the profligacy of the island rulers exceeded even his imagination: "The kings have as many women as they wish, but only one principal wife, whom all the others obey. The king of Tidore had a large house outside the city, where two hundred of his chief women lived with a like number of women to serve them. When the king eats, he sits alone or with his chief wife in a high place like a gallery where he can see all the other women who sit about the gallery; and he orders whoever best pleases him to sleep with him that night. After the king has finished eating, if he orders those women to eat together, they do so, but if not, each one goes to eat in her chamber. No one is allowed to see those women without permission of the king, and if anyone is found near the king's house by day or by night, he is put to death. Every family is obliged to give the king one or two of its daughters. That king had twenty-six children, eight sons, and the rest daughters." And on a neighboring island, Gilolo, the situation was even more extreme. Two kings shared the island; one had 600 children, the other 525.

Those were Muslim kings, Pigafetta noted. "The heathens do not have so many women; nor do they live under so many superstitions, but adore all that day the first thing they see in the morning when they go out of their houses. The king of those heathens, Rajah Papua, is exceedingly rich in gold, and lives in the interior of the island."

Once again, Pigafetta set about compiling a dictionary of words and phrases, with heavy emphasis on parts of the body and procre-ation. He worked swiftly, and his dictionary of the Malay dialect spoken in the Spice Islands blossomed into his most elaborate effort at lexicography.

*T*rading for spices got under way with astonishing speed. The king of Tidore gave orders to prepare a trading house—probably recov-ered from the days of the Portuguese occupation—to accommodate the new arrivals, and by Tuesday, November 12, four days after they had dropped anchor in Tidore harbor, the Armada de Molucca was in business. "We carried almost all our goods thither, and left three of our men to guard them. We immediately began to trade in the following manner. For ten brazas of red cloth of very good quality, they gave us one bahar of cloves, which is equivalent to four quintals and six libras." A quintal of cloves equaled one hundred pounds, and it was the most important unit for measuring the value of a spice shipment.

The men of the fleet valued their take according to the quinta-lada they received. A quintalada was a percentage of the storage space set aside for the crew members and officers. Following the instructions King Charles gave to Magellan on May 8, 1519, each significant member of the armada received a specific number of quintaladas. Once they paid one twenty-fourth of the amount to the king, they could keep the rest for themselves. Magellan, as the Cap-tain General, was naturally awarded the largest amount: sixty quin-tals plus another twenty quintaladas. The other officers received almost as much, and on down through the roster of boatswains, gun-ners, caulkers, coopers, the barber, and the master-at-arms. Even the priests received allotments.

Over the next several days, trading continued at a feverish pace. "For fifteen brazas of cloth of not very good quality, one quintal and one hundred libras; for fifteen hatchets, one bahar; for thirty-five glass drinking cups, one bahar (the king getting them all); for seventeen catis of silver, one bahar; for twenty-six brazas of linen, one bahar; for twenty-five brazas of finer linen, one bahar; for one hundred and fifty knives, one bahar; for fifty pairs of scissors, one bahar; for forty pairs of caps, one bahar; for ten pieces of Gujarat cloth, one bahar; for three of those gongs of theirs, two bahars; for one quintal of bronze, one bahar." The men of the armada traded the gongs, the knives, and other items pirated from the Chinese junks they had raided en route for the cloves. In return for these trinkets, they received a haul that a sailor might expect to see once or twice in a lifetime.

A detachment of well-armed crew members guarded the post, but as they knew from tragic experience, staying ashore overnight posed special hazards, even in a peaceful setting. Almanzor earned a measure of trust by warning them not to venture beyond the post at night, or they might encounter a renegade cult of men who appeared to be headless, and who carried with them a poison oint-ment. Anyone coming into contact with the ointment "falls sick very soon and dies within three or four days." The king explained that he had tried to discipline these menacing presences, and even had many of them hanged, but they still posed a danger. Forewarned (if scared out of their wits), the guard successfully avoided them.

As trading proceeded, Almanzor did all he could to put the armada at ease, even when the officers revealed that they were hold-ing sixteen captives, taken from islands they had visited. Perhaps their existence could no longer be concealed, or the space they occupied could be more profitably devoted to cloves or cinnamon. To the officers' surprise, the confession delighted the king, and he asked to take possession of the captives "so that he might send them back to their land with five of his own men that they might make the king of Spain and his fame known." There was also the ticklish matter of Carvalho's harem of three captive women, whom the offi-cers delivered to Almanzor for his personal use.

In return for his generous assistance, Almanzor asked only that the Europeans "kill all the swine that we had in the ships," in accordance with Muslim dietary laws, "for which he would give us an equal number of goats and fowls." Their food supply assured, the Europeans happily complied with the request. "We killed them in order to show him a pleasure and hung them up under the deck. When those people happen to see any swine they cover their faces in order that they might not look upon them or catch their odor."

If any member of the Armada de Molucca paused in the midst of his chores to reflect on these days in the Spice Islands, he could only marvel at how fortune, after punishing the fleet for months, had now chosen to favor it.

On the afternoon of November 13, Pedro Alfonso de Lorosa, Francisco Serrão's companion, hailed the fleet from a *proa*. He excitedly explained that the king of Ternate had given permission for the visit and instructed him to answer all questions truthfully, adding, in royal jest, "even if he did come from Ternate." There followed one of the more remarkable reunions in the Age of Discovery. In a time when travelers separated from their native cultures were often never heard from again, here was the Portuguese explorer standing before the armada's officers after a ten-year silence, in good humor and eager to impart vital intelligence concerning Armada de Molucca.

From Pedro Alfonso de Lorosa's detailed recollections, the officers learned that the implacable Portuguese authorities had been pursuing the armada around the globe: "He told us that he had already been sixteen years in India, and ten in the Moluccas, and that it was many years since the Moluccas had been secretly discovered, and that one year less fifteen days ago a great ship from Malacca had come there and left with a cargo of cloves." And this ship was still looking for the armada.

> Her captain was Tristão de Meneses, a Portuguese. And
> he [Pedro Alfonso] asked him what news there was in

Christendom; and he had replied that a fleet of five ships had sailed from Seville to discover the Moluccas in the name of the King of Spain, with Ferdinand Magellan, a Portuguese, as captain. And that the King of Portugal, in anger that a Portuguese should oppose him, had sent some ships to the Cape of Good Hope, and as many to Cape St. Mary, where cannibals lived, to guard and forbid the passage, and that he had not found them.

According to Pedro Alfonso de Lorosa, the Portuguese pursuit of the Armada de Molucca did not stop there, and he ended his tale with a bombshell:

A few days earlier a caravel with two junks had been there to learn news of us. But the junks went to Bacan to load cloves with seven Portuguese. And because they did not respect the king's wives and subjects, although the king had often told them not to behave thus, and since they refused to abstain and withdraw, they were put to death. And when the men of the caravel learned this, they immediately returned to Malacca, leaving the junks with four hundred bahars of cloves and as much merchandise as would purchase another hundred bahars. Moreover, he told us that every year many junks come from Malacca to Bandan to take and load mace and nutmeg, and from Bandan to Molucca to get cloves. And that these people go with their junks from Molucca to Bandan in three days, and from Bandan to Malacca in fifteen. And that the King of Portugal had already secretly enjoyed Molucca for ten years, that the King of Spain should not know.

This last piece of information explained why King Manuel had refused Magellan four times; a water route such as Magellan proposed, no matter how daring, threatened to disturb Portugal's lucrative

but clandestine trade in spices. Spain, with no such secret relationship, would naturally benefit from Magellan's plan. How strange and
wrongheaded to imagine, as did the mutineers and those whom they
influenced in Spain, that Magellan attempted to subvert the fleet to
aid Portugal. After fleeing Portugal, Magellan had been as loyal to
Spain as he claimed to be.

The officers of the armada plied Pedro Alfonso de Lorosa with
alcohol, so the revelations came thick and fast. Not until three o'clock
in the morning did the exhausted wanderer reach the end of his tale.
Amazed and persuaded by his stories, the officers begged him to join
their number by "promising him good wages and salaries." A man
without a country, he agreed. After eluding the agents of the Portuguese crown for so long, he would live to regret this decision.

On Friday the fifteenth of November," wrote Pigafetta, "the king
told us that he was going to Bacan to fetch the said cloves that those
Portuguese had left there, and he requested of us two presents to
give to the two governors of Motir in the name of the king of Spain.
And passing through our ships, he wished to see how we fired out
hackbuts, crossbows, and culverins, which are larger than an arquebus, and the king fired three shots of a crossbow, for that pleased
him more than the other weapons."

Still more gunplay ensued when Iussu, the king of Gilolo—"very
old, and feared through all those islands for the great power that he
had"—also paid a courtesy call on the armada on Saturday, prompting another exchange of gifts. "Since we were friends of the king of
Tidore," he advised, "we were also his, because he loved him like his
own son, and if any of us ever went into his country he would do
him very great honor."

He returned the next day to ask the armada to demonstrate its
firearms, and the gunners gladly complied. "He took the greatest
pleasure in it," Pigafetta noted. "He had been a great fighter in his
youth, as we were told."

*L*ater that day, Pigafetta finally had his chance to examine cloves carefully. These aromatic, humble bushes (*Syzygium aromaticum*) had inspired the voyage that had cost so many lives, and moved the destinies of empire around the world. Kingdoms in the East and West alike depended on them for economic support, and they provided the incentive for the emerging world economy. Centuries before Magellan, the Chinese had imported cloves, which were believed to have medicinal value. They were also used to flavor food and to sweeten breath. Europe found even more applications for the clove. Its essence, when applied to the eyes, supposedly improved vision. Its powder, when applied to the forehead, supposedly relieved fevers and colds. If added to food, it supposedly stimulated the bladder and cleansed the colon. If consumed with milk, it supposedly made intercourse more satisfying. It was miraculous, precious, and wonderful in all respects.

The word "clove" is derived from *clou*, the French word for nail, and the shape of its dried flowerbud is indeed reminiscent of a nail. The trees are slow to mature; from seedling to crop can take as long as seven or even eight years. Until it reaches the age of twenty-five or thereabouts, a clove tree will yield approximately eight pounds of the precious spice each year, depending on fluctuations in the climate. The ideal soil for growing cloves could be found in the Spice Islands: a deep, loamy, well-drained volcanic soil. Drenching rain is essential. The islands receive about one hundred inches of rain a year, ideal for cloves. The clove buds vary in length from one-half to three-quarters of an inch, and they contain up to 20 percent essential oil. The principal component is eugenol, an aromatic oil that imparts to cloves their distinctive, smoky flavor.

Harvesting cloves requires considerable care because the buds are fragile. The trick is to pull the buds away from the stems without damaging the branches; this was usually done by using the hand as a brush to sweep clusters of buds into waiting baskets or extended aprons. Once harvested, the buds were placed in the open for a few days to dry out. When desiccated, the stems and heads of the clove turn brown, and their weight is reduced by as much as two-thirds. Even after they are packed, they continue to lose moisture and weight, though at a much slower rate.

Now that Pigafetta was face-to-face with the source of all this wealth and struggle, he described it with obvious fascination:

> The clove tree is tall and as thick as a man's body or thereabouts. Its branches spread out somewhat widely in the middle, but at the top they have the shape of a summit. Its leaves resemble those of the laurel, and the bark is of a dark color. The cloves grow at the end of the twigs, ten or twenty in a cluster. Those trees generally have more cloves on one side than on the other, according to the season. When the cloves sprout, they are white, when ripe, red, and when dried, black. They are gathered twice per year, once at the nativity of our Savior and the other at the nativity of St. John the Baptist; for the climate is more moderate at those two seasons. . . . When the year is very hot and there is little rain, those people gather three or four hundred bahars in each of those islands. Those trees grow only in the mountains, and if any of them are planted in the lowlands near the mountains, they do not live. The leaves, the bark, and the green wood are as strong as the cloves. If the latter are not gathered when they are ripe, they become large and so hard that only their husk is good. No cloves are grown in the world except in the five mountains of those five islands. . . . Almost every day we saw a mist descend and encircle now one and now another of those mountains, on account of which those cloves become perfect.

Nutmeg was almost as important and valuable as cloves, and Pigafetta offered this description of its appearance in the wild: "The tree resembles our walnut tree, and has leaves like it. When the nut is gathered it is as large as a small quince, with the same sort of down, and it is of the same color. Its first rind is as thick as the green rind of our walnut. Under that there is a thin layer, under which is found the mace. The latter is a brilliant red and is wrapped about the rind of the nut, and within that is the nutmeg."

*I*n the early hours of Monday, November 25, Almanzor sailed out to the fleet in his *proa* to the resonant accompaniment of gongs. As he passed between the armada's ships, he announced that the cloves would be ready for delivery within four days. Overjoyed, the men of the armada fired their weapons to celebrate the event and to impress the king.

Later the same day, the men began to load what eventually amounted to 791 catis of cloves, about 1,400 pounds. "As those were the first cloves which we had laden in our ships, we fired many pieces." The more spices they took on board, the more anxious the men of the armada became to return to Spain before another disaster befell them.

*N*ow that the Europeans finally had their hands on the spices, Almanzor chose this moment to involve them in local politics, explaining that he wanted his visitors to return to the islands as soon as possible with even more ships. Even though the officers had experienced the bitter lessons of becoming ensnared in local vendettas, they blithely assured Almanzor they would help him. Content with this vague promise of assistance, the king invited everyone ashore for a banquet to celebrate the occasion.

The innocent gesture immediately sent the men of the armada into a panic because it reminded them of both the massacre at the banquet on Cebu and of Serrão's death by poisoning. Suddenly, the officers of the armada saw signs of impending doom wherever they looked; for example, "We saw those Indians speaking very low to our captives." Even the recently cleaned streets of the village, visible from the boats, appeared ominous. But they could not spurn the king's invitation because they depended on his goodwill for access to the spices. "Some of us, supposing that this was some treachery... were in great doubt and of contrary opinion to those who wished to go to the banquet, saying that we ought not to go ashore and reminding them of another such misfortune." Rather than go ashore, the officers offered to invite the king onto their ships, where they would bestow gifts on him, and even leave behind four men who wished to remain in the Spice Islands. (And good luck to those who remained in this dangerous place; they would certainly need it.)

Accepting the counteroffer, Almanzor immediately boarded *Trinidad*, boasting that he "entered there as safely as into his own houses." As the suspicious sailors listened, he said he was "greatly amazed" to hear that the armada was about to weigh anchor and sail away. "The space of time for lading the ships was thirty days," he explained. He meant no harm, or so he said, and only wanted to help them obtain their spices and journey home safely. "He besought us that we should not leave at once, seeing that it was not yet the season for navigation among those islands, and also because of the rocks and reefs that were around the island of Bandan, and also because we might easily have encountered the Portuguese." These were all persuasive arguments, as the officers realized. And he demonstrated his sincerity by saying that if the armada wanted to leave now, he would do nothing to stop it; he requested only that they take back all the gifts they had conferred on him "because the kings his neighbors would say that the king of Tidore had received so many gifts from so great a king"—that is, King Charles—"and had given him nothing, and they would think that we had departed only for fear of some deception and treachery, whereby he would always be named and reputed a traitor."

Here, at last, was the underlying reason why Almanzor wanted the armada to stay: to save face in front of the neighboring rulers. If he could maintain an alliance with his powerful visitors, he would impress and intimidate the jealous rulers of the other islands, but if he lost the visitors' favor, if they dismissed him as insignificant, he would appear vulnerable to the rival kings.

The officers began to appreciate what Magellan had always refused to acknowledge in his dealings with islanders: their presence placed both sides in peril. There were hazards for the Europeans (the islanders might massacre them), and there were hazards for the islanders themselves (the Europeans might take their women or disturb the local balance of power). Seeing himself as a savior who, in the name of Christianity and the king of Spain could do no wrong, Magellan remained blind to such nuances. But his pragmatic successors, chastened by experience, listened carefully to the king, both to protect their own lives and their precious cargo of spices.

The king became even more emotional as he sought to appeal to their hearts as well as their minds. "He had his crown brought and, first kissing it and setting it on his head four or five times," Pigafetta observed in astonishment, "he said in the presence of all that he swore by Allah, his great god, and by his crown which he had in his hand, that he desired to be forever a very loyal friend of the King of Spain. And he spoke these words almost weeping."

The king's tears softened the officers' hearts, and they decided to stay another fifteen days. To strengthen their shared bond of loyalty to the king of Spain, the officers gave the grateful Almanzor a royal banner displaying the insignia associated with Charles.

The king was apparently sincere in his goodwill toward the crew, but what about the other islanders? A few days later, the crew members heard that the lesser chiefs had urged Almanzor to kill all the Europeans because "it would give great pleasure to the Portuguese." The king sternly replied that he would not harm the visitors under any circumstances, "knowing the King of Spain and because he had made peace with us and plighted his faith." Although Almanzor had proved himself to be a man of his word, the crew members were right to be cautious. Even if he protected them, others might not follow his orders. By remaining aloof, yet carefully attuned to the king of Tidore, the armada, which had sailed into so many disasters, averted another, and perhaps final, calamity.

*W*orking feverishly throughout the last days of November and the early days of December, the men of the Armada de Molucca purchased and stored cloves until they had no more trinkets, caps, bells, mirrors, hatchets, scissors, or bolts of cloth to exchange for spices, and no more room to store the aromatic treasure. The ships reeked of the fragrant cloves; every breath the sailors drew was permeated with the scents of wealth, ease, and luxury.

The various kings of the Spice Islands paid daily visits to the ships, and the crew kept them entertained by firing off their weapons and engaging in mock swordplay. Despite the deep mistrust lingering between the islanders and the Europeans, a bond had formed between the two peoples. It was based, in part, on a mutual dislike of the Portuguese authorities (and the kings remained oblivious to the fact so many of the officers and crew happened to be Portuguese), but more than that, a genuine rapport developed between the armada's crew and the inhabitants of Tidore, which only complicated leave-taking.

*O*n Monday, December 9, Almanzor, whom Pigafetta unselfconsciously took to calling "our king," brought three betel-bearing women on board *Trinidad* to impress them with the power and glory of the king of Spain. Almanzor was followed closely by the king of Gilolo, who asked plaintively for one last blast of their guns, and, if they pleased, a final demonstration of swordplay and armor.

After the exhibition, Almanzor, who may have injected his own feelings into the matter, confided that Gilolo's king was bereft, "like a child who was taking milk and knew his sweet mother, who on departing would leave him alone; but that more especially he would remain desolate, because he had already known us and tasted some of the things of Spain." Tearfully accepting that the armada must leave, he advised the departing sailors to sail only by day to avoid the shoals strewn throughout these waters. When the officers informed him that they planned to sail "day and night," he told them he would pray daily for their safety.

The decorous leave-taking was marred only by an incident concerning Pedro Alfonso de Lorosa. Ever since his decision to return with the fleet to Spain he had remained in seclusion aboard *Trinidad*, out of harm's way. With the departure only days away, the son of Ternate's king, Chechili, traveled out to the fleet in a "well-manned *proa*," seeking to lure Lorosa into his vessel.

Fearing that he would be kidnapped and killed, Lorosa refused to go along, declaring that he was returning to Spain, "Whereupon," said Pigafetta, "the king's son tried to enter the ship, but we refused to allow him to come aboard, as he was a close friend of the Portuguese captain of Malacca, and had come to seize the Portuguese [Lorosa]."

Frustrated in his attempt to capture Lorosa, Chechili returned to his island, venting his wrath on those who had let Lorosa go.

*O*n December 15, the king of Bacan and his brother approached the fleet in the largest native vessel the crew had seen. Three tiers of oarsmen—120 men in all—propelled the craft through the water, "and they carried many banners made of white, yellow, and red parrot feathers." Its progress announced by the sound of gongs, used to synchronize the oarsmen's strokes. It was accompanied by two *proas* "filled with girls." As it happened, the king's brother was about to marry Almanzor's daughter, and the girls were intended as presents for the couple.

A summit meeting between kings unfolded with elaborate protocol. "When they passed near the ships, we saluted them with our artillery, and they in salute to us sailed round the ships and the port." Afterward, "our king," the king of Tidore, "came to congratulate him as it is not the custom for any king to disembark on the land of another king. When the king of Bacan saw our king coming, he rose from the carpet on which he was seated, and took his position at one side of it. Our king refused to sit down upon the carpet, but on its other side, so no one occupied the carpet. The king of Bacan gave our king five hundred *patols,* because the latter was giving his daugh-ter as wife to the former's brother. The said *patols* are cloths of gold and silk manufactured in China, and are highly esteemed among them. Whenever one of those people died, the other members of his family clothe themselves in those cloths in order to show him more honor."

The festivities resumed the next day, when Almanzor dispatched fifty women "all clad in silk garments from the waist to the knees" with a banquet for the king of Bacan. "They went two by two with a man between each couple. Each one bore a large tray filled with other small dishes which contained various kinds of food. The men carried nothing but the large wine jars. Ten of the oldest women acted as macebearers. Thus did they go to the *proa,* where they pre-sented everything to the king, who was sitting upon the carpet under a red and yellow canopy." The crew members watched this ceremony with fascination and longing, because during their weeks in the Spice Islands, they had refrained from the orgies that high-lighted their earlier layovers. Catching sight of the yearning sailors, the women decided to have a little fun and boarded one of the ships, where they "captured" them; in all likelihood, the hostages did not put up much resistance. The flirtatious game continued until "it was necessary to give them"—the women—"some little trifle in order to regain their freedom," Pigafetta commented.

More industrious crew members busied themselves in bending and decorating the sails for the ships, restoring the rigging, and mak-ing sure the vessels would be able to withstand the rigors of the jour-ney home. When hoisted, the sheets revealed a freshly painted design: an elaborate cross and beneath it the legend, "This is the sign of our good fortune."

As that bold legend indicated, the officers and crew of the armada were proud of their accomplishments. Their voyage finally demonstrated what Columbus and so many other explorers had failed to show, that a water route to the Moluccas existed, and that it was possible to reach the East by sailing west. Those who had survived the grueling journey could look back on countless moments of courage and even heroism that helped to bring them to this place, and they could console themselves with dreams of glory and avarice.

*A*s the hour of departure approached, the pace of activity quickened. The fleet took on board eighty casks of water and a supply of wood cut by one hundred laborers assigned to the task by the king of Bacan, who rallied to the cause of the armada and Spain. To seal the alliance, he arranged a meeting on the neighboring island of Mare with representatives of the armada (including Pigafetta) and Almanzor. The ceremony was impressive: "Before the king walked four men with drawn daggers in their hands. In the presence of our king and of all the others he said that he would always remain in the service of the king of Spain, and that he would save in his name the cloves left by the Portuguese until the arrival of another of our fleets, and he would never give them to the Portuguese without our consent."

To demonstrate his good faith, he gave the armada a slave as a present for the king of Spain; two additional bahars of cloves (he would have sent ten, but the ships were so heavily laden with spices that there was no room); and "two extremely beautiful dead birds," which caught Pigafetta's imagination. "The people told us that those birds came from the terrestrial paradise, and they call them *bolon diuata*, that is to say, 'birds of God.' " The birds of paradise, as they came to be known throughout Europe, were as celebrated as the cloves, a token of heaven on earth. Maximilian of Transylvania reported that the Moors believed the birds were born in Paradise, spent the entire lives aloft, never falling from the sky until they died. Anyone who retrieved their skins and wore them in battle was supposed to be protected from harm. So these were extremely valuable presents, as Pigafetta realized at the time.

On the day of departure, the kings of all the Spice Islands assembled on the island of Mare to see the fleet off. *Victoria* weighed anchor and set sail, standing off the harbor awaiting *Trinidad,* the flagship, to join her. The ships' gunners fired their artillery one more time, but in the midst of the excitement, *Trinidad*'s cables fouled and to the dismay of everyone, she began taking on water. None of the eyewitnesses supplied a reason for the near-disaster; most likely, the ship had not been adequately repaired during the long layover on Cimbonbon. But the leak was worse than ever, and she was in danger in losing her cargo of spices.

With her sister ship in distress, "*Victoria* returned to her anchorage, and we immediately began to lighten *Trinidad* to see whether we could repair her. We found that the water was rushing in as through a pipe, but we were unable to find where it was coming in. All that and the next day we did nothing but work the pump." The work was grueling, but necessary. The loss of *Trinidad* would have been a disaster, depriving the armada of the rewards of its long-sought-after spices. Even worse, *Victoria* lacked room to hold the crews of both vessels. The arduous pumping continued until the men were exhausted, "but we availed nothing." Laden with spices, the flagship of the fleet was on the verge of sinking at her mooring.

After all the pomp and circumstance, not to mention the gongs, girls, and parrot feathers surrounding the fleet's departure, the situation was humbling, indeed. And it was just the sort of mishap that Magellan would likely have prevented, because he had always been meticulous about the condition of his ships and saw to it that they were seaworthy at all times. *Trinidad* had fallen into disrepair from sheer neglect, and with that ship disabled, the officers' hasty decision to burn *Concepción* returned to haunt them. Not even Magellan would risk taking one, and only one, ship all the way from the Spice Islands back to Spain.

*A*s soon as Almanzor—"our king"—heard about the plight of *Trinidad,* he sprang into action, boarding the afflicted ship and prowling below deck, trying to locate the source of the damnable leak, but without success. Then, "He sent five men into the water to see whether they could discover the hole. They remained more than one half hour under water, but were quite unable to find the leak." The ship was listing badly, and desperate measures were required. "Seeing that he could not help us and the water was increasing hourly, [he] said almost in tears that he would send to the head of the island for three men, who could remain under water a long time." Almanzor went in search of them, as the ship slowly but unmistakably settled into the water.

After an anxious night, Almanzor reappeared with the men by the first light of dawn. "He immediately sent them into the water with their hair hanging loose so that they could locate the leak by that means." Water entering into ship would draw strands of their hair into its current. But even these men failed to locate the leak, and when they emerged, grim-faced, from the water, the king finally broke down in tears. Who among them, he pleaded, would be able to return to Spain now and tell King Charles about the loyalty of the king of Tidore?

Pigafetta and the others tried to calm the distraught ruler by describing their new plan for returning to Spain. "We replied to him that *Victoria* would go there in order not to lose east winds that were beginning to blow, while the other ship, until being refitted, would await the west winds and go then to Darién, which is located in another part of the sea in the country of Yucatán." In other words, Elcano would take *Victoria* on a westerly course, which was the most direct route back to Spain. But it brought special dangers because it cut a swath through the Portuguese hemisphere, as defined by the

Treaty of Tordesillas. If Portuguese navigators captured a Spanish ship loaded with spices in their waters, they would be merciless. *Trinidad*'s course home promised even greater risks. Once she was repaired, she would try to catch favorable winds carrying her along an easterly course to the American continent. Her cargo of spices would then be transferred to mules, and the beasts would carry the spices to another Spanish fleet heading for Seville.

As devoted and helpful as ever, Almanzor pledged no fewer than 250 carpenters to perform "all the work" required to return *Trinidad* to seaworthiness, and he promised to treat all the sailors who remained behind as if they were his own sons, vowing that "they would not suffer any fatigue beyond two of them to boss the carpenters in their work." The king's sincerity and generosity finally wore away the officers' skepticism: "He spoke these words so earnestly that he made us all weep."

*T*he unsuccessful efforts to repair *Trinidad*'s mysterious leak and the deliberations leading to the decision that *Victoria* would return alone consumed five days. Just before *Victoria* left Tidore, the crew members loaded her with as many cloves as they could salvage from *Trinidad*, but once they saw *Victoria* riding low in the water, "mistrusting that the ship might open," they lightened her by removing sixty quintals of cloves and storing the spices in the trading house.

Victoria was so dilapidated that many crew members refused to board her. They preferred to remain with *Trinidad* in Tidore until she was repaired. Still others stayed behind because they feared that those aboard *Victoria* would "perish of hunger" long before they reached Spain. So the crew divided itself between the two ships, each man seeking the lesser of two evils: *Victoria*, the flimsy vessel that would depart for Spain immediately, or the much larger *Trinidad*, which needed weeks if not months of repairs before she could begin her journey home. Dangers abounded both on land and at sea; starvation and shipwreck imperiled those who sailed, while headless marauders or poison might fell those who remained behind.

In the end, Carvalho was designated captain of *Trinidad,* and Elcano took over the command of *Victoria.* Among the fifty-three men who cast their lot with *Trinidad* were Ginés de Mafra, the pilot; Gonzalo Gómez de Espinosa, the master-at-arms (and second in command to Carvalho); and Hans Vargue, a German gunner. Pigafetta faced the most critical decision of the entire journey: Which ship would he join? His instinct for survival had stood many tests, and he elected to go along with Elcano aboard *Victoria;* that ship would carry him among her crew of about sixty men, including sixteen Indians. Although he detested the Basque mariner, he clearly had more confidence in Elcano's seamanship than in Carvalho's.

Each ship in the divided fleet contained a memoirist, Pigafetta aboard *Victoria* and de Mafra aboard *Trinidad.* The Venetian resumed his passionate, eloquent descriptions of the Indies, while de Mafra—"a man of few but true words," by his own account—stuck to a more practical report of what he perceived as bad judgment and missed opportunities.

*E*arly on the morning of December 21, Almanzor, ever helpful, came aboard *Victoria* for the last time, delivering two pilots, paid for by the crew, to guide the ship safely through the maze of islands and shoals. The king then took his leave. Familiar with the tides, the pilots insisted that early morning was the most advantageous time to depart, but the men who remained behind persuaded *Victoria* to delay a few hours while they wrote long letters for her to carry home to Spain. Finally, at noon, it was time to leave the Spice Islands. "When that hour came," Pigafetta recalled, "the ships bid one another farewell amid the discharge of the cannon, and it seemed as though they were bewailing their last departure. Our men [remaining behind] accompanied us in their boats a short distance, and then with many tears and embraces we departed."

This should have been a festive occasion, the ships bulging with spices, heading for home port and the prospect of a grand reception from King Charles, but the damage to *Trinidad* dramatically altered the final leg of this voyage around the world and mocked the proud legend painted on her sails. The crew faced more than the ordinary pangs of leaving port for another long journey at sea, although those pangs—the monotony of life at sea, the nights interrupted by watches, the gradual diminution of their fresh food to a diet of salted dried meat, salted biscuit, and salted dried fish—were hard enough to bear, but now, in addition to all that, they knew their lives would be at risk the moment they were out of sight of the Spice Islands.

Despite the obstacles they had faced, the men of the armada had always taken comfort in the knowledge that they had extra ships at their disposal. Even two ships had a reasonable chance of making it back to Seville, but one ship was hardly equal to the task, no matter how skillful the crew's seamanship, or how favorable the winds. One ship, alone on the high seas, was always at the mercy of storms, shoals, pirates, termites, or faulty navigation. On the high seas, no king could protect them, and at least one sovereign, the king of Portugal, wanted them dead. (Of all the captains in the armada, only Magellan had fully appreciated the full extent of Portuguese malice toward him.) Yet they had no choice but to face the tests presented by the ten-thousand-mile-long route home.

CHAPTER XIV

Ghost Ship

O Wedding-Guest! this soul hath been
Alone on a wide wide sea:
So lonely 'twas, that God himself
Scarce seeméd there to be.

Laden with cloves and about sixty survivors, *Victoria* left the island of Tidore on December 21, 1521. Heading southwest, she called at a small island nearby to load firewood, and resumed her southerly course toward one of the most dangerous stretches of ocean in the world: the Cape of Good Hope.

Embarking on the final leg of the unprecedented journey around the world should have been an occasion for relief among the homeward-bound crew members, but it was not. The character of the expedition had changed completely; the Armada de Molucca finally had its spices, but it had lost its soul. The absence of Magellan's guiding hand, his fierce discipline, even his quixotic delusions of grandeur, left the two remaining ships and their crew members without a sense of overriding purpose. Only survival mattered now.

Even if the crew survived the voyage home, they were anxious about the reception they would receive in Spain. Although they had no knowledge of *San Antonio's* arrival in Seville seven months earlier, they suspected that the mutineers aboard that ship might have made it back and succeeded in discrediting Magellan. Elcano and *Victoria's* crew feared they would be arrested and jailed for treason the moment they tied up at the dock. Desertion might have been an appealing option among the grim choices facing the sailors, except for their fear of cannibals inhabiting the islands surrounding them. In the end, staying aboard ship served as the best strategy to fore-stall disaster. They found themselves prisoners of peculiar circum-stances, hostages to a situation created largely by those who had predeceased them.

Even Antonio Pigafetta, so determined to bring the news of Ma-gellan's accomplishments back to Europe, was at a loss for words, con-tent simply to note the islands *Victoria* passed: Caion, Laigoma, Sico, Giogi, and Caphi, all part of the Moluccas. On the advice of local pi-lots, he recorded, "We turned toward the southeast, and encountered an island that lies in a latitude of two degrees toward the Antarctic Pole, and fifty-five leagues from Maluco. It is called Sulach [later called Xulla], and its inhabitants are heathens." Here Pigafetta briefly resumed his amateur anthropology: "They have no king, and eat human flesh. They go naked, both men and women, only wearing a bit of bark two fingers wide before their privies." Cannibals seemed to be everywhere; Pigafetta listed ten islands to be avoided at all costs.

Two days after Christmas, the ship found anchorage in Jakiol Bay, where the crew obtained fresh, and much needed, supplies, along with an Indonesian pilot who knew his way around these islands. Under his guidance, the crew sailed on as if in a trance, heading south, narrowly avoiding Moors and cannibals, coral reefs and hid-den sandbars. Eventually, *Victoria* put the Indonesian islands astern, passing through the Alor Strait and eluding pirates. As a lone vessel laden with spices, *Victoria* was especially vulnerable to predators.

On January 8, 1522, *Victoria* entered the Banda Sea, extending west of the Moluccas, and the torpid weather suddenly changed.

"We were struck by a fierce storm," Pigafetta reported, "which caused us to make a pilgrimage to Our Lady of Guidance. Running before the storm, we landed at a lofty island, but before reaching it we were greatly worn out by the violent gusts of wind that came from the mountains of that island, and the great currents of water." The squall nearly shattered the ship, but Elcano avoided the rocks and reefs, and when the seas moderated, *Victoria* limped to an anchorage close to shore. The next day, divers inspecting the hull discovered extensive damage, and the men gingerly hauled the vessel onto a beach to commence repairs and caulking.

The inhabitants of this island, known as Malua, shocked even these hardened sailors. They were, said Pigafetta, "savage and bestial, and eat human flesh," and their appearance combined the frightening and the outlandish. They went naked, or nearly so, "wearing only that bark as do the others, except when they go to fight, they wear certain pieces of buffalo hide behind, and at the sides, which are ornamented with small shells, boars' tusks, and tails of goat skins fastened before and behind." They lavished most of their attention on their hair, "done up high and held by bamboo pins which they pass from one side to the other." Completing this curious picture, "They wear their beards wrapped in leaves and thrust into small bamboo tubes—a ridiculous sight." All in all, Pigafetta judged them to be "the ugliest people in the Indies."

Despite the inhabitants' bizarre appearance, the sailors, by this time old hands in such transactions, bestowed trinkets on them, and both sides quickly made peace. As the sailors set to work repairing the ship, the aristocratic Pigafetta, spared the indignity of physical labor, roamed the island, studying its flora and fauna, noting an abundance of fowl and goats and coconuts and pepper: "The fields in those regions are full of this pepper, planted to resemble arbors." He was speaking of black pepper, which had been introduced to the island some time before the Europeans' arrival, and which the inhabitants carefully cultivated.

$\mathcal{T}$wo weeks later, with repairs to the hull completed, Elcano gave the order to resume their voyage home, and the crew set sail on Saturday, January 25. *Victoria*, having sailed five leagues or so, called at the island of Timor, towering nearly ten thousand feet above the shimmering surface of the Pacific. Everyone aboard her looked forward to a luxurious, satisfying time ashore, because food, spices, almonds, rice, bananas, ginger, and fragrant wood were all said to grow there in abundance.

Pigafetta's linguistic skills gave him a prominent part to play in the dealings with the locals to obtain provisions. "I went ashore alone to speak to the chief of a city called Amaban to ask him to furnish us with food. He told me that he would give me buffaloes, swine, and goats, but we could not come to terms because he asked many things for one buffalo." Assessing his surroundings, Pigafetta realized the chief lived in luxury, attended by numerous naked serving women, all of them adorned with gold earrings "with silk tassels pendant from them, as well as amulets of gold and brass." And the men displayed even more gold jewelry than the women.

While Pigafetta was negotiating, two young crew members deserted; Martín de Ayamonte, an apprentice seaman, and Bartolomé de Saldaña, a cabin boy, swam ashore under cover of darkness and were never heard from again. They were exceptions to the generally cautious behavior of *Victoria*'s crew in Timor. For example, they refrained from enjoying the charms of the local women, believing they were infected with syphilis—"the disease of St. Job." They had seen evidence of what they assumed to be syphilis all over the Moluccas, according to Pigafetta, but the greatest concentration occurred here, on this island. The origins of syphilis in this part of the world are a mystery. Portuguese traders or sailors might have carried it with them (syphilis was also known as "the Portuguese disease"), but it is worth noting that the disease was reported in China centuries earlier than in Europe, and that junks regularly plied these waters. It is also possible that the sailors' diagnosis was mistaken, and they had come across islanders affected with leprosy.

To guarantee cooperation with the islanders, Elcano ordered a party of sailors ashore in search of a bargaining chip: "Since we had but few things, and hunger was constraining us, we restrained in the ship a chief and his son from another village." With their hostages in hand, the armada's officers proceeded to negotiate for the food they so desperately needed. The strategy worked exactly as planned, and the recalcitrant islanders delivered a ransom of six buffalo, a dozen goats, and as many pigs to the grateful yet rapacious sailors in exchange for the hostages' freedom.

*O*nce the slaughtered beasts had been loaded, *Victoria* prepared to set sail once more, this time heading for the island of Java, the largest and, to Europeans, the best-known destination in the Indies. Among the crew members, Java possessed a mysterious allure, if only because the Javanese reputedly practiced exotic customs such as *palang*. Pigafetta relished telling the tales he heard of Java, beginning with its funeral rites. "When one of the chief men of Java dies, his body is burned," he wrote. "His principal wife adorns herself with garlands of flowers and has herself carried on a chair through the entire village by three or four men. Smiling and consoling her relatives who are weeping, she says, 'Do not weep, for I am going to sup with my dear husband this evening and to sleep with him this night.' Then she is carried to the fire, where her husband is being burned. Turning toward her relatives, and again consoling them, she throws herself into the fire, where her husband is being burned. If she did not do that, she would not be considered an honorable woman or a true wife to her dead husband." For all its melodrama, this was a fairly accurate account of a funeral ceremony as practiced on the island of Bali, located little more than a mile east of Java, and in India.

And then there was the role *palang* played in Javanese courtship rites. Magellan's relative Duarte Barbosa, in his account of the region, had described Javanese *palang* in excruciating detail. "They are very voluptuous," he wrote of the inhabitants, "and have certain round hawk's bells sewn and fastened in the head of their penis between the flesh and the skin in order to make them larger. Some have three, some five, and others seven. Some are made of gold and silver and others of brass, and they tinkle as the men walk. The custom is considered quite the proper thing. The women delight greatly in the bells, and do not like men who go without them. The most honored men are those who have the most and largest ones."

Pigafetta observed that the custom still formed a vital part of Javanese life. "When the young men of Java are in love with any gentlewoman, they fasten certain little bells between their penis and foreskin. They take a position until their sweetheart hears the sound. The sweetheart descends immediately, and they take their pleasure; always with those little bells, for their women take great pleasure in hearing those bells ring from the inside of their vagina. Those bells are all covered, and the more they are covered, the louder they sound."

Normally a careful observer, Pigafetta could not resist telling tales when the mood came over him. In the same breath as his description of *palang*, he conjured Amazons, among the most persistent of all the illusions of Neverland, and perhaps the hardest for the lonely sailors who roamed the world to give up. Pigafetta lent at least partial credence to an account he heard about Amazons on a neighboring island kill their male offspring and raise only females. And any man found exploring the island would be attacked instantly. Needless to say, the survivors of so many shipwrecks, mutinies, ambushes, and other disasters elected not to risk the wrath of the Amazons they believed to be in their midst.

Although *Victoria* remained hundreds of miles distant from the southernmost point of China, Pigafetta heard dramatic stories of the

Middle Kingdom from local traders. "The king," as Pigafetta referred to the emperor, "never allows himself to be seen by anyone. When he wishes to see his people, he rides about the palace on a skillfully made peacock, a most elegant contrivance, accompanied by six of his principal women clad like himself; after which he enters a serpent called a *nagha*"—the name given to a mythical dragon— "which is as rich a thing as can be seen, and which is kept in the greatest court of the palace. The king and the women enter it so that he may not be recognized among his women. He looks at his people through a large glass which is in the breast of the serpent. He and the women can be seen, but one cannot tell which is the king. The latter is married to his sisters, so that the royal blood may not be mixed with others."

The emperor, it seemed, had absolute power over all his subjects, and he wielded it with impressive, if fiendish, enthusiasm. "When any seigneur is disobedient to the king, he is ordered to be flayed, and his skin dried in the sun and salted. Then the skin is stuffed with straw or other substance, and placed head downward in a prominent place in the square, with hands clasped above the head, so that he may be seen to be performing *zonghu*, that is, obeisance."

Pigafetta's vivid evocation of Chinese customs reveals his yearning to visit the Middle Kingdom and play the role of diplomat and translator as he had throughout the voyage. Perhaps Magellan, had he been alive, would have made a detour and allowed Pigafetta to fulfill his dream, but Elcano had no such ambitions. China remained tantalizingly remote.

*I*n the early hours of Wednesday, February 11, *Victoria* weighed anchor and put the island of Timor astern, sailing along a southwesterly course. With Java and, later on, Sumatra barely visible to starboard, she headed for her meeting with destiny at the Cape of Good Hope.

The struggle with the elements was joined within days of leaving Timor as *Victoria* became the plaything of the unstable weather systems of the southern latitudes. "In order that we might double the Cape of Good Hope, we descended to forty-two degrees on the side of the Antarctic Pole. We were nine weeks"—nine weeks!—"near that cape with our sails hauled down because we had the west and northwest winds on our bow quarter and because of a most furious storm," Pigafetta explained. He went on to warn, "It is the largest of and most dangerous cape in the world." And he was right.

Although the Cape of Good Hope was first rounded in 1488 by Bartolomeu Dias and nine years later by Vasco da Gama—both major accomplishments in Portuguese exploration history—it was still considered extremely hazardous and barely navigable even by the most seaworthy of ships and the most experienced of captains. It occupied a nearly mythical place in the Portuguese consciousness as the most fearsome place in the entire world.

Sebastián Elcano had never experienced anything like the fierce, confused winds and riptides of Cabo Tormentoso; doubling it would tax his navigational skills, his patience, and his daring to the utmost. Many of the crew wanted to jump ship at the island of Madagascar rather than risk doubling the cape, said Pigafetta, "because the ship was leaking badly, because of the severe cold, and especially because we had no other food than rice and water; for as we had no salt, our provisions of meat had putrefied." Doing so meant a life of exile and slavery, because Madagascar was a Portuguese stronghold, with ships flying the Portuguese colors calling there on their way to and from the Indies.

A few brave souls on board *Victoria* had no use for Madagascar. They retained their principles and allegiance to King Charles, and preferred death to spending the rest of their days marooned off the coast of Africa. They were, said Pigafetta, "more desirous of their honor than of their own life, determined to reach Spain, dead or alive."

*H*alfway between Australia and Africa, *Victoria* began to leak dangerously. Deliverance seemed at hand on March 18, when the crew sighted the prominent hump of what is now known as Amsterdam Island. Elcano hoped to perform urgently needed repairs on the shores of this small volcanic landmass, but after four days of tacking in rough weather and surging seas, he was unable to find a secure anchorage. "We saw a very high island, and we went towards it to anchor, and we could not fetch it; and we struck the sails and lay to until the next day," Albo recorded in frustration.

Elcano eventually gave up on the idea of reaching Amsterdam Island, and repairs took place in the ocean swells. As the men worked, they might have seen the killer whales or elephant seals, and if they lifted their gaze, they would have seen several species of albatross circling above them, the same benignly smiling bird that Samuel Taylor Coleridge's imagination transformed into a symbol of hope and innocence corrupted by thoughtless violence.

Once the repairs were completed, *Victoria* resumed her westerly course. Over the following days and weeks, the crew, hovering on the verge of starvation and dreading the onset of scurvy, steadily ate their way through their supply of rice and awaited whatever destiny had in store.

*F*ifteen hundred miles east of Amsterdam Island, *Trinidad* prepared to leave the island of Tidore. On April 6, after more than three months of repairs, she finally weighed anchor and unfurled her sails. The ship carried a full load of spices, one thousand quintals of cloves—fifty tons!—more than enough to justify the expense of the entire voyage.

Magellan's former flagship was commanded by Gonzalo Gómez de Espinosa, and her pilot was Juan Bautista Punzorol, known to history as the "Genoese pilot," after the name of the short memoir of the voyage he left behind. Sorely missed was Juan Carvalho, the capable pilot who had become a corrupt Captain General; he had died of unknown causes on February 14.

As the fleet's *alguacil,* or master-at-arms, Espinosa had performed as a loyal servant to King Charles, and he had helped Magellan maintain authority over his often rebellious crew. During the mutiny in Port Saint Julian, when Magellan lost control of three of his ships, Espinosa had come to his aid, and as a career soldier, he discharged his dangerous duties without fuss or complaint. But as a captain, Espinosa was hopelessly out of his element. Without Magellan to advise and protect him, it became apparent that he lacked the navigational skills to take his ship through rough weather; beyond his lack of expertise, his character, seemingly so straightforward and loyal, turned ambivalent when he should have been resolute, and naïve where he should have been canny. It was not that he lacked discipline, or the support of the men; the problem was that Espinosa, a soldier, was simply not qualified to command a ship. The challenge of guiding *Trinidad* halfway around the world, often against the prevailing winds, was beyond him, as it might have been beyond even Magellan, had he lived to face it.

Espinosa decided to leave behind four men to operate a trading post on the island of Tidore. The post would store cloves and serve as a symbol of Spanish rule in the Spice Islands. The four men stationed there were, Ginés de Mafra recalled, "Juan de Campos and Luis de Molino and a Genoese and a certain Guillermo Corco." While serving time at their remote outpost, they picked up alarming intelligence: "Some Indian merchants who had come there to buy cloves told them that a Portuguese armada was coming from India to the Moluccas because they had learned of the Castilians' presence there." They, too, wanted to establish an outpost, but more than that, they planned to seize control of the spice trade. The four men left behind suddenly found themselves vulnerable to both Portuguese marauders and to the island's residents, whose loyalties could be purchased or transferred with a show of force.

Setting sail, Espinosa backtracked and followed an easterly

course through waters the fleet had already explored, past Gilolo and Morotai, and into the Philippine Sea, all the way to the island of Komo, where *Trinidad* took on more provisions. From this point on, stout easterly headwinds got the better of his navigational skills, and he took a more northerly course. The choice proved disastrous. Although he now understood how large the Pacific Ocean was, his ideas about the location of landmasses in the Northern Hemisphere were deeply flawed. He mistakenly believed that Asia was connected to the American continent, and that misunderstanding led him to assume that if he sailed far enough north, he would catch benign westerly winds. But soon after his departure, the monsoon season started in earnest, bringing with it a seemingly endless succession of storms and drenching rains.

"After ten days of sailing," according to de Mafra, "we arrived at one of the Islands of the Thieves." Their position was uncannily close to the armada's first landfall after the ninety-eight-day ordeal of crossing the Pacific during the voyage out. "There Gonzalo de Vigo stayed, much tired of the travails." Nor was he the only one to desert—in all, three crew members fled, preferring to take their chances on a remote Pacific island rather than remain aboard Espinosa's ship of doom. (De Vigo remained in the Philippines for the rest of his life; the other two deserters were killed by islanders.)

De Mafra wrote that *Trinidad* "sailed to the northeast until she reached 42 degrees North." Espinosa faced winds of ever-increasing intensity, and soon storms overwhelmed the isolated ship. A more ill-advised global detour cannot be imagined. One can only wonder what he was thinking as he sailed as far north as Japan, into ever more frigid waters, because this course took him away from his goal of reaching Darién.

Scurvy returned to plague the men, and its miseries made the living envy the dead. "At this point many began dying," said de Mafra, "and one of them was opened to see what it was that they were dying of, and his body was found to be as if all its veins had burst open because all the blood had spread all over the interior of his body. Henceforth, whenever anyone fell sick he was bled because it was thought that the blood was suffocating him, but they kept dying all the same and did not elude death, so thenceforward the sick men were considered helpless and left untreated." Scurvy ultimately claimed the lives of thirty men, leaving only twenty to carry on. In their frail and bewildered state, the handful of survivors sought an explanation for their suffering. "Some claimed that it was because of the venom poured by the Ternate Indians into the well where they had collected water for the voyage," de Mafra suggested.

Even Espinosa admitted that his course placed the ship in peril, first from the weather, and then from illness: "It became necessary for me to cut the castles and quarter-deck because the storm was so big and the weather so cold that aboard the ship that we could not cook any food. The storm lasted twelve days and because the people did not have any bread to eat, most of them lost weight and when the storm had passed and the people could once again cook food, on account of the many worms we had, it gave them nausea, which affected most people."

Finally, Espinosa came to his senses. "When I saw the people suffering, the contrary weather, and [realized] that I had been at sea for five months, I turned back to the Moluccas, and by the time we got to the Moluccas . . . it had been seven months at sea without taking [on] any refreshments."

After a brief respite at the Islands of the Thieves to collect water, Espinosa commanded *Trinidad* to resume her retreat toward Tidore, but as he approached his goal, he received shocking news. On May 13, five weeks after *Trinidad*'s departure from Tidore, a fleet of seven Portuguese ships, all looking for Magellan and the Armada de Molucca, had arrived at the island. Their leader was António de Brito, bearing a royal appointment as governor of the Spice Islands.

His Portuguese soldiers, heavily armed, imprisoned the four crew members Espinosa had left behind to maintain a trading post. Then

Brito turned his attention to Almanzor, the king of Tidore, demanding to know how he could have allowed the Spanish to maintain a
post on his island. Almanzor pleaded for mercy, explaining that the
Spanish had forced him to yield, but now that Captain Brito had
come to rescue Almanzor from the Spanish, he would gladly switch
his allegiance back to the Portuguese. Captain António de Brito,
whose cynicism concerning Almanzor's protestations can be imagined, reclaimed the Spice Islands in the name of Portugal.

Espinosa dispatched a boat bearing a letter for Captain Brito,
begging for sympathy. He told a pathetic tale. His ship was in bad
condition, down to its last anchor; one storm could send her to the
bottom. And he was in desperate need of supplies. Had Magellan
been alive, he would never have been so foolish as to write a letter to
the Portuguese captain charged with capturing him, and the last
thing he would have done was to reveal his whereabouts and weak-
nesses to the enemy. He would have known there was no chance of
mercy from the Portuguese.

Rather than the compassion Espinosa expected, the letter only
made Brito gloat. After searching the Indies for three years, the Portuguese governor now knew exactly where the Armada de Molucca
was located, and once he had captured the crew, he would treat
them as cruelly as he wished.

A few days later, a Portuguese caravel with twenty armed men
stormed Benaconora, the harbor where Espinosa had sought refuge.
The soldiers boarded *Trinidad*, expecting to overwhelm the crew,
but were repelled by the grievous spectacle of men near death, a
foul and unhealthy stench that no one dared to brave, and a ship on
the verge of sinking. Everything Espinosa had said in his letter to
Brito was true; *Trinidad* and her crew were in desperate condition
and offered no threat to the Portuguese.

Unmoved, the Portuguese soldiers arrested Espinosa and sailed
Magellan's fetid and decrepit flagship to Ternate. There Brito took
possession of *Trinidad*'s papers, logbooks, quadrants, and astrolabes.
Included in the haul were the diary of Andrés de San Martín and, it

is said, Magellan's personal logbook. Brito ordered the ship stripped
of all her sails and rigging, and in this condition, she rode helplessly
at anchor until a severe storm hit the island. The winds smashed
apart the remains of the once-proud ship, her precious cargo of
cloves sank, and the splintered remnants of her hull washed ashore.
The flagship of the Armada de Molucca ended up as driftwood.

Espinosa had squandered his chance for glory. If he had succeeded in guiding *Trinidad* home, he would have earned a place in
history and a fortune for himself. Instead, his indecision claimed the
lives of over a score of men, half the remaining assets of the armada,
a valuable cargo of cloves, and the records maintained by *Trinidad*'s
officers, including Magellan himself.

When Brito perused the logbooks, he became incensed because
they contained damning evidence of the armada's route through
Portuguese waters and its attempts to snatch the Spice Islands away
from Portugal. The source of the intelligence was impeccable: the
records of the fleet's official astronomer, Andrés de San Martín. To
make matters worse, Brito discovered that the astronomer had
secretly altered the location of various lands to obscure the embar-
rassing fact that the ships had wandered into the Portuguese hemisphere, at least as it was defined by the Treaty of Tordesillas. With
this information, Brito had his motive for revenge.

*H*is first victim was Pedro Alfonso de Lorosa, the Portuguese
renegade who had joined the fleet when it first called at the Spice
Islands. He was beheaded.

Brito then considered executing several sailors and pilots, but preferred that they die a slow death in the tropical heat. He later
reported to the king of Portugal, "So far as concerns the master, clerk,
and pilot . . . it would be more to your Highness's service to order
their heads to be struck off than to send them [to India]. I kept them
in the Moluccas because it is a most unhealthy country, in order that
they might die there, not liking to order their heads to be cut off,

since I did not know whether your Highness would be pleased or not." Brito based his judgment of the climate on his own troops' suffering; of the two hundred under his command, only fifty survived. The Portuguese governor did spare the lives of two men, a boatswain and carpenter, but he did so only to press them into service for the Portuguese. He sent the rest of the crew to a fortress under construction on the island of Ternate, with orders to help build it. The timber used to construct the Portuguese fort, and the cannon to protect it, came from the wreck of *Trinidad,* formerly Magellan's flagship and the symbol of Spanish sea power in the Indies.

Espinosa, now just another prisoner, at first refused to comply with Brito's humiliating dictates, but eventually he was forced to go along: "I was rewarded for my labor by threats of being hanged from the yardarms and the seizure of the ship loaded with cloves and all of the equipment." The Portuguese clapped several of his men into leg irons, and even Espinosa himself, "dishonoring me and saying that I was a thief in front of all the native people and not paying respect to me at all, and saying"—and this was the ultimate insult—"'Now we'll see [who will prevail], the King of Spain, or that of Portugal.'"

Espinosa was forced to admit that the Portuguese, not the Spanish, remained firmly in control of the Spice Islands.

*T*rinidad's voyage came to its heartbreaking end in October 1522. Now there was only one ship left of the five comprising the original Armada de Molucca. This was *Victoria,* under Elcano's command, and her prospects of returning to Seville appeared even less certain than *Trinidad's.*

Six months earlier, Elcano had tried repeatedly to set a course around the Cape of Good Hope, each time without success, but without serious damage either. After weeks of failed attempts, *Victoria* finally sought refuge in a harbor located in South Africa, perhaps Port Elizabeth. More disappointment ensued when a scouting party found no helpful natives, in fact, no people of any kind; and no food.

Burning precious calories, the explorers climbed a hill to survey the landscape only to realize that, after all their attempts, they had yet to double the cape. It still lay ahead of them, far to the east.

With the greatest of reluctance, *Victoria* put to sea once more, battling a set of weather conditions found nowhere else on earth, the result of the interaction between the Agulhas current and ever-changing winds. The Agulhas current runs from the northeast to southwest, following the contour of the continental shelf, often at speeds of up to six knots. As if the current did not pose a sufficient threat, the ship also had to battle giant waves and gales that can change from northeasterly to southwesterly in a matter of minutes.

The wind was an even more dangerous force than the current. The major wind belts around southern Africa are influenced by two high-pressure systems, the South Atlantic High and the Indian Ocean High, which form part of the so-called subtropical ridge. The Coriolis effect deflects these winds to the left in the Southern Hemisphere, and they blow around in a counterclockwise direction. Such systems are also called "anticyclones." Winds can reach up to one hundred miles an hour, and *Victoria* experienced blasts powerful enough to sheer away her fore-topmast and main yard.

Sixty-foot-high rogue waves, monstrous walls of water, inflicted additional misery on the crew. Each upsurge threatened to swallow the fragile little ship, but somehow she managed to emerge from the churning troughs in one buoyant piece and to surge forward into the next wall of water. After a while, the mauling *Victoria* received came to seem, if not routine, then predictable. The sea had its own patient rhythm of destruction.

Given the wretched and chaotic existence the men endured, the logs and diaries covering this segment of the journey are understandably sparse and occasionally in conflict with one another. Albo, the pilot, and Pigafetta, whose records are generally in close agreement, diverge over milestones they reached by as much as two weeks. Apparently, they were too preoccupied, and the ship too unsteady, to make detailed note-taking possible.

The constant pummeling exhausted the crew, and simply finding a quiet moment to consume a few handfuls of barely edible food, usually rice, came to seem a major accomplishment, and getting through the day a miracle of sorts. Of course, the weather continued to batter the boat by night as well, so there was no rest for the crew, nor safe harbor, nor cooking fire, nor soft dry blanket, nor guarantee that their misery would end anytime soon. They might double the cape in a matter of days, but then again they might never be able to accomplish the feat. And if they were forced to turn back, the prospect of starvation in the open stretches of the Indian Ocean or death at the hands of the Portuguese awaited them. And so they tried again and again, fleeing for their lives, hoping to cheat death just one more time.

*J*ust when it seemed that the cape was impassable, the wind shifted slightly and the storms relented briefly. Elcano seized the moment to round Cape Agulhas, the point farthest south on the African continent, with the Cape of Good Hope coming up quickly, almost easy to handle in comparison.

Fighting churning waters, sailing as close to the wind as he dared, Elcano finally drove his ship around the Cape of Good Hope. Pigafetta wrote, with evident relief, "Finally, by God's help, we doubled that cape . . . at a distance of five leagues." It was only a guess, for the cape lay shrouded in fog and mist, an invisible, menacing presence now falling behind. They had survived one more ordeal, and that was enough to give thanks to a merciful Lord.

*B*y now it was May 22, 1522, the winds had abated, and *Victoria* was at last able to proceed on a northerly course. Elcano led the weather-beaten ship and her worn-out crew into what is now called Saldanha Bay, just north of Cape Town, where the men rested. There is no record that they thought of themselves as heroic for having outlasted the storms surrounding the Cape of Good Hope;

there was no longer any boldness or swagger about them. They had suffered too much for that; the sea had not killed them, but it had humbled them, and they were simply grateful to be alive. Nothing else mattered in comparison with that singular fact.

When the men recovered a bit of their strength, there was work to be done. They occupied themselves loading enough water and wood to see them home. For once, they were not alone because they shared the bay with a Portuguese ship plying the India route. Elcano imprudently risked making his presence known to the Portuguese captain, who saluted and sailed away, two ships at the end of the world pursuing their disparate goals.

Although *Victoria* had passed the supreme navigational test, the torments afflicting her crew were not over yet. On June 8, 1522, she crossed the equator again; this was the fourth time since the departure from Seville. "Then we sailed northwest for two months continually without taking on any fresh food or water," Pigafetta reported. Inevitably, scurvy returned to devastate the crew. "Twenty-one men died during that short time. When we cast them into the sea, the Christians went to the bottom face upward, while the Indians always went face downward." The victims included Martín de Magallanes, Magellan's young nephew, who had sailed as a passenger. Despite everything he had endured, Pigafetta retained his touching faith. "Had not God given us good weather, we would all have perished from hunger." The survivors summoned the strength to go on.

"Finally, constrained by our great extremity, we went to the islands. On Wednesday, July 9, we reached one of the Saint Jacob islands"—by which Pigafetta meant Santiago, the largest of the Cape Verde Islands, off the coast of West Africa, the very same islands that had served as the marker for the line of demarcation under the Treaty of Tordesillas. The islands remained a Portuguese stronghold, a center for commerce in materials and in men. The seas surrounding the Cape Verde Islands were familiar to Portuguese mariners, too familiar, in fact, for *Victoria's* safety. The farther north she journeyed, the more likely she was to encounter vindictive Portuguese authorities.

As soon as *Victoria* dropped anchor in the port of Ribeira Grande, on Santiago Island, Espinosa dispatched a longboat for food needed by the starving crew. Fearing that the Portuguese would likely pounce, the men crafted a story designed to elicit sympathy and avoid uncomfortable facts: "We had lost our foremast under the equinoctial line (although we had lost it under the Cape of Good Hope), and when we were restepping it, our Captain General had gone to Spain with the other two ships."

The cover story omitted any mention of their visit to the Spice Islands, the precious cloves they were carrying, Magellan's death, the mutinies, their doubling of the Cape of Good Hope, among other incursions into Portuguese waters, and, most important of all, their nearly complete circumnavigation of the globe. Instead, they posed as an unlucky, storm-battered Spanish cargo ship, hardly worth troubling over. The ruse seemed to work, and Pigafetta exulted, "With those good words, and with our merchandise, we got two boatloads of rice."

As an afterthought, Elcano told his men to confirm the date with the Portuguese, just to make sure the ship's log remained accurate after nearly three years' record-keeping. The reply—Thursday— baffled the sailors. "We were greatly surprised for it was Wednesday with us, and we could not see how we had made a mistake; for I had always kept well, and had always set down every day without interruption." How could they have omitted a day? As they learned later, "It was no error, but as the voyage had been made continually toward the west, and we had returned to the same place as does the sun, we had made a gain of twenty-four hours." But this miscalculation meant that they violated their faith by eating meat on Fridays, and celebrating Easter on a Monday.

This was no mere bookkeeping oversight: Albo, Pigafetta, and the rest of the survivors erred because the international date line did not yet exist. No Western cosmologist or astronomer, not even Ptolemy, had anticipated that a correction would be necessary to compensate for sailing around the globe. It took the first circumnavigation to demonstrate the need for a twenty-four hour gain. By general agreement, the international date line now extends west from the island of Guam, in the Pacific Ocean.

*A*s *Victoria* was about to slip away from Santiago Island, Elcano made a serious mistake. "On Monday, the fourteenth [of July]," wrote Albo, "we sent the ship's boat ashore for more rice. It returned the next day, and went back for another load. We waited until night, but it did not return. Then we waited until the next day, but it never returned." Something had gone awry, but no one aboard the ship knew what it was. One possibility was that the four Indians who had gone ashore to fetch rice tried to purchase food with cloves. When the Portuguese authorities saw this contraband, which could only have come from the Spice Islands, they became deeply suspicious of *Victoria.*

That was not all. While on the island of Santiago, one of the sailors let slip that their Captain General, Ferdinand Magellan, was dead. Pigafetta's all-too-brief mention of the incident suggests that whoever revealed Magellan's death also revealed that Elcano and the others were afraid to return to Spain, a remark calculated to raise suspicions. The sailor suspected of betraying secrets was Simón de Burgos, a Portuguese who had passed himself off as a Castilian to join the armada. His concealed identity might have had an innocent explanation—he simply wanted to find work, and with restrictions on the number of Portuguese crew members, pretending to be Spanish was the only way around the problem—or it might have been more sinister. It is possible that once he was among his fellow Portuguese in Santiago, he felt free to reveal his identity and betray his long-suffering crew members in exchange for favors. The severity of the subsequent Portuguese reaction to Burgos's admissions—assuming he was the source—suggests that he exposed still more about the expedition, including its visit to the Moluccas and incursion into Portuguese waters—all inflammatory matters.

Burgos was not the only crew member who tried to seek asylum from the Portuguese. Elcano had revealed the true nature of the expedition to a Portuguese captain shortly after *Victoria* doubled the Cape of Good Hope, and in the distant Spice Islands, Espinosa had also implored the Portuguese to come to his rescue. Assuming that many of the crew felt the urge to surrender to the Portuguese for the sake of survival, Burgos's admission might be seen as a diplomatic feeler rather than a betrayal of men who had suffered and died for each other. The crew, near death after three years of incessant journeying, deserve a measure of sympathy. To these beaten-down men, throwing themselves on the mercy of the Portuguese seemed a reasonable strategy for survival.

In practice, however, their attempt to disclose the true nature of the expedition as a prelude to defection failed miserably. "We went nearer the port," Albo continues, "to discover the reason of the delay, whereupon a vessel came out and demanded our surrender, saying that they would send us with the ship that was coming from the Indies, and that they would place their men in our ship, for thus had their officials ordered."

Victoria's officers stoutly resisted. "We requested them to send us our men and the ship's boat. They replied that they would bear our request to their officials. We answered that we would take another tack and wait. Accordingly, we tacked about and set all our sails full, and left with twenty-two men, both sick and well." The number probably included eighteen Europeans and four captives acquired en route. Twenty-two men: all that remained of the approximately 260 who had left Seville with the armada three years earlier. Twenty-two survivors of an endless succession of calamities, storms, scurvy, drowning, torture, execution, war, desertion, and now this final indignity: capture by the Portuguese. The prisoners included

Martín Méndez, the fleet's accountant; Ricarte de Normandia, a carpenter; Roland de Argot, a gunner; four sailors; an apprentice seaman, Vasquito Gallego; and two passengers who had avoided misfortune until this point in the journey.

"Fearing lest we also be taken prisoner by certain caravels," Pigafetta recorded, "we hastily departed."

It was July 15, 1522.

With barely enough men to handle the ship, Elcano took *Victoria* along a northerly course to her rendezvous with destiny in Spain. The diarists' silence concerning the final weeks of the circumnavigation suggests both their distaste for Elcano's barely legitimate authority and the suffering they endured from scurvy, other forms of malnourishment, depression, and exhaustion. Each day, familiar, well-charted landmarks along the coast of North Africa slid past, bringing no cheer, markers on a voyage to disgrace and prison—or so it seemed to the handful of men occupying their ramshackle ship.

Leaks constantly threatened to scuttle *Victoria*, and the men, in their exhausted condition, were forced to work the pumps night and day, simply to stay afloat. Their incessant labor paid off, and by July 28, Tenerife swung into view, signaling the beginning of a new course toward the Azores to negotiate the northerly winds. Elcano, still in command, approached the Azores, hoping to take on the fresh provisions they desperately needed and depart before the Portuguese, who claimed these islands, pursued them, but he wisely judged the maneuver too dangerous to attempt.

As they worked the pumps, the ship's crew discerned Cape Saint Vincent to the north on September 4. It would be the last important landmark they observed before reaching their goal, and it was a fitting sight, for Sagres, the location of Prince Henry the Navigator's academy, was located right on the cape; the developments that he had pioneered there a century before had culminated in this strange, difficult, and heroic voyage. Cape Saint Vincent disappeared in the mists as the "Portuguese trades" bore *Victoria* and her skeleton crew east toward the mouth of the Guadalquivir River, its waters churning just as they had three years earlier, when the ship, part of the proud Armada de Molucca, began the expedition to the Spice Islands.

*O*n Saturday, September 6, 1522, we entered the bay of San Lúcar with only eighteen [European] men, the majority of them sick, all that were left of the sixty men who had left the Moluccas. Some died of hunger; some deserted at the island of Timor; and some were put to death for crimes." So wrote Antonio Pigafetta, in an elegiac mode.

His cryptic reference to "crimes" has given rise to speculation that Elcano had to endure a mutiny during the final weeks of the voyage, and might have sunk to the same level of cruelty as Magellan had in quelling the uprising. Yet the mutiny, if there was one, must have been pathetic and halfhearted, because no other diarist has a word to say on the subject. More likely, the crimes mentioned by Pigafetta were the mundane deeds of desperate men, crimes such as the theft of *Trinidad*'s cloves or the dwindling food supply. Or the malefactors might have been one of the Indians still aboard the ship. The armada had captured a number of prisoners during its travels through Indonesia, some of them pilots, others hostages to be used as bargaining chips, and still others women whose chief role was to serve in a harem. The fleet's roster, so scrupulous and detailed concerning European crew members, offers little help in tracking the Indians taken aboard during the voyage. Even Pigafetta, who recorded the sad history of John the Giant with great interest and compassion, evinces little interest in later captives and offers no hint concerning their fates, but such prisoners would likely be the first to desert or to be condemned to death for their transgressions.

At last, Pigafetta allowed himself a moment of pride concerning the chief accomplishment of the Armada de Molucca. "From the time we left that bay until the present day, we sailed fourteen thousand four hundred and sixty leagues"—nearly sixty thousand miles—"and furthermore completed the circumnavigation of the world from east to west." The distance the armada traveled was fifteen times longer than that covered by Columbus's first voyage to the New World, and correspondingly more dangerous.

*T*o complete her journey around the world, *Victoria* and her decimated crew had to make one last passage, from the harbor of Sanlúcar de Barrameda along the Guadalquivir River into Seville. Elcano sent for a small boat to tow the battered craft and her exhausted crew to the teeming city, now abuzz with talk and excitement concerning the unprecedented voyage. Although her hull was in such poor condition and leaking so copiously that the men had to keep pumping all the way just to stay afloat, *Victoria* completed her journey along the river to Seville and tied up at a quay on September 10.

Under the scrutiny of representatives of the king and his financiers, dock workers unloaded the precious cargo that *Victoria* had traveled around the world to collect: cloves. Even without the other four ships, the amount of cloves in *Victoria*'s hold was sufficient to turn a profit for the expedition's backers. The king's agents were pleased to note that the cloves were of first quality, far exceeding those obtained from merchants who had acquired them in the traditional manner, from middlemen using land routes. The cloves filled no less than 381 sacks, weighing 524 quintals. Their value came to 7,888,864 *maravedís*. By royal order, the cargo passed directly to the expedition's backer, Cristóbal de Haro. Within weeks, he was in possession of the precious cloves, which he dispatched to his brother Diego in Antwerp for sale. The profits were divided between the Haros and the nearly insolvent Spanish crown.

Beyond the profits from spices, the completion of Magellan's voyage finally gave the Spanish a water route to the Spice Islands, if they wanted it. In terms of prestige and political might, the achievement was the Renaissance equivalent of winning the space race—a competition between the world's two great maritime superpowers,

Spain and Portugal, for territory of vital economic and political importance. In a sudden reversal of the balance of power, Spain was poised to control the spice trade and, by extension, global commerce.

*T*he day after arriving in Seville, the eighteen European survivors, attired only in their ragged shirts and breeches, did penance. Their number included Elcano; Francisco Albo, the pilot; Miguel Rodas, a ship's master; Juan de Acurio, the boatswain; Hernando Bustamente, the barber and medic; Antonio Pigafetta, whose eloquent, occasionally X-rated journal became the primary source of information about the entire voyage; and twelve seamen who, through luck and caution, had managed to survive where so many of their cohorts had perished. Walking barefoot, holding a candle, each world traveler slowly walked, still getting accustomed to the unusual feeling of solid, unshakable land beneath his feet. Elcano led the gaunt, weary pilgrims through Seville's narrow, winding streets to the shrine of Santa María de la Victoria, where they knelt to pray before the statue of the blessed Virgin and Child. They returned to Seville as sinners and penitents rather than conquerors. Their voyage had commenced as a Shakespearean drama, bristling with significance and passion, starring the heroic Magellan, but three years had taken a dreadful toll and the journey was ending as a play by Samuel Beckett. The survivors were shell-shocked, tentative, and chastened by all they had seen and experienced.

As curious onlookers watched, they rose and hobbled in bare feet over a wooden pontoon bridge across the Guadalquivir River and proceeded to another shrine, Santa María del Antigua, in Seville's massive cathedral. The grandeur of the cathedral dwarfed the little band of mariners as they trudged through the square to the chapel.

Their prayers concluded, the remnants of the first crew to circle the globe dispersed. They shed the rags they had brought with them from the sea, donned new clothing, and sought out their modest homes.

*I*n a bustling square in Sanlúcar de Barrameda, there is today a small marble plaque mounted high on the stone façade of a well-worn building. The plaque's tarnished inscription commemorates the eighteen survivors of the first-ever circumnavigation of the globe:

Juan Sebastián Elcano	Captain
Francisco Albo	Pilot
Miguel de Rodas	Master
Juan de Acurio	Boatswain
Martín de Judicibus	Sailor
Hernando Bustamente	Barber
Hans of Aachen	Gunner
Diego Carmona	Sailor
Nicholas the Greek, of Naples	Sailor
Miguel Sánchez, of Rodas	Sailor
Francisco Rodrigues	Sailor
Juan Rodríguez de Huelva	Sailor
Antonio Hernández Colmenero	Sailor
Juan de Arratia	Sailor
Juan de Santandres	Ordinary seaman
Vasco Gomes Gallego	Ordinary seaman
Juan de Zubileta	Page
Antonio Pigafetta	Passenger

In the entire list, only Elcano, the captain; Albo, the pilot; Busta-mente, the barber; and Pigafetta, Magellan's chronicler, could be considered notable members of the armada's original roster. The others, for the most part, were ordinary men, many still in their twenties or even younger, the overlooked servants of more powerful officers and specialists. No matter what their status, they had sur-veyed more of the world than anyone else before them; by accident or design, their names belong among history's great explorers.

They had seen a great deal, and although they failed to understand much of what they experienced, they had made records for
others to study, enlarging the Europeans' knowledge of the world.
They had circled the globe, only to demonstrate that the world was
now a larger place than previously imagined, not smaller. Seven
thousand miles had been added to the globe's circumference, as well
as an immense body of water, the Pacific Ocean. They had learned
that beyond Europe, people existed in astonishing profusion and
variety, as tall as the giants of Patagonia and as short as the pygmies
of the Philippines, as generous as the courtiers of Brunei, and as violent as the inhabitants of Mactan. Banished were phenomena such
as mermaids, boiling water at the equator, and a magnetic island
capable of pulling the nails from passing ships. All these discoveries
came at the cost of over two hundred lives and extreme hardship.
No other voyage had been as prolonged and complicated as this
one; no other voyage during the Age of Discovery would ever equal
it for ambition and daring.

The expedition had ended, but its effects on Spain, and on world
history, were just beginning.

CHAPTER XV

After Magellan

The Mariner, whose eye is bright,
Whose beard with age is hoar,
Is gone: and now the Wedding-Guest
Turned from the bridegroom's door.

As the skeleton crew guided the weather-beaten *Victoria* along the Guadalquivir River to her mooring in Seville, Juan Sebastián Elcano employed his considerable skills of persuasion in a letter to King Charles to boast of the voyage's multifaceted accomplishments and to justify his assumption of command after Magellan's death.

By the florid and long-winded standards of the era, the dispatch was a marvel of concision:

Most High and Illustrious Majesty:

Your high Majesty will learn how we eighteen men only have returned with one of the five ships which Your Majesty sent to discover the Spice Islands under Captain Ferdinand Magellan (to whom glory); and so that Your Majesty may have news of the principal things which we have passed through, I write to say briefly this:

First, we reached 54° S of the Equator where we found a Strait which passed through Your Majesty's mainland to the Sea of India, which Strait is of 100 leagues and from which we debouched; and, in the time of three months and twenty days, encountering highly favorable winds, and finding no land save two small and uninhabited islands; afterward we reached an archipelago of many islands quite abundant in gold. We lost by his death Captain Ferdinand Magellan, with many others, and unable to sail for want of people, very few having survived, we dismantled one of the ships and with the two remaining sailed from island to island, seeking how to arrive, with God's grace, at the Isles of Maluco [Spice Islands], which we did eight months after the death of the Captain; and there we loaded the two ships with spices. . . .

Having departed the last of these islands, in five months, without eating anything but wheat and rice and drinking only water, we touched at no land for fear of the King of Portugal, who had given orders in all his dominions to capture this fleet. . . . We arrived at the islands of Cape Verde, whose governor seized my boat with thirteen men, and sought to throw me and all my men into a ship which was sailing from Calicut to Portugal charged with spying . . . but we resolved, with common accord, to die before falling into the hands of the Portuguese. And so with very great labor at the pumps, which we had to work day and night to free her of water, and as exhausted as any man ever was, with the aid of God and of Our Lady, and after the passage of three years, we arrived. . . .

Your Majesty will know best that what we should esteem and admire most is that we have discovered and made a course around the entire rotundity of the world—that going by the occident we have returned by the orient.

After boasting of his feats of discovery, Elcano turned to the commercial aspects of the expedition and petitioned the king to excuse the men who had suffered so greatly for him from having to pay duties on profits from their personal store of spices:

> I beg Your Majesty, in view of the many travails, sweats, famine, and thirst, cold and heat that these people have endured in the service of Your Majesty, to give us grace for the fourth and twentieth of their property and of what they brought with them. And with this I close, kissing the feet and hands of Your high Majesty.
>
> Written on board the ship *Victoria*, in Sanlúcar, on the 6th day of September of 1522.
>
> The Captain
Juan Sebastián Elcano

The first account of the first journey around the world, Elcano's letter was dispatched from Sanlúcar de Barrameda even before the ship reached Seville—an indication of Elcano's eagerness to offer explanations. But his letter did little to clear up the mystery of how Magellan died; nor did Elcano explain how he, a Basque mariner, emerged as the fleet's Captain General. And any connection between the two events—Magellan's fall and Elcano's rise—was similarly obscured. The letter concealed more than it revealed. Many serious questions loomed concerning the voyage: mutinies, the sailors' licentious behavior and outright orgies with women in distant lands, which had been expressly forbidden by the king; and, most important of all, Magellan's conduct at sea and accusations of torture.

King Charles never mourned the loss of the Captain General, even though Magellan had always regarded the young sovereign as the paragon of all virtues, the recipient of all his loyalties and efforts, the justification of all his suffering. And Magellan's fanatical devotion was not returned in kind. Charles felt no sense of kinship with the ardent Portuguese mariner who had presented himself at

Valladolid four years before, pleading for royal backing of an expedition. The armada's many scientific and geographic discoveries and
the claiming of dozens of islands and lands for Spain made little
impression on this preoccupied sovereign, who, through lifelong
habit, merely considered such tributes his due. King Charles barely
acknowledged that, thanks to Magellan's efforts, he now laid claim
to much of the known world, at least for a short time. Eventually, he
took to boasting about the expedition because it had returned with
a shipload of cloves, the aromatic equivalent of gold. He counted
the number of bahars of cloves aboard the battered *Victoria* and
ignored the number of souls Magellan and the priests had converted to Christianity. For Charles, the Armada de Molucca could
be considered a commercial success; that was all, and that was
enough.

King Charles proudly wrote to his Flemish aunt, the Archduchess Marguerite of Austria, the Netherlands' regent, to proclaim
the arrival of the prized cargo transported against all odds from
halfway around the world. "The armada that three years ago I sent
to the Spice Islands has returned and has been to the place where
the said spices grow, where the Portuguese or any other nation has
never been . . ."—that was manifestly untrue, but Charles had to
maintain the fiction that Spain reached the Moluccas first in order
to claim them—"and the captain of the said armada asserts that on
this voyage they went so far that they roamed around the entire
world." These boasts reveal a twenty-one-year-old king attempting
to assert his legitimacy and authority, and he asked his aunt to help
bring the spices to market, "as if it were my own affair." He
reminded her that he had "borne great expenses for this new and
untried effort, in addition to the work and care my people gave to
it," and he reminded her that he expected the entire empire over
which he ruled, from Spain to the Netherlands, to profit, that is to
say, get out of debt to the Haro family: "I hope that certainly my
realms on this side and also my said countries on that side, and the
subjects of each, will receive great benefit, convenience, and profit

in the future, as you may well expect. And as to the value of the spices that the ships brought, what will come of them . . . will serve to furnish the preparation of a larger armada that I have decided to send to these Spice Islands as soon as possible."

Excited by the thought of these riches, the archduchess requested her nephew to designate Bruges, the flourishing Flemish city in her realm, as the new center of the European spice trade, but Charles, thinking he had found a surefire way out of debt, insisted on keeping it in Spain "because this merchandise was first found at the expense of this realm."

Still gloating over this unexpected success, King Charles summoned Elcano and two men of his choosing to visit the royal residence at Valladolid to provide a full account of their exploits. Elcano selected the pilot, Albo, and the barber-medic, Bustamente, to back up his account. Significantly, he excluded Pigafetta, whom he knew to be a Magellan loyalist. As a sign of royal favor, Elcano's delegation received a lavish disbursement for formal clothes and traveling expenses to Valladolid; they could be assured of making an impressive appearance before their sovereign.

The city, in north central Spain, was a time capsule of the Spanish past, held for centuries by the Moors, who named it. Christians conquered the city in the tenth century, and it became a stronghold of commerce, its citizens renowned for speaking the purest Spanish anywhere, so important to the kingdom as a whole that by the dawn of the Renaissance the kings of Castile made it their official seat. For this reason, Valladolid exerted its bureaucratic influence over a substantial part of the world. By the time King Charles took up residence in Valladolid, the city was at its zenith.

Charles received the three world travelers on October 18 with apparent warmth and congratulated them on having reached the Spice Islands through a water route and claiming them for Spain. Keenly aware of what was expected, Elcano solemnly presented His

Majesty with samples of the spices brought back from the Moluccas, as well as letters from the island chieftains swearing loyalty to
the unknown ruler of the distant land. All that was very impressive,
but just for show.

Clouds of suspected disloyalty, even mutiny, hung over the survivors' heads. Just before their arrival in Valladolid, disquieting
rumors had reached King Charles. It was whispered that Magellan
had not been killed by warriors on Mactan but by the members of
the fleet. Could Elcano have been among them? And there were
conflicting accounts of the bitter mutiny at Port Saint Julian, some
blaming the Spanish officers for the uprising and others holding the
Portuguese contingent responsible.

To get to the bottom of these stories, the three men—Elcano,
Albo, and Bustamente—faced an inquiry conducted by Valladolid's
mayor, acting on orders from King Charles himself. The proceeding, which began on October 18, consisted of thirteen questions put
to the men. The questions concentrated on two themes, the mutiny
and the commercial aspects of the voyage. Elcano had given considerable thought to the charge of disloyalty that he was bound to face,
and during his examination he carefully explained his way out of
the mutinies that had occurred on the ships by condemning Magellan. Elcano rearranged events to make it sound as though he had
been invited by the Spanish captains to serve as the Captain General, that Magellan had favored his relatives on board the ships at
the expense of all others, especially the Spanish captains, and that
Magellan had defied the king's explicit orders. "Elcano declared
that Magellan said that he did not wish to . . . carry out the instructions entrusted to him by His Majesty," read the transcript of the
proceedings.

By portraying himself as a humble defender of Spanish honor,
Elcano skillfully played up to King Charles, but he was less successful in his defense of the expedition's commercial aspects. Why, his
inquisitors demanded, were there only 524 quintals of cloves on
board *Victoria* when she tied up at the quay in Seville, but the ship's

own register clearly showed she had taken on no less than 600 quintals in the Spice Islands?

In his response, Elcano carefully explained that he had relied on the weight given by the islanders from whom he had purchased the cloves, that he personally supervised weighing the cargo in Seville, and that any discrepancy could be accounted for by drying during the long voyage home.

Next, Elcano was asked why he had failed to keep accounts. According to the transcript, "Elcano was asked to declare all that was done on the voyage to the disservice of His Majesty and to defraud him of his property."

Again, the Basque-born mariner tried to shift the blame to Magellan, claiming that as long as Magellan was alive, he had written nothing "because he dared not to do so," while after Magellan's death, he did record transactions. This explanation made no sense because Magellan was scrupulous about recording the fleet's activities, whether in Pigafetta's diary or in Albo's pilot's log. Ignoring those inconvenient pieces of evidence, Elcano instead spoke grandly and vaguely about Magellan's "disservice" to the king and the fleet, which he recklessly "abandoned to its fate." His indictment of Magellan was as damning as it was unsupported by the events.

Finally, Elcano was forced to confront the disquieting rumors surrounding Magellan's death. In his brief reply, Elcano held the Mactanese islanders completely responsible. By burning their hamlet, Elcano implied, Magellan had goaded them into taking revenge. His explanation went unchallenged, and served as the basis of the official determination of the cause of Magellan's death.

Elcano's testimony was sufficiently dexterous to exculpate himself from royal disfavor or worse. And his two companions, giving answers remarkably similar to Elcano's, achieved the same result. By the time the inquiry ended, King Charles and his advisers were reminded that the survivors had brought them a fortune in spices, a claim to the Spice Islands themselves, a new water route to the islands, and an unequaled mastery of the ocean—all of it priceless, no matter how underhanded they had been in getting it.

*I*n the end, King Charles waived the royal duties on the spices the men brought home for their personal enrichment and offered a quarter of his own proceeds from the voyage to the three survivors who had testified in Valladolid. Elcano's bonus included even more: an annual pension of five hundred ducats, a knighthood, and a coat of arms befitting the mariner who had sailed around the world. It depicted a castle, spices, two Malay kings, a globe, and the legend:

Primus circumdedesti me

Thou first circled me.

Of equal importance, Elcano received a royal pardon for his role in the failed mutiny against Magellan's command. Elcano insisted on having the document published, making his exoneration complete. He would now be qualified to lead future expeditions for Castile.

With all his new riches, Elcano acquired two mistresses, one of whom bore him a daughter, the other a son, but he lived with neither.

*T*he other survivors of the expedition received similar marks of royal favor. Martín Méndez, *Victoria's* accountant; Hernando Busta- mente, the barber; Miguel de Rodas, the master of *Victoria*; and Espinosa each received individualized coats of arms commemorat- ing their accomplishments. (Meanwhile, the coat of arms for the Magellan family remained defaced and dishonored, as it had been since Magellan left Portugal to serve the king of Spain—the king who had all but forgotten him now.)

The men who had mutinied against Magellan—an entire ship filled with them—were freed from prison and absolved of their crimes. Álvaro de Mesquita, who had served as captain of *San Anto- nio* until the mutineers overwhelmed him, had languished in jail ever

since 1521, when his ship returned to Seville. With *Victoria's* survivors corroborating his story, the diehard Magellan loyalist was also
freed in a general amnesty designed to end lingering controversy
about the voyage. Having had enough of Spanish justice, he fled
home to Portugal.

*D*espite Elcano's skill at self-promotion, and King Charles's
endorsement, a different interpretation of the voyage emerged soon
after *Victoria's* return. Maximilian of Transylvania, a secretary to
King Charles, pounced on Elcano, Albo, and Bustamente at Valladolid, interviewed them all at length, and very likely talked to
Antonio Pigafetta, Magellan's official chronicler, as well. Within a
month of *Victoria's* return to Seville, he delivered his lengthy report
to King Charles.

In his account, Maximilian saw past the expedition's internal
power struggles to emphasize how it changed the way the entire
world would be seen from this time forth. "I have resolved to write as
truly as possible," he remarked. "I have taken care to have everything
related to me exactly by the captain and by the individual sailors
who have returned with him." These men were so sincere that it was
apparent to Maximilian that "they seemed not only to tell nothing
fabulous themselves, but by their relation to disprove and refute all
the fabulous stories which had been told by ancient authors."

*B*y far the most authoritative and eloquent chronicle of the first
voyage around the world flowed from the pen of Antonio Pigafetta,
who had faithfully maintained his diary throughout the entire
expedition. To counter what he expected would be Elcano's self-
serving distortions of the events that had occurred at sea, Pigafetta
immediately set about writing his own impassioned plea for Magellan's valor and loyalty to the king and the Church. He provided
eloquent eyewitness testimony about how Magellan had died and

more important, how he had lived. He revealed Magellan as the fearless disprover of long-standing myths and overturner of tenacious fallacies.

Leaving Seville, Pigafetta headed directly for Valladolid, where he presented the young monarch with "neither gold nor silver, but things very highly esteemed by such a sovereign. Among other things, I gave him a book, written by my hand, concerning all the matters that had occurred day to day during our voyage"—the most important account of distant lands to appear since *The Travels of Marco Polo.*

Pigafetta's diplomatic background served him well, because he then succeeded in giving his account to sovereigns who were often bitter enemies of one another: "After this, I left for Portugal, where I gave an account to King João of all that I had seen. Passing again through Spain, I also went to present some rare objects from the other hemisphere to the very Christian King François. Finally, I went to Italy, to the very industrious lord, Philippe Villiers de l'Isle-Adam, a worthy grandmaster of Rhodes, and placed at his disposal my person and services of which I would be capable." Pigafetta's thorough and even-handed distribution of his account ensured Magellan's leading role in the adventure for posterity—and, not so incidentally, his own. "I made the voyage and saw with my eyes the things hereafter written," Pigafetta vowed, "that I might win a famous name."

After traveling across Europe, Pigafetta arrived in Venice, his home, and immediately caused a stir. "There came into the college a Venetian who had been appointed a knight errant, a brother of [the Order of] Rhodes, who has been three years in India," wrote Marin Sanudo on November 7, 1523, of Pigafetta's visit. "And all the college listened to him with great attention, and he told half of his voyage ... and after dinner also he was with the doge, and related those things in detail, so that his Serenity and all who heard him were rendered speechless over the things of India."

In August of the following year, Pigafetta, by this time settled in Venice, requested that the doge and city council allow him to print his sensational account; he supplied two reasons, the overwhelming importance of the events recorded, and Pigafetta's singular authority in relating them:

> Most Serene Prince and your Excellencies:
> Petition of me, Antonio Pigafetta, Venetian knight of Jerusalem, who, desiring to see the world, have sailed, in past years, with the caravels of his Caesarean Majesty [Charles V], which went to discover the islands of the new Indies where the spices grow.
> On that voyage, I circumnavigated the whole world, and since it is a feat that no man had accomplished, I have composed a short narration of the entire voyage, which I desire to have printed. For that purpose, I petition that no one may print it for twenty years, except myself, under penalty to him who should print it, or who should bring it here if printed elsewhere, of a fine of three lire per copy, besides the loss of the books. [I petition] also that the execution [of the penalty] may be imposed by any magistrate of this city who shall be informed of it; and that the fine be divided as follows: one-third to the arsenal of your Highness, one-third to the accuser, and one-third to those who shall impose it.
> I humbly commend myself to your kindness.

Pigafetta's request met with a favorable response, and he was granted the privilege that "no other except himself be allowed to have it printed for twenty years."

The first copies of Pigafetta's "relation," the ones he brought with him to the courts of Europe, were lavish handwritten manuscripts illustrated with maps of his own devising, items literally fit for a king. It is believed that Pigafetta wrote his "relation" in the Venetian dialect, mixed with Italian and Spanish, but the original has been lost. Instead, four early versions produced by expert scribes have come down over the centuries, one in Italian and three in French. By

general agreement, the most handsome, complete, and extravagantly
illustrated version resides today in the collection of the Beinecke
Rare Book and Manuscript Library at Yale University. To read this
memoir, and to turn its ancient vellum pages, is to be transported
instantly five hundred years into the past. Although Pigafetta tells
his story more or less in chronological order, he has not constructed
a linear narrative; rather, it is a compilation of events, illustrations,
translations of foreign tongues, prayers, descriptions, epiphanies,
and bawdy asides, all of them heavily cross-referenced and color-
coded in brilliant inks of black, blue, and red. Yet it is also a personal document, unusual for that time, when the idea of an
individual consciousness was just beginning to take root. The reader
of Pigafetta's chronicle hears his voice, alternately bold, astonished,
devastated, fascinated, and, in the end, amazed to be alive in the
cruelly beautiful world of his time.

Although a few influential voices celebrated Magellan's extraordinary accomplishments and appreciated the extent of his ordeal, he
was despised or discounted by most authorities and observers from
Seville to Lisbon; in both countries he was considered a traitor, and
court historians everywhere prepared to blacken pages with their
indictments of his nefarious deeds and treachery. Ironically, Magellan's most fervent admirers were in England, where political commentators urged their island nation to emulate his daring example.
Closer to home, King João III of Portugal (the son of the monarch
who had spurned Magellan) seethed at the news that one of the
ships of the Armada de Molucca had returned to Seville with a full
load of cloves. He hotly protested to King Charles, insisting that the
Spice Islands actually belonged to Portugal. Charles, for his part,
patiently but insistently pressed for the release of the men taken
prisoner by the Portuguese in the Cape Verde Islands, and they
trickled in to Spain in small groups throughout the following year.

The additional survivors of *Victoria* included Roland de Argot, a gunner; Martín Méndez, the fleet's accountant; Pedro de Tolosa, a steward; Simón de Burgos, suspected of betraying the other crew members in the Cape Verde Islands; and one Moluccan, who went by the name of Manuel.

Victoria's two groups of survivors, for all the hardships they had endured since leaving the Spice Islands, enjoyed far better fortune than the sixty men who had chosen to sail home aboard *Trinidad.* Only four of that number ever returned to Spain or Portugal. A deaf seaman named Juan Rodríguez, at forty-eight the oldest survivor, stowed away on a Portuguese ship bound for Lisbon. He spent a short time in jail, won his release, made his way back to Seville, and, despite his age, his infirmity, and the hardships he had endured during his years at sea, applied to the Casa de Contratación to sail to the Indies once again.

Enduring months of hard labor and humiliation in the Moluccas, Espinosa was transported along with several of his crew members to Cochin, a Portuguese outpost on the west coast of India. Refusing a Portuguese invitation to fight the Arabs, he wrote to King Charles, complaining that the viceroy, Vasco da Gama, was busy "menacing me and telling me that my head would be cut off and dishonoring me with many evil words, saying that he would hang the others."

*I*n 1526, after four miserable years in captivity, Gonzalo Gómez de Espinosa, the former captain, and Ginés de Mafra, the garrulous pilot, joined the crew's gunner, Hans Vargue, aboard a ship bound for Lisbon. Freedom would elude them a while longer, though; on arrival, the heroic circumnavigators were thrown into jail. Vargue died there, leaving all his worldly possessions—his back pay and a package of cloves—to Espinosa.

Toughened by years of adversity, de Mafra and Espinosa survived their time in a Lisbon prison as they had survived everything else, and upon their release they returned to Seville, only to be jailed again. Their case came to trial in 1527; at last they were acquitted and finally released.

The harsh fate of these two men, who had been loyal to Magellan and King Charles, stood in marked contrast to the mutineers who had returned to Seville aboard *San Antonio*; all of them had been freed, except for the true loyalist among them, Álvaro de Mesquita, whom they had taken hostage during their mutiny. The injustice was particularly striking in Espinosa's case, because whatever his failings as a captain, he had performed effectively as the *alguacil* in moments of crisis, and had played a crucial role in helping Magellan regain control of the fleet after the mutiny in Port Saint Julian. Magellan's father-in-law, still living in Seville, took up the cause of these unjustly punished survivors, and risked all to write in their defense to King Charles. Rather than being punished for their acts of disloyalty, the mutineers had been "very well received and treated at the expense of Your Highness," Barbosa remarked, "while the captain and others who were desirous of serving Your Highness were imprisoned and deprived of all justice. From this, so many bad examples arise—heartbreaking to those who try to do their duty."

Both men found their homecoming to be bitter, indeed. De Mafra, for one, learned that his wife, assuming that he was dead, had remarried; not only that, she had spent his entire fortune with her new husband. Disgusted with his lot, de Mafra returned to the life he knew best, that of a pilot in the Pacific; by 1542, he was back in the Philippines in the service of Spain.

Espinosa faced a more ambiguous destiny. On August 24, 1527, King Charles granted him an enormous pension—112,500 *maravedís*—but Espinosa never received it. The Casa de Contratación, as mean-spirited as ever, withheld the salary he earned during his years in jail, arguing that he was not actually "in the service of Spain" at the time. Outraged by the treatment he had received at the hands of unfeeling bureaucrats, he sued for twice the amount, settled for half of the original pension, and, in the end, received only a fraction of the settlement, and even that modest amount was contingent on his participating in another expedition to the Moluccas. (The king did allow Espinosa to keep the 15,000 *maravedís* left to him by Hans Vargue.)

Understandably, Espinosa refused to return to the lands that had claimed so many Spaniards' lives, and where he had suffered in prison for four long years. In 1529, King Charles decided to bestow another pension on his loyal servant, this time in the amount of 30,000 *maravedís*, and he received a comfortable job as an inspector, at an annual salary of 43,000 *maravedís*. He lived out his days in Seville.

Spain and Portugal agreed to hold another conference to determine the locations of the line of demarcation and the Spice Islands. The Spanish delegates included experts such as Sebastián Elcano, Giovanni Vespucci (Amerigo's sibling), and Sebastian Cabot. Despite the good intentions of the two nations, and the credentials of the delegates, the proceedings quickly degenerated into farce.

To symbolize the strict impartiality of the deliberations, the summit was held on a bridge spanning the Guadiana River, along the Spanish-Portuguese border, but the location nearly undid the conference. As the distinguished members of the Portuguese delegation happened to be walking across the bridge, they were stopped by a small boy, who asked if they were carving up the world with King Charles. The former governor of India, Diogo Lopes de Sequeira, acknowledged that indeed they were. At that, the boy lifted his shirt, turned to reveal his bare bottom, and with his small finger traced the line between his buttocks.

"Draw your line right through this place!" he declared.

The parties adjourned to towns on either side of the river. The cosmologists and astronomers continued to argue over longitude, and could not even agree on the length of a degree, so the question of where to place the Moluccas remained unresolved. Magellan had

traversed the Pacific, it was true, but no one yet knew how to measure the distance he had traveled, except by dead reckoning, of limited value over long distances.

For all these reasons, the attempt to redefine the line of demarcation ended in failure. As might be expected, both sides claimed
victory—and possession of the Spice Islands.

*B*lithely ignoring the conference, King Charles splurged on lavish follow-up expeditions to the Moluccas, heedless of the cost and
the risks involved in these tragic enterprises.

In 1525, the Casa de Contratación commissioned a well-
connected officer, Francisco García Jofre de Loaysa, to lead the next
Armada de Molucca. Sebastián Elcano, honored as the first circumnavigator, received an appointment as second-in-command. The
voyage's primary goal—to build a fully staffed Spanish trading post
and fort in the Spice Islands—demonstrated how hollow King
Charles's conciliatory language actually was. Spain remained determined to break the Portuguese monopoly on the spice trade and to
claim the islands, no matter what.

The second Armada de Molucca left Seville with the hope that
retracing Magellan's route would make the voyage to the Spice
Islands far safer and faster, but just the opposite occurred. Without
Magellan's navigational genius to guide the ships, the second armada
met with an even harsher fate than the first. Elcano, despite his
experience, made one navigational error after another during the
journey, and he arrived at the strait discovered by Magellan only
after dangerous delays. Storms in these low latitudes battered the
ships of the armada, reducing their number from five to just two. In
the Pacific, scurvy broke out among the officers and crew, just as it
had during Magellan's crossing, and this time, no one had quince
preserves to protect them from its ravages, not even the Captain
General.

Loaysa died, leaving behind an envelope prepared by King

Charles himself, naming a successor. When the seal was broken, the letter appointed Elcano as the next admiral. The Basque mariner had finally reached the summit of ambition, but the time remaining to him proved to be cruelly brief, because he was already suffering from scurvy. Retreating to his small cabin, he drew up his will; the document carefully inventoried all his worldly possessions down to the last article of clothing and ream of paper; it listed his many charitable bequests; it specified gifts to his two mistresses; and it requested that his funeral take place in his hometown, Guetaria. The will was witnessed by seven others, each one a Basque. Five days after assuming command, on August 4, 1526, Sebastián Elcano died at sea, another casualty of the Age of Discovery. His body was committed to the deep amid the rolling blue expanses of the Pacific.

In an eerie recapitulation of Magellan's voyage, just one of the five original ships of the second Armada de Molucca reached the Spice Islands. And of the 450 men who set sail from Spain aboard these ships, only 8 lived to see Spain again, an even greater loss of life than Magellan's crew suffered.

The extraordinary death rate, to say nothing of the expense involved, did nothing to deter King Charles from trying to reach the Spice Islands again—and again. He sent Sebastian Cabot, then the *piloto mayor,* or chief pilot, of Seville, in search of the Indies, but the hapless mariner only got as far as the Río de la Plata, the false strait on the east coast of South America. After a time, he led the third Armada de Molucca, back to Spain, where he was charged with failing to complete his mission because he was afraid to enter the real strait and face its dangers.

Soon after that debacle, Hernándo Cortés, the conqueror of Mexico, dispatched his own expedition to the Moluccas from his outpost in Aguatanejo, Mexico. Although this expedition promised to be shorter, and did not have to pass through the strait, it, too, met with disaster. Only one ship reached the Spice Islands, and the Portuguese captured her crew and seized her cargo, abruptly aborting the mission.

With each failure, the dream of establishing a Spanish outpost in the Spice Islands and bringing the wealth of the Indies into Spanish coffers faded, and the scope of Magellan's superhuman accomplish- ment and fierce determination came to seem greater and greater.

Despite all the setbacks, King Charles refused to let the dream of dominating the world's economy die. He backed plans for a fifth armada, led by Simón de Alcazaba, another Portuguese sailing for Spain, and this one promised to be the most ambitious—and aggressive—of all. The fleet consisted of eight ships capable of transporting a large garrison of Spanish soldiers to the Spice Islands. They were to drive out the Portuguese and claim the islands for the crown once and for all. But before these ships put to sea, King Charles found himself in desperate financial straits. Fighting off the French had drained his coffers, and his longtime financial backers, Cristóbal de Haro and the Fugger dynasty, refused to back another expedition in search of the elusive goal that had claimed the lives of so many brave men. For the next two decades, the House of Fugger tried to recover its huge investment in the failed armadas, but the Spanish crown, teetering on the brink of insolvency, failed to repay the debt.

Desperately short of cash, Charles was unable to send any more expeditions to the Spice Islands. But he did not give up his goal; instead, he sought a diplomatic solution to thwart or slow Portugal's imperial ambitions. He invited Portugal to join a commission to study the Spice Islands quandary, and he asked the Vatican to arbi- trate in case of disagreement. In the end, João III had no choice but to agree to the plan, or risk seeming bellicose and heedless of papal authority. In this way, King Charles maintained his interest in the Spice Islands through diplomacy—but not for long.

Unable to raise money from his usual backers, Charles was forced to turn to Portugal for aid. In 1529, he borrowed 350,000 ducats from João III, and as security for the loan he pledged the Moluccas and all the islands lying to the east. Both nations signed the Treaty of Saragossa, ending the epic struggle for control of the global economy. Just seven years after Magellan's voyage and three unsuccessful follow-up expeditions to the Spice Islands, King Charles, facing bankruptcy, gave up and returned the islands to the Portuguese. In matters of empire, everything had its price.

Not until 1580, fifty-eight years after *Victoria* returned to Seville, did another explorer, Sir Francis Drake, complete a circumnavigation. His voyage took him through the Strait of Magellan. To accomplish the feat, Drake relied on the knowledge so painfully and heroically acquired by the Captain General and his crew.

Little *Victoria,* the first ship to complete a circumnavigation, had her own curious epilogue. No one thought to preserve the battered vessel as a testament to Magellan's great achievement. Instead, she was repaired, sold to a merchant for 106,274 *maravedís,* and returned to service, a workhorse of the Spanish conquest of the Americas. As late as 1570, she was still plying the Atlantic. En route to Seville from the Antilles, she disappeared without a trace; all hands on board were lost. It is assumed that she encountered a mid-Atlantic storm that sent her to the bottom, her wordless epitaph written on the restless waves.

In 1531, one of the first accurate maps of the Strait of Magellan appeared. Oronce Finé's representation placed the strait in its proper position in South America, and although the map does not name the strait, it does call the Pacific "Magellanicum." The name Magellanica, or Magellanic Land, would appear on many later maps of South America, usually indicating Patagonia or Chile. Gerardus Mercator, the Flemish cartographer, canonized the strait on his famous globe in 1536. In time, the name Magellan came to designate only the strait—no lands, in fact, none of the territories that he once dreamed of bequeathing to his heirs. At least, such was the case on earth. In the heavens, his name came to be associated with the two dwarf galaxies he had discovered, the Magellanic Clouds, visible in the Southern Hemisphere.

Although no continent or country was named after him, Magellan's expedition stands as the greatest sea voyage in the Age of Discovery. In its epic, world-straddling scope, his voyage harkened back to Greek and Roman antiquity, which had been rediscovered and embraced with such conviction during the Renaissance. "Worthier, indeed, are our sailors of eternal fame than the Argonauts who sailed with Jason," wrote Peter Martyr, Magellan's contemporary and the first historian of the New World. "And much more worthy was their ship of being placed along the stars than that old Argo; for they only sailed from Greece through Pontus, but ours through the whole west and southern hemisphere, penetrating into the east, and again returned to the west."

By confronting the intellectual and spiritual limitations of the ancient view of the world, by subjecting its assumptions to the ultimate reality check—traveling around the globe—Magellan looked ahead of his time to the Age of Reason and beyond, to the present. In their lust for power, their fascination with sexuality, their religious fervor, and their often tragic ignorance and vulnerability, Magellan and his men epitomized a turning point in history. Their deeds and character, for better or worse, still resonate powerfully.

Notes on Sources

Ferdinand Magellan remains controversial even today, considered a tyrant, a traitor, a visionary, and a hero by various chroniclers. As befits an explorer who led a multinational crew on a voyage around the world, accounts of his life and circumnavigation have been heavily influenced by divergent manuscript traditions arising from a rich store of primary and important secondary sources in Spanish, French, Portuguese, Latin, and Italian. In re-creating Magellan's epic voyage, I have generally relied on these diverse primary sources—diaries, journals, contemporaneous accounts, royal warrants, and legal testimony. Some important early Magellan sources have been translated into English for the first time for use in this book. These include a lengthy memoir by Ginés de Mafra, who was one of the survivors; early histories by João de Barros, António de Herrera y Tordesillas, and Gonzalo Fernández de Oviedo y Valdez; and legal documents pertaining to the voyage now archived at Brandeis University, in Massachusetts.

The most important (though not the only) source of primary information about the voyages of Magellan and other explorers is the Archive of the Indies in Seville. Martín Fernández de Navarrete edited a multivolume compilation of the archive's chief holdings, published in Spanish in 1837, which advanced understanding of Magellan and his era; most of the archive's records pertaining to Magellan's voyage are in Volume 4.

As a result of this wealth of primary sources, Spanish historians have tended to feed off earlier works in Spanish, but they are not the only important Magellan chroniclers. Portuguese historians have emphasized Portuguese sources and attitudes, often sharply critical of Magellan. More recently, English-language historians, who generally portray Magellan in a heroic light, have drawn on a wider variety of sources and languages; but as the decades have passed, they, too, have become another manuscript tradition. In particular, the naval historian Samuel Eliot Morison wrote several heavily documented chapters on Magellan in his classic work, *The European Discovery of America: The Southern Voyages*

(1974), to which I happily acknowledge my debt. Curiously, F.H.H. Guillemard's *Life of Ferdinand Magellan* (1890) remains the standard biography more than one hundred years after its publication; since then, new sources and approaches to the era have emerged, making it possible to give a more three-dimensional account of the voyage, including graphic and intimate details that custom prevented Guillemard from mentioning, except, perhaps, in a Latin whisper. Also worthwhile is Tim Joyner's *Magellan* (1992), a concise biography buttressed by a generous selection of primary sources. Martin Torodash's "Magellan Historiography," published in *The Hispanic American Historical Review*, surveys the entire field, offering reliable if occasionally heavy-handed assessments.

The best and most affecting eyewitness account of Magellan's circumnavigation was written by Antonio Pigafetta, the young Venetian scholar and diplomat who was among the handful of survivors. His chronicle remains one of the most significant documents of the Age of Discovery. The best and fullest English translation, by James A. Robinson, an American scholar, was published in three substantial volumes in 1906. Robinson worked from a Portuguese translation of the original, which meant the occasional blurring of Pigafetta's distinctive humor and irony. In 1969, R. A. Skelton's new translation managed to convey a sense of Pigafetta's voice and sensibility, and includes a facsimile of the Pigafetta manuscript in the Beinecke Rare Book and Manuscript Library at Yale University. I am indebted to both of these scholars for their diligent work, as is anyone who wants to learn more about Magellan and Pigafetta. My quotations from Pigafetta's diary are drawn largely from Robinson's translation, but where possible I have checked it against the original and other sources, and silently corrected a number of slips and archaic usage or euphemisms.

Pigafetta was not a disinterested source. He was, touchingly, a Magellan loyalist, and as a result, made only the briefest mention of the various mutinies during the voyage and Magellan's drastic efforts to quell them. To present a fuller account of these events, I have turned to the testimony of other sailors who witnessed or participated in them, including de Mafra and Vasquito Gallego. In addition to the diaries, Francisco Albo's pilot's log gives a day-by-day record of the voyage.

CHAPTER I: THE QUEST

Concerning the Treaty of Tordesillas, Samuel Eliot Morison's *The European Discovery of America: The Southern Voyages* and Tim Joyner's *Magellan* both contain valuable analyses of the treaty as it affected Magellan's proposed expedition, as does Jean Denucé's *Magellan: La Question des Moluques et la Première Circumnavigation du Globe* (pp. 46–47).

Pedro de Medina's *A Navigator's Universe*, ed. Ursula Lamb, sheds light on the subject of Renaissance cosmology, as does Alison Sandman's accomplished thesis, *Cosmographers vs. Pilots.* Pablo Pérez-Malláina's *Spain's Men of the Sea* mentions "coarse" pilots (p. 233).

For more on spices and the spice trade throughout history, see notes to chapter 13. Maximilian of Transylvania's remark about spices comes from Charles E. Nowell's *Magellan's Voyage Around the World* (p. 275), a convenient if not definitive anthology of several accounts.

Prince Henry the Navigator's remark about peril and reward can be found in John Noble Wilford's *The Mapmakers* (pp. 67–69). And J. H. Parry's *The Discovery of the Sea* contains a sweeping summary of Portuguese ocean exploration.

For extended discussions of Magellan's ancestry, see Manuel Villas-Boas, *Os Magalhães;* Joyner (p. 309); and Morison (pp. 327–329).

The lives and influence of Spanish and Portuguese Jews have been written about by many scholars, including Jane Gerber, *The Jews of Spain;* Frederic David Mocatta, *The Jews of Spain and Portugal and the Inquisition;* and Ruth Pike, *Linajudos and Conversos in Seville.*

Among the many accounts of Magellan's early career are those by Morison; Charles Parr, *So Noble a Captain;* and F.H.H. Guillemard. Joyner's *Magellan* (pp. 33–57) is especially robust.

Leonard Y. Andaya's *The World of Maluku* mentions the extreme sensitivity of Portuguese maps (p. 9). Magellan's dealings with the Barbosa clan are described by Morison (p. 333) and Denucé (p. 168). Roger Craig Smith's thesis, *Vanguard of Empire,* offers background about the Casa de Contratación (pp. 32–33), and Denucé (p. 175) quotes Peter Martyr, as well as describing Ruy Faleiro's decline (pp. 169–171).

Guillemard's assessment of Adrian (p. 101) is quoted.

Donald Brand's articles in the *The Pacific Basin* and Mairin Mitchell's *Elcano* (p. 69) discuss Serrão, whose correspondence with Magellan was lost in the Lisbon earthquake of 1755; all that survives is accounts of it in the records of early Portuguese historians.

Las Casas's account of Magellan's plan can be found in Morison (p. 319), and the royal replies come from Martín Fernández de Navarrete, *Colección de los viajes y descubrimientos que hicierón por mar los españoles desde fines del siglo XV* (vol. 4, pp. 11–12, 113–116), which are available in an English translation in Rodrigue Lévesque's *History of Micronesia* (vol. 1, pp. 119–121, 123–125).

Denucé (pp. 172, 210, 214–218) describes Haro's financial arrangements for Magellan's voyage.

Navarrete (vol. 4, pp. 121–122) contains the document formally authorizing Magellan. An English translation can be found in Blair and Robertson's *The Philippine Islands: 1493–1898* (vol. 1, pp. 271–275).

CHAPTER II: THE MAN WITHOUT A COUNTRY

Magellan's distraught letter to King Charles can be found in Licuanan and Mira, *The Philippines Under Spain* (pp. 11–13). The original mentions placing four flags on the capstan, but it was unlikely that piece of machinery would be used for that purpose. A mast was far more likely. For more, see Morison (pp. 340–341).

King Charles's correspondence about Magellan's voyage is reproduced in Blair and Robertson (vol. 1, pp. 277–279 and 280–292). For Magellan's sailing orders, see Navarrete (vol. 4, pp. 130–152), and Blair and Robertson (vol. 1, pp. 256–259).

Documents pertaining to Ruy Faleiro's role in the expedition are reproduced in Navarrete (vol. 4, p. 497) and in Ignacio Vial and Guadalupe Morente, *La Primera Vuelta al Mundo: La Nao Victoria* (pp. 44–45).

The list of navigational supplies carried by the fleet comes from Vial and Morente (pp. 85–86).

Morison (pp. 338–339) is particularly blunt on the subject of Fonseca, as is Joyner, *passim*. Documents concerning Fonseca's dealings with the armada are contained in Navarrette (vol. 2). Although it lacks source notes, Charles Parr's biography, *So Noble a Captain* (p. 230), is strong on preparations for the voyage, including Fonseca's machinations.

Vial and Morente (pp. 95–96) discuss the Seville waterfront and the armada's provisions (p. 128).

The Casa de Contratación's efforts to reign in Magellan are detailed in Vial and Morente (p. 51) and documented in Navarrete (vol. 5). Denucé discusses Magellan's packing the roster with his relatives (pp. 236–239) and the solemn mass at Santa María de la Victoria (pp. 241–246).

Joyner (pp. 286–287) has the complete text of Magellan's will, and Denucé (p. 255) tells of Sabrosa's sad decline after Magellan fled Portugal.

CHAPTER III: NEVERLANDS

The prayerful commands are recorded by Pérez-Mallaína (p. 69).

The literature of early cartography is vast. A good place for general readers to start is Lloyd A. Brown's *The Story of Maps*, along with Rodney Shirley's *The Mapping of the World: Early Printed World Maps, 1472–1700*. John Noble Wilford's *The Mapmakers*, now in a revised edition (2000), is another valuable summation.

Stephen Frimmer's *Neverlands* offers a diverting introduction to the subject of mythical kingdoms. The quotations from Pliny the Elder are found in the Penguin edition of *Natural History* (pp. 76, 81). John Livingston Lowes's *The Road to Xanadu* (pp. 117–118) catalogs some colorful monsters of the deep. Accounts of the Prester John phenomenon are drawn from Robert Silverberg's

The Realm of Prester John (pp. 41–45, 63). Marco Polo's words come from the Penguin edition of *The Travels* (pp. 96, 106), and Mandeville's fanciful descriptions can be found in the Penguin edition of *Sir John Mandeville* (pp. 117, 122, 129, 130). John Larner's *Marco Polo and the Discovery of the World* (p. 166) has also been consulted. Finally, Rabelais's satirical skewering of Hearsay can be found in the Penguin edition of *Gargantua and Pantagruel* (p. 679).

CHAPTER IV: "THE CHURCH OF THE LAWLESS"

Details of the contretemps concerning the proper form of address to Magellan come from Morison (p. 358), and the incident involving António Ginovés is told most persuasively by Vial and Morente (p. 111).

For more on the social and political aspects of homosexuality in Spain, see Roger Bigelow Merriman, *The Rise of the Spanish Empire* (p. 53). It was common practice for homosexuals and even those suspected of homosexual practices to be denounced and punished in public. In August 1519, at the time of Magellan's departure from Spain, a clergyman in Valencia used the public punishment of a number of homosexuals as the occasion for a hysterical sermon condemning the accused, and his listeners cried out for the death of those who had escaped with lesser punishments. The hysteria boiled along as the populace took up arms; the uprising appeared to end when the authorities confiscated the weapons and demanded that the protestors confine themselves to their homes, but even then the controversy continued as the protestors formed a fraternity and insisted on bearing arms.

Albo's account of the fleet's arrival in Rio de Janeiro can be found in Lord Stanley, *First Voyage* (p. 212), and Morison (p. 299) discusses early Portuguese efforts to exploit the region's natural resources. Joyner (p. 125) offers details of Carvalho's past.

Vespucci's ripe description of Brazilian Indians is reproduced by Morison (pp. 285–286).

Morison (p. 362) discusses Magellan's efforts to calculate latitudes.

Details of the sailor's existence aboard ship are drawn from Pérez-Malláina (pp. 135–159) and Morison (pp. 165–171). Joyner (p. 250) has an interesting discussion of the *ampolletas*. And only Morison (p. 171), it seems, would trouble to explain the difficulties sailors faced when they had to relieve themselves at sea. Roger Craig Smith's thesis (pp. 175–176) and the *Colección General de Documentos Relativos a las Islas Filipinas Existentes en el Archivo de Indias de Sevilla* (vol. 2, pp. 165–168) describe Bustamente's limited store of medical supplies.

Information about the saints in the ships' rosters comes from Pérez-Malláina (p. 238) and from Louis Reau, *Iconographie de l'Art Chrétien* (vol. 3, pp. 115–122, 169–177, 804).

For more on the *Consulado,* see Paul S. Taylor, "Spanish Seamen in the New World During the Colonial Period," *The Hispanic American Historical Review.*

Early conceptions of the strait are discussed by Guillemard (pp. 191–193), who quotes Galvão about the "Dragon's taile"; by Justin Winsor, *Narrative and Critical History of America* (p. 107); and by Morison (pp. 301–302). See also Mateo Martinic Beros, *Historia del Estrecho de Magallanes* (1977).

CHAPTER V: THE CRUCIBLE OF LEADERSHIP

The quotation from Albo's diary is drawn from Stanley (p. 217). Morison (p. 365) provides details of Magellan's reconnaissance during the waning days of February. Guillemard, normally scrupulous, mentions only one island discovered on February 27, but as Pigafetta makes clear, there were two. See Skelton's translation of Pigafetta's *Magellan's Voyage* (p. 46).

The actual animals that Magellan and his crew saw in this part of the world are open to debate because Pigafetta did not provide enough details for exact identification. Guillemard and his followers labeled the "sea wolves" that Magellan's men saw as "fur seals," but that is probably not correct. In general, fur seals do not live in this part of the world, but are found in Australia or more northerly waters, around the Bering Strait, for example. It is more likely that Pigafetta was describing the sea lion or sea elephant (sometimes called elephant seal), which is far more common in these latitudes.

The case for Magellan's deliberately obscuring the location of Port Saint Julian is made by Denucé, whose Portuguese sources might have imputed sinister motives to Magellan and his pilots where none existed. Nevertheless, there are a number of strong hints that as the voyage proceeded Magellan came to realize he had sailed into Portuguese waters, and it was too late for him to do anything about it except hope he was not caught.

Accounts of Magellan's motivational speech are found in Guillemard (p. 163) and Antonio Herrera, *The General History* (vol. 2, pp. 357ff, and vol. 3, p. 14).

Since Pigafetta is silent on the subject of the mutiny out of loyalty to his Captain General, de Mafra's recollections, found in his *Relation,* are particularly useful, but he was not writing about events at the time they occurred; rather, he was reminiscing—to a scribe—some years after the fact. Nevertheless, de Mafra, unlike Pigafetta, was able to speak freely about controversial matters. See Blázquez and Aguilera's *Descripción* for the complete de Mafra account. The translation is by Víctor Ubéda. Useful, if predictable, eyewitness accounts of the mutiny in Port Saint Julian can also be found in Navarrete (vol. 4); Elcano's comment can be found on p. 288. See also Joyner (pp. 284, 291). Finally, Gaspar Correia's brief but jumbled account of the voyage (found in Lord Stanley of Alderly, ed., *The First Voyage Round the World* and in Charles E.

Nowell, ed., *Magellan's Voyage Around the World*) supplies details of Magellan's use of trickery in regaining control over his ships. Unfortunately, Correia, one of the earliest historians of the voyage, confuses Cartagena with Quesada and relates that Magellan had Cartagena drawn and quartered, when it was Quesada who suffered that fate. Guillemard (pp. 165–170) makes sense of the chaotic set of events surrouding the mutiny.

Descriptions of torture procedures are drawn from Henry Lea, *Torture* (p. 116); Philippus Limborch, *The History of the Inquisition* (vol. 1, pp. 217–220); and John Marchant et al., *A Review of the Bloody Tribunal* (pp. 357–358). The final stage of the *strappado* can be found on pp. 219–220 of Limborch. The punctuation has been modernized. Denucé names the victims of torture and declares Magellan's deeds to be illegal (pp. 272–280).

CHAPTER VI: CASTAWAYS

Morison (p. 374) quotes praise for Serrano's industriousness. An account of *Santiago's* ill-fated reconnaissance mission appears in Stanley (p. 250), presenting Correia, who appears to confuse *Santiago's* final voyage with the crew's subsequent journey over land. (Correia states that the ship returned "laden with the crew," which was not the case.) For Charles Darwin's description of the Santa Cruz region, plus many other detailed natural descriptions, see *Voyage of the Beagle* (p. 167).

Pigafetta's sketchy description of *Santiago's* crew efforts to survive the trek back to Port Saint Julian is ably supplemented by Guillemard and especially Herrera (pp. 17–18), who writes about the frozen fingers.

Concerning the first signs of Indians in Port Saint Julian, Pigafetta describes the unexpected appearance of a "giant" on the beach, but de Mafra more plausibly recalls the appearance of smoke prior to the giant's arrival. Pigafetta's account of these Indians and guanacos appear in Skelton's translation (pp. 47–50), Guillemard (p. 183), and Herrera (p. 19).

Ginés de Mafra recalled that first encounter with the Indians of the region quite differently. In de Mafra's unsentimental account, no friendly giant danced on the shore and pointed heavenward, no religious conversion occurred, and no banquet for the Indians took place aboard the flagship.

After two months in Port Saint Julian, he wrote, "One night the night watchman said there were fires on shore." On hearing the news, Magellan sent a party ashore to find them and, if they were fortunate, a new source of food beyond their steady diet of salted sea elephant and shellfish. He forbade his men to harm the Indians, if they found any. When the men reached the fires, de Mafra recalled, they "found a hut that was like the small lean-to of a wine grower, covered with animal skins. Our men surrounded the lean-to so carefully that none of the seven people inside left." The Europeans saw that the lean-to was divided into two sections, one for men, the other for women and children. Just outside the lean-to were "five sheep of a very good size and shape never seen before." These were guanacos. The landing party camped near the lean-to, the sailors shivering under the borrowed skins and maintaining a vigil over the Indians in case they attacked in the dead of night, but the precaution was unnecessary because the Indians slept deeply and snored loudly until the morning.

The next day, the Europeans feasted on tough, stringy, and relatively taste-less guanaco meat with the Indians. Only drink was missing; the thirsty sailors craved wine, or even water, to chase the sinewy guanaco meat.

When the landing party returned to the flagship and told Magellan of their find, the Captain General sent them back ashore with orders to return with an Indian, but the crew members found the huts deserted, apparently on short notice. "Our men spotted the tracks in the abundant snows and followed them," said de Mafra. "It was late when they found them in another hut erected in a different valley." The Indians fled, the Europeans gave chase, and a skirmish ensued. "Our men tried to capture them, and when they rushed at them, the Indians wounded a certain Barassa"—an apprentice seaman aboard *Victoria*— "in his groin, as a result of which he later died. The Indians escaped, and our men could do nothing to prevent them."

The crew members spent the night ashore, "Where they made a fire and roasted some of the meat that they had taken and drank melted snow from bowls, and with no protection other than their spears, they passed the night, even though it was very cold." In the morning, they broke camp and returned to the waiting ships, where they made their report. Magellan "ordered thirty men to go ashore and kill whomever they found to avenge for the dead one, and since the first men had not buried him, to bury him." As ordered, the war party went ashore and buried their fallen comrade, but they failed to find "anybody on whom to avenge their anger and rage." After a fruitless search lasting eight days, they returned to the ships, exhausted and frustrated.

Joyner (p. 150) describes the plight of Cartagena and Pero Sánchez de la Reina, as does Morison (p. 375).

CHAPTER VII: DRAGON'S TAIL

The original details concerning the eclipse most likely came from the fleet's astronomer, San Martín, whose records are described by Guillemard (p. 187). He draws on Herrera, who had access to the actual papers, which are lost.

Gallego's remarks are from the Leiden Narrative, translated by the indefati-gable Morison (p. 12) and Albo's entry about the strait appears in Stanley (pp. 218–219). The "later explorer" is quoted by Morison (p. 380). For an analysis of Pigafetta's use of the word *carta,* see Morison (p. 382), and for early misconceptions of the strait, see Parry, *The Discovery of the Sea* (p. 248) and Morison (pp. 382–383). Guillemard cites *"terra ulterior incog."* on p. 192.

Although Magellan staked the success of the expedition on navigating the strait, he reluctantly revealed that he had an alternate plan. "Had we not discovered the Strait," Pigafetta informs us, "the Captain General had determined to go as far as seventy-five degrees toward the Atlantic Pole. There in that latitude, during the summer season, there is no night, or if there is any night it is but short, and so in the winter with the day."

Albo is quoted in Stanley (p. 219), and Francis Pretty in Charles William Eliot, ed., *Voyages and Travels.* Other descriptions of the strait are drawn from Morison (pp. 390–391), Darwin's *Voyage of the Beagle* (pp. 196–197, 203), and Herrera (chap. 14). Parr (pp. 317–318) relates a dramatic encounter between "a half dozen naked Indians" paddling a canoe and Magellan's fleet. But none of the diarists mention it (Pigafetta, fascinated by indigenous tribes, surely would have); nor do other historians. In the absence of sources, this incident lacks a basis in fact.

Magellan's desire to persist in the voyage is related by Denucé (p. 288) and by Herrera (chap. 15). Joyner (p. 276) discusses Gomes's resentment. Denucé (pp. 287–288) provides details concerning the supposed placement of papers in the strait. Herrera says the mutineers killed Mesquita, but as numerous other accounts demonstrate, that was not the case.

Magellan's and San Martín's important missives appear in João de Barros's *Da Asia: Decada Terceira,* translated for this book by Víctor Úbeda. See also Stanley (pp. 177–178). Barros retrieved the documents from the papers of San Martín, later seized by the Portuguese. In Barros's words, "We do not deem it unfitting to include here the contents of such orders, as well as San Martín's reply, so that it can be seen, not by our words, but by their own, the condition in which they found themselves, and also Magellan's purpose with regard to the route that he planned to follow in case the way he wished to find should fail him."

Pigafetta and Albo disagree on the precise date the armada sailed from the western mouth of the strait. Pigafetta gives the date as November 28, and Albo the twenty-sixth. The discrepancy could be explained in various ways; for example, Pigafetta and Albo could have selected different landmarks to mark the strait's end. See Morison (pp. 400–401).

CHAPTER VIII: A RACE AGAINST DEATH

For more on Setebos in the English literary tradition, see Robert Browning's long poem "Caliban upon Setebos or, Natural Theology in the Island" (1864), a philosophical rumination by Caliban on his plight.

Magellan's failure to make a landfall in the Pacific before Guam has long prompted questions. One school of thought holds that he was actually farther north than his chroniclers indicated, and distant from all islands. Although all the eyewitnesses—Albo, Pigafetta, and de Mafra—agree that the armada headed west into the Pacific at the approximate latitude of Valparaiso, Chile, others have suggested that the diarists falsified their accounts to conceal the true location of the Spice Islands, in case they were found in the Portuguese part of the world rather than the Spanish. The assumption makes little sense because they wrote their accounts for different purposes; Pigafetta wrote to glorify Magellan and ingratiate himself with European nobility, Albo to keep track of their whereabouts, and de Mafra dictated his account years later, when the location of the Spice Islands was no longer controversial.

Information on little San Pablo is drawn from Samuel Eliot Morison's unpublished *Life* article (February 24, 1972).

CHAPTER IX: A VANISHED EMPIRE

Much of the information in this chapter is drawn directly from Pigafetta's account, which eloquently describes the armada's Pacific passage.

For an extended and valuable discussion of Magellan's first landfall in the Pacific, see Robert F. Rogers and Dirk Anthony Ballendorf, "Magellan's Landfall in the Mariana Islands," in the *Journal of Pacific History* (vol. 24, October 1989). The authors re-created the landfall to be precise about the fleet's movements; however, alterations wrought by erosion can compromise the value of such exercises. Also worth consulting is Rogers's book *Destiny's Landfall* for details of Chamorran culture. Guillemard (p. 226) and Joyner (p. 269) discuss Master Andrew.

For a fascinating account of island navigation systems in theory and practice, see Ben Finney, *Voyage of Rediscovery,* especially pp. 56–64.

Louise Levathes's *When China Ruled the Seas* (1996) is the one reliable guide to the subject written in English. Also eminently worthwhile is Ma Huan's diary of one expedition, *The Overall Survey of the Ocean's Shores* (1433). Gavin Menzies's recent book *1421* suggests that the Treasure Fleet reached the Caribbean and perhaps completed a circumnavigation one hundred years before Magellan. However, hard evidence to prove these tantalizing assertions is still lacking.

As his candidate for the first person to complete a circumnavigation, Morison (p. 435) nominates Magellan's slave Enrique. Morison argues that Magellan's voyage brought Enrique back to his point of origin. Even if this assumption is correct, Enrique traveled around the world only because Magellan took him along.

For a discussion of the armada's weaponry, see Charles Boutell, *Arms and Armour* (p. 243) and Parr (p. 383). Roger Craig Smith's 1989 thesis, *Vanguard of Empire: 15th- and 16th-Century Iberian Ship Technology in the Age of Discovery,* offers more specialized information on the subject. Also recommended are Courtlandt Canby's *A History of Weaponry* (vol. 4) and John Hewitt's *Ancient Armour and Weapons in Europe* (vol. 3), as well as *The Penguin Encyclopedia of Weapons and Military Technology.*

Guillemard (p. 235) mentions the bats seen by the sailors. Albo's description of Cebu comes from Navarrete (vol. 4, pp. 219–221).

CHAPTER X : THE FINAL BATTLE

Two very different facets of Pigafetta's wide-ranging interests are on display in his account of Magellan's visit to Cebu. As a former papal diplomat, he was duty-bound, but also genuinely moved, by the Captain General's efforts to convert the Filipinos. In addition, Pigafetta is virtually the only source on the subject. See Robertson's translation of Pigafetta (pp. 133–169) for more details of Magellan's religious convictions.

Pigafetta also dwells at length on *palang,* which fascinated him. The subject frequently appears in accounts of Pacific and Eastern cultures during the Age of Discovery, and even the Chinese sailors with the Treasure Fleet came across a variant of *palang,* and, like Magellan's men, were both fascinated and appalled by the practice. In this instance, *palang* took the form of small sand-filled beads inserted into the scrotum, and when men who were thus adorned moved or walked, they made a faint noise, reminiscent of bells ringing. It was, said Ma Huan, "a most curious thing."

In his assessment of *palang,* Pigafetta was unusually tolerant, at least by European standards. Other European visitors wrote about *palang* in censorious tones. Andrés Urdaneta, the capable Spanish navigator, visited the region several times, beginning in 1525, four years after the Armada de Molucca, and he left an account of *palang* in which the Indians of Borneo fasten a "few small round stones" to the penis with a leather sleeve, while others are pierced with "a tube of silver or tin . . . and on those tubes, they put thin sticks of silver or gold at the time they want to engage with women in coitus." In practice, the bearer of *palang* often inserted a range of objects into the tube; pig's bristles were employed, as were bamboo shavings, beads, and even shards of glass. Urdaneta was appalled, and missionaries in the Philippines preached against it.

Antonio Morga, a Spanish historian who wrote one of the first accounts of the Philippines, was also revolted by the practice, which he considered highly immoral, but he provided a detailed description of *palang* as practiced elsewhere in the Philippine archipelago. By the time Morga got around to describing *palang,* in 1609, it was clearly a practice on the way out, thanks to the strenuous efforts of the Catholic clergy to discourage it: "The natives...especially the women, are very vicious and sensual, and their wickedness has devised lewd ways of intercourse between men and women, one of which they practice from their youth onwards. The men skillfully make a hole near the head of the penis into which they insert a small serpent's head of metal or ivory. Then they secure this by passing a small peg of the same material through the hole, so that it may not work loose. With this device they have intercourse with their wives and for long after the copulation they are unable to withdraw. They are so addicted to this, and find such pleasure in it, that although they shed a great deal of blood, and receive other injuries, it is a common practice among them. These devices are known as *sagras,* and there are very few of them left, because after they become Christians, care is taken to do away with such things and not permit their use."

For more on the subject, see Morison (p. 435). The two articles by Tom Harrison listed in the bibliography contain the quoted descriptions.

Juan Gil, in his recent *Mitos y Utopiás del Descubrimiento* (1989), is one of the few commentators to consider the possibility that Magellan's disaffected officers let the Mactanese slaughter him.

Simon Winchester describes the reenactment of the battle between Magellan and Lapu Lapu in "After Dire Straits, an Agonizing Haul Across the Pacific," *Smithsonian* (pp. 84–95).

CHAPTER XI : SHIP OF MUTINEERS

In addition to Pigafetta and other accounts mentioned in the text, details concerning Enrique's treachery are drawn from Gonzalo Fernández de Oviedo y Valdez's *Historia General y Natural de las Indias* (pp. 13ff). Denucé (pp. 323–326) adds to the picture of the massacre's aftermath. See also Morison (pp. 438–441) and Navarrete (vol. 4).

Concerning *San Antonio*'s return to Spain, Guillemard (p. 215) remarks that Argensola, an early and occasionally inaccurate historian, states that Cartagena and the priest were rescued by *San Antonio,* but no records support this claim. Although Guillemard (p. 216) believes *San Antonio* ran low on food during the return journey, that was likely not the case, for she carried the entire fleet's provisions. It is possible that those aboard *San Antonio* invented this story to gain sympathy. Skelton provides the date of the ship's arrival (p.156).

The official reports and orders concerning the mutiny of *San Antonio* and her paltry contents can be found in Lucuanan and Mira, *The Philippines Under Spain* (pp. 17, 24–28, 43–44). See also Denucé (p. 293). Joyner (p. 159) says Mesquita had to pay for his trial-related costs.

Roger Merriman offers much more on King Charles's astonishing ascent to power in *The Rise of the Spanish Empire* (vol. 3, 1925).

For accounts of daily life in sixteenth-century Seville, see Pike's "Seville in the Sixteenth Century," *The Hispanic American Historical Review*, vol. 41, no. 3, August 1961.

CHAPTER XII: SURVIVORS

Elcano's ascent and the problems facing the Armada de Molucca after Magellan's death are ably set forth in Mitchell. See especially pp. 42–48 and 63–64.

Robertson's translation of Pigafetta carries forward into vol. 2 at this point in the narrative.

Morison's description of Palawan appears on p. 442, and Albo's exasperation while trying to reach Brunei can be found in Stanley (pp. 226–227).

Jones's 1928 translation of *The Itinerary of Ludovico de Varthema of Bologna* has been quoted. Varthema's description of the Spice Islands (pp. 88–89) offers a fairly exact preview of scenes the armada later encountered. And for more on the Bajau, see Harry Nimmo's *The Sea People of Sulu* (1972).

Argensola's description of the Moluccas comes from Stevens's 1708 translation of *The Discovery and Conquest of the Molucco and Philippine Islands* (p. 7).

The quotation of *The Lusíads* comes from Landeg White's translation (p. 223).

CHAPTER XIII: ET IN ARCADIA EGO

Barros's harsh view of the inhabitants of the Spice Islands is cited in Charles Corn's charming and evocative study, *The Scents of Eden* (p. 58), and in Andaya (p. 16). Given the reputation of the Spice Islands' inhabitants, it is surprising that the armada treated them with as much civility as they did.

António Galvão's useful and vivid description of the Spice Islands' volcanoes and rainfall can be found in his *Treatise on the Moluccas*, tr. Hubert Jacobs (1971), and *The Discoveries of the World*, tr. Richard Hakluyt (1862, originally published in 1601). Barbosa's descriptions of cloves and Almanzor's family are drawn from his *Description of the Coasts of East Africa and Malabar*, tr. Henry E. J. Stanley, 1866 (pp. 201–202). Andaya discusses the primacy of oral over written agreements (p. 61).

On the subject of Serrão's curious odyssey in the Spice Islands, Guillemard offers several unsubstantiated theories. According to one scenario, he was "poisoned by a Malay woman who acted under Portuguese orders." But Guillemard also cites Argensola's assertion that Serrão was not poisoned at all; rather, he was sent back to India and he died aboard ship (p. 281).

Anyone wanting to learn more about cloves should start by consulting Frederic Rosengarten's *Book of Spices* (rev. ed., 1973), especially pp. 200–204. Much of the information about spices in this chapter is drawn from this comprehensive and entertaining reference work. Other useful works on the subject include Parry's *The Story of Spices* (1953) and Larioux Bruno's "Spices in the Medieval Diet: A New Approach," *Food and Foodways*, vol. 1, no. 1, 1985. Also of interest is M. N. Pearson, ed., *Spices in the Indian Ocean World* (1996).

CHAPTER XIV: GHOST SHIP

It is still possible that syphilis in Timor—if that was what the sailors saw—originally came from Portugal, because the Portuguese went to China as early as 1513; the Chinese might then have carried it to Timor.

Pigafetta's elaborate account of China relies on stories he gathered in Indonesia from a well-traveled Arab merchant. Pigafetta sketched a convincing description of the emperor's seat, Peking: "Near his palace are seven encircling walls, and in each of those circular places are stationed ten thousand men for the guard of the place [who remain there] until a bell rings, when ten thousand other men come for each circular space. They are changed in this manner each day and night. Each circle of the wall has a gate. At the first stands a man with a large hook in his hand, called *satu horan* with *satu bagan;* in the second, a dog, called *satu hain;* in the third, a man with an iron mace, called *satu horan* with *pocum becin;* in the fourth, a man with a bow in his hand, called *satu horan* with *anat panam;* in the fifth, a man with a spear, called *satu horan* with *tumach;* in the sixth, a lion, called *satu horiman;* in the seventh, two white elephants called *gagua pute.*

"The palace has seventy-nine halls which contain only women who serve the king. Torches are always kept lighted in the palace, and it takes a day to go through it. In the upper part are four halls, where the principal men go sometimes to speak to the king. One is ornamented with copper, both below and above; one all with silver; one all with gold; and the fourth with pearls and precious gems. When the king's vassals take him gold or any other precious things as tribute, they are placed in those halls, and they say, 'Let this be for the honor and glory of our Santhoa Raia.' "

Concerning the Cape of Good Hope: in Canto Five of *The Lusíads*, Luis de Camões personified it as a mighty giant named Adamastor, who resented the intrusion of mere humans, even audacious Portuguese navigators, into his domain.

> *I am that vast, secret promontory*
> *You Portuguese call the Cape of Storms,*
> *Which neither Ptolemy, Pompey, Strabo,*

> *Pliny, nor any authors knew of.*
> *Here Africa ends. Here its coast*
> *Concludes in this, my vast, inviolate*
> *Plateau, extending southwards to the Pole*
> *And, by your daring, stuck to my very soul.*

Espinosa's sad comment about turning back is in Lévesque (p. 306).

Much of what is known about *Trinidad*'s tragic end comes from Barros, whose account is skewed in favored of the Portuguese. Barros (Chapter 10) states that Brito discovered the armada's attempts to alter the locations of various lands, and Guillemard (p. 303) approvingly quotes Brito's callous report to the Portuguese crown about the armada's survivors. *Trinidad*'s tragic end inspired Barros to twist events so that Brito emerges as the savior of Magellan's men, when in fact he was happy to let them die. "The first thing he did," Barros writes of Brito, "on request of a certain Bartolomé Sánchez, clerk of that ship, whom Gonzalo Gómez de Espinosa had sent for help due to their sorry condition, was to dispatch a caravel with plenty of provisions and anchors for the ship. . . . António de Brito had the crew cured and tended as carefully as they had been natives of this kingdom, and not gone to those lands to cause us trouble." In conclusion, Barros writes, "We are free of any suspicion."

The description of Espinosa's travails in the Portuguese penal colony is drawn from Lévesque (p. 306), Guillemard (p. 304), and Navarrete (vol. 4, pp. 378ff).

The vignette of *Victoria*'s encountering an indifferent Portuguese vessel at the Cape of Good Hope is related by Joyner (p. 231), and Morison (p. 461) mentions *Victoria*'s multiple crossings of the equator. Joyner (p. 234) explores Burgos's character and motives for betraying his crew members.

CHAPTER XV: AFTER MAGELLAN

The full text of Elcano's evasive letter to King Charles is reprinted by Morison (pp. 471–473).

Mitchell (pp. 178–182) provides the relevant documents pertaining to the subsequent inquiry into the expedition, and Joyner (p. 242) tells of Juan Rodríguez's fate. Espinosa's last days are accounted for by Joyner (pp. 265, 277–278) and Morison (p. 456). Morison (p. 477) also discusses the ill-fated Guadiana River conference. Those seeking still more detailed information on lawsuits arising from the loss of the armada's ships should consult the Special Collections Department of the Brandeis University Libraries; documents there shed light on Cristóbal de Haro's efforts to recover the cost of backing the expedition.

Accounts of Elcano's last voyage and death can be found in Mitchell (pp. 148–157), Morison (pp. 475–483), Parry, *The Discovery of the Sea* (p. 257), and Nowell (p. 338).

Lawrence C. Wroth's *Early Cartography of the Pacific* (pp. 149–150) details the ultimate outcome of the struggle between Spain and Portugal for the control of the Spice Islands.

The mournful tale of *Victoria's* final voyage is told by Parry (p. 261), Joyner (p. 243), and Mitchell (pp. 106–107).

Bibliography

BOOKS

Alcocer Martínez, Mariano. *Don Juan Rodríguez Fonseca: Estudia Crítica Biográfica*. Valladolid: Imprenta de la Casa Católica, 1923.

Andaya, Leonard Y. *The World of Maluku*. Honolulu: University of Hawaii Press, 1993.

Andrews, William. *Bygone Punishments*. London: William Andrews & Co., 1899.

Arber, Edward. *The First Three English Books on America*. New York: Kraus. Reprint, 1971.

Baker, J. *The History of the Inquisition*. Westminster, England: O. Payne, 1736.

Baker, J.N.L. *A History of Geographical Discovery and Exploration*. Boston: Houghton Mifflin, 1931.

Barbosa, Duarte. *The Book of Duarte Barbosa*, tr. Mansel Longworth Dames. New Delhi: Asian Educational Society, 1989. (Originally published in 1812.)

———. *A Description of the Coasts of East Africa and Malabar*, tr. Henry E. J. Stanley. London: Hakluyt Society, 1866.

Barros, João de. *Da Asia: Decada Terceira*. Lisboa: Na Régia Officina Typografica, 1777.

Barros Arana, Diego. *Vida i viajes de Hernando de Magallanes*. Santiago de Chile: Imprenta Nacional, 1864.

Bates, Robert L., and Julia A. Jackson, eds. *Dictionary of Geological Terms*. New York: Anchor Books, 1984.

Benson, E. F. *Magellan*. London: John Lane, 1929.

Birmingham, Stephen. *The Grandees: America's Sephardic Elite*. New York: Harper & Row, 1971.

Blair, Emma Helen, and James Alexander Robertson, eds. *The Philippine Islands: 1493–1898*, vol. 1. Mandalyong, Rizal: Cachos Hermanos, 1973.

Blázquez, Antonio, and Delgado Aguilera, eds. *Descripción de los reinos, costas, puertos e islas que hay desde el Cabo de Buena Esperanza hasta los Leyquios, por*

Fernando de Magallanes; Libro que trata del descubrimiento y principio del Estrecho que se llama de Magallanes, por Ginés de Mafra; y Descripción de parte del Japón. Madrid: Publicaciones de la Real Sociedad Geográfica, 1920.

Boorstin, Daniel J. *The Discoverers.* New York: Random House, 1983.

Bourne, Edward Gaylord. *Discovery, Conquest, and Early History of the Philippine Islands.* Cleveland: Arthur H. Clark, 1907.

————. *Spain in America: 1450–1580.* New York: Harper & Brothers, 1904.

Boutell, Charles, ed. *Arms and Armour in Antiquity and the Middle Ages.* New York: D. Appleton & Co., 1870.

Brand, Donald D. "Geographical Exploration by the Spaniards" and "Geographical Exploration by the Portuguese." In *The Pacific Basin: A History of Its Geographical Exploration,* ed. Herman R. Friis. New York: American Geographical Society, 1967.

Braudel, Fernand. *The Mediterranean and the Mediterranean World in the Age of Philip II,* 2 vols. New York: Harper & Row, 1972–1973.

————. *The Structures of Everyday Life: The Limits of the Possible.* London: Collins, 1981.

Brown, Lloyd A. *The Story of Maps.* New York: Dover Publications, 1977. (Originally published in 1949.)

Buehr, Walter. *Firearms.* New York: Thomas Y. Crowell, 1967.

Bueno, José María. *Soldados de España: El Uniforme Militar Español Desde los Reyes Católicos hasta Juan Carlos I.* Self-published, Málaga, 1978.

Cabrero Fernández, Leoncio, ed. *Historia general de Filipinas.* Madrid: Ediciónes de Cultura Hispánica, 2000.

Camões, Luíz de. *The Lusíads,* tr. Landeg White. Oxford: Oxford University Press, 1997.

Campbell, John. *The Spanish Empire in America.* London: M. Cooper, 1747.

Canby, Courtlandt. *A History of Weaponry,* vol. 4. New York: Hawthorn Books, 1963.

Carpenter, Kenneth, J. *The History of Scurvy and Vitamin C.* Cambridge: Cambridge University Press, 1986.

Cipolla, Carlo M. *Guns and Sails in the Early Phase of European Expansion, 1400–1700.* London: Collins, 1965.

Colección General de Documentos Relativos a las Islas Filipinas Existentes en el Archivo de Indias de Sevilla (5 vols). Barcelona: L. Tasso, 1918–1923.

Corn, Charles. *The Scents of Eden.* New York: Kodansha International, 1998.

Crane, Nicholas. *Mercator: The Man Who Mapped the Planet.* London: Weidenfeld & Nicholson, 2002.

Crow, John A. *Spain: The Root and the Flower.* New York: Harper & Row, 1975.

Dalrymple, Alexander. *An Historical Collection of the Several Voyages and Discoveries in the South Pacific Ocean*, vol. 1. London: J. Nourse, 1770.

Darwin, Charles. *Voyage of the Beagle*. London: Penguin Books, 1989. (Originally published in 1839.)

Denucé, Jean. *Magellan: La Question des Moluques et la Première Circumnavigation du Globe*. In *Mémoires, Académie Royale de Belgique*, vol. 4. Brussels: Hayez, 1908–1911.

DeVries, Kelly. *Medieval Military Technology*. Lewiston, N.Y.: Broadview Press, 1992.

Diamond, Jared. *Guns, Germs, and Steel*. New York: W. W. Norton, 1997.

Diffie, Bailey W., and George D. Winius. *Foundations of the Portuguese Empire, 1415–1580*. Minneapolis: University of Minnesota Press, 1977.

Eliot, Charles William, ed. *Voyages and Travels: Ancient and Modern*. New York: P. F. Collier & Son, 1910.

Faria y Sousa, Manuel de. *The Portugues [sic] Asia*, tr. John Stevens. Westmead, England: Gregg International Publishing, 1971. (Originally published in 1695.)

Finney, Ben. *Voyage of Rediscovery*. Berkeley: University of California Press, 1994.

Frimmer, Steven. *Neverland*. New York: Viking Press, 1976.

Galvano, Antonio, tr. Richard Hakluyt. *The Discoveries of the World*. London: Hakluyt Society, 1862. (Originally published in 1601.)

Galvão, Antonio. *A Treatise on the Moluccas*, tr. Hubert Jacobs. Rome: Jesuit Historical Institute, 1971.

Gerber, Jane S. *The Jews of Spain*. New York: The Free Press, 1992.

Gil, Juan. *Mitos y Utopías del Descubrimiento*. Madrid: Alianza Editorial, 1989.

Guillemard, F.H.H. *The Life of Ferdinand Magellan*. London: George Philip & Son, 1890.

Haliczer, Stephen. "The Expulsion of the Jews as Social Process." In *The Jews of Spain and the Expulsion of 1492*, ed. Moshe Lazar and Stephen Haliczer. Lancaster, Calif.: Labyrinthos, 1997.

Haring, Clarence Henry. *The Spanish Empire in America*. New York: Harcourt Brace & World, 1973. (Originally published in 1947.)

———. *Trade and Navigation Between Spain and the Indies*. Cambridge: Harvard University Press, 1918.

Harvey, Miles. *The Island of Lost Maps*. New York: Random House, 2000.

Hawthorne, Daniel. *Ferdinand Magellan*. Garden City, N.Y.: Doubleday, 1964.

Henisch, Bridget Ann. *Fast and Feast*. University Park: Pennsylvania State University Press, 1976.

Herrera y Tordesillas, Antonio de. *The General History of the Vast Continent and Islands of America*, vols. 2 and 3. London: Jer. Batley, 1725. (Originally published in 1601–1615.)

————. *Historia general de los hechos de los Castellanos en las islas i tierra firme del mar océano*, vol. 5. Madrid: Tipografía de Archivos, 1936.

Hewitt, John. *Ancient Armour and Weapons in Europe*, vol. 3. Graz, Austria: Akademische Druck, 1967.

Hildebrand, Arthur Sturges. *Magellan*. New York: Harcourt, Brace & Co., 1924.

Joyner, Tim. *Magellan*. Camden, Me.: International Marine, 1992.

Kimble, George H. T. *Geography in the Middle Ages*. London: Methuen, 1938.

Lagôa, João António de Mascarenhas Judice, Visconde de. *Fernão de Magalhãis: a sua vida e a sua viagem* (2 vols.). Lisboa: Seara Nova, 1938.

Lalaguna, Juan. *A Traveller's History of Spain*, 4th ed. Brooklyn, N.Y., and Northampton, Mass.: Interlink Books, 1999.

Larner, John. *Marco Polo and the Discovery of the World*. New Haven: Yale University Press, 1999.

Lea, Henry Charles. *A History of the Inquisition of Spain*, vol. 3. New York: AMS Press, 1988. (Originally published in 1906–1907.)

————. *Torture*. Philadelphia: University of Philadelphia Press, 1973. (Originally published in 1866.)

Leonardo de Argensola, Bartolomé. *Conquista de las islas Malucas*. Madrid: A Martín, 1609.

————. *The Discovery and Conquest of the Molucco and Philippine Islands*, tr. John Stevens. London, 1708.

Levathes, Louise. *When China Ruled the Seas*. New York: Oxford University Press, 1996.

Lévesque, Rodrigue, ed. *History of Micronesia*, vol 1. Gatineau, Québec: Lévesque Publications, n.d.

Licuanan, Virginia Benitez, and José Llavador Mira, eds. *The Philippines under Spain: A Compilation and Translation of Original Documents*, book 1. Manila: National Trust for Historical and Cultural Preservation of the Philippines, 1990.

Limborch, Philippus van. *The History of the Inquisition* (2 vols), tr. Samuel Chandler. London: J. Gray, 1731.

López de Gómara, Francisco. *Historia general de las Indias*. Madrid: Amigos del Círculo del Bibliófilo, 1982.

Lord Stanley of Alderly, ed. *The First Voyage Round the World, by Magellan*. London: Hakluyt Society, 1874. (Reprinted in 1964.)

Lothrop, Samuel Kirkland. *The Indians of Tierra del Fuego*. New York: Museum of the American Indian, 1928.

Lowes, John Livingston. *The Road to Xanadu*. Boston: Houghton Mifflin, 1927.

Macksey, Kenneth. *The Penguin Encyclopedia of Weapons and Military Technology*. London: Viking, 1993.

Manchester, William. *A World Lit Only by Fire: The Medieval Mind and the Renaissance.* New York: Little, Brown, 1993.

Marchant, John, et al. *A Review of the Bloody Tribunal; Or the Horrid Cruelties of the Inquisition.* Perth: G. Johnston, 1770.

Markham, Clements, tr. *Early Spanish Voyages to the Strait of Magellan.* London, Hakluyt Society, 1911.

———, tr. *The Letters of Amerigo Vespucci.* London: Hakluyt Society, 1894.

Martinic Beros, Mateo. *Historia del Estrecho de Magallanes.* Santiago, Chile: Editorial Andrés Bello, 1977.

Medina, José Toribio, ed. *Colección de documentos inéditos para la historia de Chile,* vols 2, 3. Santiago, Chile: Imprenta Ercilla, 1888.

———. *Colección de Historiadores de Chile y de Documentos Relativos,* vol. 27. Santiago, Chile: Imprenta Elzeviriana, 1901.

———. *El descubrimiento del Océano Pacífico: Vasco Nuñez Balboa, Hernando de Magallanes y sus compañeros.* Santiago, Chile: Imprenta Elzeviriana, 1920.

Medina, Pedro de. *A Navigator's Universe.* Chicago: University of Chicago Press, 1972. (Originally published in 1538.)

Melón y Ruiz de Gordejuela, Amando. *Magallanes-Elcano; o, La primera vuelta al mundo.* Zaragoza: Ediciones Luz, 1940.

Menzies, Gavin. *1421: The Year China Discovered the World.* London: Bantam Press, 2002.

Merriman, Roger Bigelow. *The Rise of the Spanish Empire in the Old World and the New.* New York: Macmillan, 1925.

Milton, Giles. *Nathaniel's Nutmeg.* New York: Farrar, Straus & Giroux, 1999.

Mitchell, Mairin. *Elcano: The First Circumnavigator.* London: Herder Publications, 1958.

Mocatta, Frederic David. *The Jews of Spain and Portugal and the Inquisition.* New York: Cooper Square Publishers, 1973. (Originally published in 1933.)

Molina, Antonio de. *Historia de Filipinas* (2 vols.). Madrid: Ediciones Cultura Hispánica del Instituto de Cooperación Iberoamericana, 1984.

Morga, Antonio de. *Sucesos de las Islas Filipinas,* tr. J. S. Cummins. Cambridge: Hakluyt Society, 1971.

Morison, Samuel Eliot. *Admiral of the Ocean Sea* (2 vols.). Boston: Little, Brown, 1942.

———. *The European Discovery of America: The Northern Voyages.* New York: Oxford University Press, 1971.

———. *The European Discovery of America: The Southern Voyages.* New York: Oxford University Press, 1974.

Morris, John G. *Martin Behaim.* Baltimore: Maryland Historical Society, 1855.

Navarrete, Martín Fernández de. *Colección de los viajes y descubrimientos que*

hicieron por mar los españoles desde fines del siglo XV, vol 4. Madrid: Imprenta
    Nacional, 1837.
Nimmo, Harry. *The Sea People of Sulu.* San Francisco: Chandler, 1972.
Nowell, Charles E., ed. *Magellan's Voyage Around the World.* Evanston: North-
    western University Press, 1962.
Nunn, George E. *The Columbus and Magellan Concepts of South American Geog-
    raphy.* Glenside, 1932. (Privately printed.)
Obregón, Mauricio. *From Argonauts to Astronauts: An Unconventional History of
    Discovery.* New York: Harper & Row, 1980.
————. *La Primera Vuelta al Mundo.* Bogotá: Plaza y Janés, 1984.
Oviedo y Valdez, Gonzalo Fernández de. *Historia General y Natural de las
    Indias.* Asunción: Editorial Guaranía, 1944–1945.
Oxford Companion to Ships and the Sea, The, ed. Peter Kemp. London: Oxford
    University Press, 1976.
Parr, Charles McKew. *So Noble a Captain.* New York: Thomas Y. Crowell, 1953.
Parry, J. H. *The Age of Reconnaissance.* Berkeley: University of California Press,
    1963.
————. *The Discovery of South America.* New York: Taplinger, 1979.
————. *The Discovery of the Sea.* Berkeley: University of California Press, 1981.
————. *The European Reconnaissance: Selected Documents.* New York: Walker &
    Co., 1968.
————. *The Spanish Seaborne Empire.* New York: Knopf, 1970.
Parry, John W. *The Story of Spices.* New York: Chemical Publishing Co., 1953.
Pearson, M. N., ed. *Spices in the Indian Ocean World.* Aldershot, Hampshire:
    Variorum, 1996.
Peillard, Leonce. *Magallanes.* Barcelona: Círculo de Lectores, 1970.
Penrose, Boies. *A Link to Magellan: Being a Chart of the East Indies, c. 1522.*
    Philadelphia: Wm. F. Fell, 1929.
————. *Travel and Discovery in the Renaissance.* Cambridge: Harvard Univer-
    sity Press, 1955.
Pérez-Mallaína, Pablo E. *Spain's Men of the Sea,* tr. Carla Rahn Phillips. Balti-
    more: Johns Hopkins University Press, 1998.
Pfitzer, Gregory M. *Samuel Eliot Morison's Historical World.* Boston: Northeast-
    ern University Press, 1991.
Pigafetta, Antonio. *Magellan's Voyage Around the World* (3 vols.), tr. James
    Alexander Robinson, Cleveland: Arthur H. Clark, 1906.
————. *Magellan's Voyage: A Narrative Account of the First Circumnavigation* (2
    vols.), tr. R. A. Skelton. New Haven and London: Yale University Press, 1969.
————. *The Voyage of Magellan,* tr. Paula Spurlin Paige. Englewood Cliffs, N.J.:
    Prentice-Hall, 1969.

Pike, Ruth. *Linajudos and Conversos in Seville*. New York: Peter Lang, 2000.

Pliny the Elder. *Natural History: A Selection,* tr. John F. Healy. New York: Penguin Books, 1991.

Polo, Marco. *The Travels,* tr. Ronald Latham. London: Penguin Books, 1958.

Prestage, Edgar. *The Portuguese Pioneers*. London: Adam & Charles Black, 1966. (Originally published in 1933.)

Rabelais, François. *The Histories of Gargantua and Pantagruel,* tr. J. M. Cohen. Baltimore: Penguin Books, 1955.

Ravenstein, E. G. *Martin Behaim: His Life and His Globe*. London: George Philip & Son, 1908.

Reau, Louis. *Iconographie de l'Art Chrétien* (3 vols.). Paris: Presses Universitaires de France, 1955–1959.

Reyes y Florentino, Isabelo de los. *Las islas Visayas en la época de la conquista.* Manila: Tipo-Litografía de Chofré, 1889.

Riling, Ray. *The Powder Flask Book*. New Hope, Pa.: Robert Halter, 1953.

Roditi, Edouard. *Magellan of the Pacific*. London: Faber & Faber, 1972.

Rodríguez, Marco, and María del Rosario. *Catálogo de Armas de Fuego*. Madrid: Patronato Nacional de Museos, 1980.

Rogers, Robert F. *Destiny's Landfall: A History of Guam*. Honolulu: University of Hawaii Press, 1995.

Rosengarten, Frederic, Jr. *The Book of Spices,* rev. ed. New York: Pyramid Books, 1973.

Sagarra Gamazo, Adelaida. *La Otra Versión de la Historia Indiana: Colón y Fonseca.* Valladolid: Universidad de Valladolid, 1997.

Schivelbusch, Wolfgang. *Tastes of Paradise*. New York: Vintage Books, 1992.

Sharp, Andrew. *The Discovery of the Pacific Islands*. London: Oxford University Press, 1960.

Shirley, Rodney. *The Mapping of the World: Early Printed World Maps, 1472–1700,* rev. ed. London: New Holland Press, 1993.

Silverberg, Robert. *The Realm of Prester John*. Athens: Ohio University Press, 1996.

Slocum, Joshua. *Sailing Alone Around the World*. New York: Barnes & Noble Books, 2000.

Sobel, Dava. *Longitude*. New York: Penguin Books, 1996.

Torres y Lanzas, Pedro. *Catálogo de los documentos relativos a las islas Filipinas existentes en el Archivo de Indias de Sevilla,* vol. 1. Barcelona: L. Tasso, 1925.

The Travels of Sir John Mandeville, tr. C.W.R.D. Moseley. London: Penguin Books, 1983.

Ulman, R.B., and D. Brothers. *The Shattered Self: A Psychoanalytic Study of Trauma*. Hillsdale, N.J.: Analytic Press, 1988.

Varthema, Ludovico di. *The Itinerary of Ludovico di Varthema of Bologna*, tr. John Winter Jones. London: The Argonaut Press, 1928.

Vial, Ignacio Fernández, and Guadalupe Fernández Morente. *La Primera Vuelta al Mundo: La Nao Victoria*. Sevilla: Muñoz Moya Editores, 2001.

Vigón, Jorge, *Historia de la Artillería Española*. Madrid, 1947.

Villas-Boas, Manuel. *Os Magalhães: Sete Séculos de Aventura*. Lisboa: Estampa, 1998.

Wilford, John Noble. *The Mapmakers*. New York: Knopf, 2000.

Winsor, Justin. *Narrative and Critical History of America*, vol. 2. Boston: Houghton Mifflin, 1884.

Wionzek, Karl-Heinz, ed. *Another Report About Magellan's Circumnavigation of the World: The Story of Fernando Oliveira*. Manila: National Historical Institute, 2000.

Wroth, Lawrence C. *The Early Cartography of the Pacific*. The Papers of the Bibliographical Society of America, vol. 38. New York: The Bibliographical Society of America, 1944.

Zweig, Stefan. *Conqueror of the Seas*. New York: The Viking Press, 1938.

PERIODICALS

Harrison, Tom. "The 'Palang.'" *Journal of the Malaysian Branch of the Royal Asiatic Society*, vol. 37, 1964, pp. 162–174.

————. "The 'Palang': II. Three Further Notes." *Journal of the Malaysian Branch of the Royal Asiatic Society*, vol. 39, 1966, pp. 172–174.

Larioux, Bruno. "Spices in the Medieval Diet: A New Approach." *Food and Foodways*, vol. 1, no. 1, 1985.

Nunn, George E. "Magellan's Route in the Pacific." *Geographical Review*, vol. 24, 1934.

Pike, Ruth. "Seville in the Sixteenth Century." *Hispanic American Historical Review*, vol. 41, no. 3, August 1961.

Rogers, Robert F., and Dirk Anthony Ballendorf. "Magellan's Landfall in the Mariana Islands." *Journal of Pacific History*, vol. 24, October 1989.

Taylor, Paul S. "Spanish Seamen in the New World During the Colonial Period." *Hispanic American Historical Review*, vol. 5, 1922.

Torodash, Martin. "Magellan Historiography." *Hispanic American Historical Review*, vol. 51, no. 2, May 1971.

Villiers, Alan. "Magellan: A Voyage into the Unknown Changed Man's Understanding of His World." *National Geographic*, June 1976.

Winchester, Simon. "After Dire Straits, an Agonizing Haul Across the Pacific." *Smithsonian*, April 1991.

Unpublished Materials

Gallego, Vasquito. "The Voyage of Fernão de Magalhães Written by One Man Who Went in His Company," tr. Samuel Eliot Morison. Harvard University Archives.

Morison, Samuel Eliot. Unpublished article for *Life*, February 24, 1972. Harvard University Archives.

Sandman, Alison. *Cosmographers vs. Pilots: Navigation, Cosmography, and the State in Early Modern Spain*, Ph.D. dissertation. University of Wisconsin, 2001.

Smith, Roger Craig. *Vanguard of Empire: 15th- and 16th-Century Iberian Ship Technology in the Age of Discovery*. Ph.D. dissertation. Texas A&M University, 1989.

Acknowledgments

$\mathcal{S}$uzanne Gluck, my literary agent, provided invaluable assistance with the development of this book every step of the way; it is a privilege to have the benefit of her keen insights and good judgment. At William Morrow, I owe a vast debt of gratitude to my editor, Henry Ferris, for his steadfast belief in this book and editorial expertise. I am also grateful to Trish Grader for her enthusiasm and guidance, and I wish to extend additional thanks to Juliette Shapland and Sarah Durand. At HarperCollins UK, I must acknowledge Val Hudson, whose editorial contributions and friendship I have long prized, as well as the support of Arabella Pike.

Magellan's circumnavigation concerns many different fields, and I conducted research in a wide variety of institutions. In New York, I was fortunate to be able to use the resources of the following institutions: Butler Library, Columbia University; the Center for Jewish History Genealogy Institute; the New York Society Library, where I wish to thank Mark Piel and Susan O'Brien for help with the interlibrary loan program; the Hispanic Society of America; the New York Academy of Medicine Library; and the New York Public Library. I also want to express my appreciation to Columbia University's John Jay Colloquium, led by the inspirational Peter Pouncey, where I had the opportunity to study classical approaches to writing history with numerous distinguished colleagues.

I owe special thanks to the John Carter Brown Library at Brown University, where Richard Ring, reference librarian; Susan Danforth; and Norman Fiering, director, offered assistance, encouragement, and a sustaining belief in the importance of discovery and exploration as the engine of history. I also received assistance at the Harvard University Archives from Melanie M. Halloran, reference assistant, and Harley P. Holden, university archivist, in researching the papers of Samuel Eliot Morison. My appreciation goes to Mrs. Emily Beck Morison for granting me access to the papers. I must also mention the Beinecke Rare Book and Manuscript Library, Yale University, the repository of the Antonio Pigafetta manuscript; the Library of Congress, Manuscript Division, Washington, D.C.; and the Special Collections Department, Brandeis University Libraries, where Susan C. Pyzynski, Eliot Wilczek, and Lisa Long guided me through their documents pertaining to lawsuits arising from Magellan's voyage; the Peabody Library, Johns Hopkins University; and John Hattendorf of the Naval War College in Newport, Rhode Island.

My thanks go also to the NASA scientists who provided up-to-date satellite images of Magellan's route and a better understanding of the physical nature of the globe. They include my good friends James Garvin, NASA's lead scientist for Mars exploration; and Claire Parkinson, principal investigator for the AQUA mission. Thanks as well to Marshall Shepherd, research meteorologist, and Chester Koblinsky, head of the Oceans and Ice Branch, for their assistance.

Many other individuals generously offered guidance. In New York, I wish to thank my son Nick for his sailing expertise and my mother, Adele, and my daughter, Sara, for their encouragement; Wilma and Esteban Cordero; Ed Darrach of Bristed-Manning for travel-related services; Daniel Dolgin, for his unstinting advice and patience; Darrell Fennell; Sloan Harris; Emily Nurkin; Roberta

Oster; Meredith Palmer; Natalia Tapies; Susan Sparrow; Susan Shapiro; Joseph Thanhauser III; and the gang at Byrnam Wood. Thanks also to Jennifer O'Keeffe for research assistance in New York. Others who helped in various ways include Alexandra Roosevelt, Martha Saxton, and Robert Schiffman.

Because primary sources about Magellan exist in many languages, especially sixteenth-century Spanish and Portuguese, I am indebted to several translators for bringing these occasionally difficult texts to light, in some cases translating them into English for the first time. They include Isabel Cuadrado, Laura Kopp, Rosa Moran, and Víctor Úbeda.

In the course of my research trips to Spain, I received assistance from Kristina Cordero, my able researcher; Javier Guardiola; and Víctor Úbeda. In Madrid I conducted research at the Museo Naval and the Biblioteca Nacional, and in Seville I consulted the Archive of the Indies, where I am grateful for the assistance of Pilar Lazaro, chief of the Reference Division. Thanks to Francisco Contente Domingues in Portugal; and in Brazil, I extend appreciation to Alessandra Blocker and Elisabeth Xavier, my editors at Objetiva.

One of the highlights of research for this book was my trip to South America in January 2001 to travel along Magellan's route through the strait that bears his name. In Patagonia I wish to thank the captain and crew of M/V *Terra Australis,* on which I sailed, and Jon V. Diamond, my traveling companion.

I also owe a debt to specialist readers of the manuscript for their perceptive comments and corrections. They include Dr. Bruce Charash; Daniel Dolgin; Professor Peter Pouncey of Columbia University; Patrick Ryan S.J.; Samuel Scott of the Peabody Essex Museum in Salem, Massachusetts; and Patricia Telles.

By way of personal thanks, I must acknowledge the contribution of my wife (and first reader) Betsy, who made it possible for me to undertake occasionally demanding travel that was an integral part of the research. During the time I worked on this book, I lost my brother and my father. They enjoyed hearing about it while it was in progress, and I wish they could have seen the finished product. For this reason, and for others that are far more important, I wish to dedicate it to their memory.

Index

Sanlúcar de
Barrameda

PERSIA

ARABIA

INDIA

CHINA

AFRICA

Malacca

South
China
Sea

Philippines

Ladrones
(Guam)

EQUATOR

Sumatra

Borneo

Moluccas

MOZAMBIQUE

Java

Timor

AUSTRALIA

Cape of
Good Hope

MAGELLAN'S
CIRCUMNAVIGATION
1519—1522